Current Topics in Behavioral Neurosciences

Volume 74

Current Topics in Behavioral Neurosciences provides critical and comprehensive discussions of the most significant areas of behavioral neuroscience research, written by leading international authorities. Each volume in the series represents the most informative and contemporary account of its subject available, making it an unrivalled reference source. Each volume will be made available in both print and electronic form.

With the development of new methodologies for brain imaging, genetic and genomic analyses, molecular engineering of mutant animals, novel routes for drug delivery, and sophisticated cross-species behavioral assessments, it is now possible to study behavior relevant to psychiatric and neurological diseases and disorders on the physiological level. The Behavioral Neurosciences series focuses on translational medicine and cutting-edge technologies. Preclinical and clinical trials for the development of new diagnostics and therapeutics as well as prevention efforts are covered whenever possible. Special attention is also drawn on epigenetical aspects, especially in psychiatric disorders.

CTBN series is indexed in PubMed and Scopus.

Sahib Khalsa • Albert Powers
Editors

Perceptual Dysregulation in Psychiatric Nosology

Springer

Editors
Sahib Khalsa
Department of Psychiatry
University of California Los Angeles (UCLA), UCLA Semel Institute for Neuroscience and Human Behavior
Los Angeles, CA, USA

Albert Powers
Department of Psychiatry
Yale University School of Medicine
The Connecticut Mental Health Center
New Haven, CT, USA

ISSN 1866-3370 ISSN 1866-3389 (electronic)
Current Topics in Behavioral Neurosciences
ISBN 978-3-032-02575-3 ISBN 978-3-032-02576-0 (eBook)
https://doi.org/10.1007/978-3-032-02576-0

This Springer imprint is published by the registered company Springer Nature Switzerland AG
The registered company address is: Gewerbestrasse 11, 6330 Cham, Switzerland

Preface

Since the advent of systems neuroscience, it has been difficult to deny the intuitive appeal of a simple feed-forward model of perception. Beginning in the retina, a stepwise cascade of feature processing seems to call into being an increasingly complete picture of our external world: contrasts become edges, edges become shapes, and shapes, wedded to movement and color, become detailed representations of the contents of our environment. Through this organized serial extraction of information, sensory input seems capable of providing everything we need for successful navigation and decision-making in the visual world. This example of visual perceptual processing is simple and elegant, but wrong. Or, more charitably, it is only a partial description of perception. Responding to the limitations of feed-forward models, contemporary computational neuroscience has proposed the existence of an evolutionarily conserved system for establishing and updating perceptual representations of the environment through a sophisticated series of reciprocal messaging involving the issuing and processing of prediction signals.

In this volume, we consider perceptual dysregulation as a transdiagnostic feature of psychiatric illness. Using the Bayesian brain as a guiding principle for perception, we examine how exteroceptive, interoceptive, and proprioceptive anomalies shape the lived experience of psychiatric symptoms. Our aim is to illuminate how computational principles unify otherwise disparate symptom presentations, and how a deeper understanding of perceptual inference may catalyze the evolution of psychiatric nosology itself. This inferential framework offers a transdiagnostic lens for understanding both normative and pathological perceptual phenomena.

Throughout this collection, we elaborate how perceptual dysregulation—defined as maladaptive learning and inference within or across sensory systems—provides a powerful unifying lens for understanding how dysfunction in discrete sensory systems can propagate into the complex and heterogeneous symptom profiles seen across psychiatric diagnoses. Bayesian frameworks like those highlighted in this collection are immensely powerful because they provide a common method for linking normative processing to pathophysiology.

This volume originated from a study group convened at the annual meeting of the American College of Neuropsychopharmacology. Arising from the ensuing discussion, our compilation advances the hypothesis that disruptions in perceptual inference across the exteroceptive, proprioceptive, and interoceptive realms lie at the heart of a diversity of neuropsychiatric illnesses. This offers not only the promise of novel treatments but also the opportunity to establish causal links between latent computational processes and observable symptoms. In doing so, these efforts help lay the groundwork for a revised nosology—one rooted not in symptom checklists or clusters but in empirically grounded models of perceptual inference, brain-body interaction, and bodily regulation.

Los Angeles, CA, USA Sahib Khalsa
New Haven, CT, USA Albert Powers

Contents

Disturbances in Auditory and Visual Perceptual Function in Schizophrenia: Patterns, Causes, and Consequences

Daniel C. Javitt

Contents

Abstract Schizophrenia is a severe mental disorder and a leading cause of global disability. Increasing evidence ties impairments in both auditory and visual perceptual processing to impaired functional outcomes. In the auditory system, deficits are observed in the ability to process changes in tonality, leading to deficits in the detection of nonverbal aspects of auditory communication, including emotion and sarcasm. In the auditory sytem, these deficits are indexed by impaired generation of mismatch negativity (MMN), which has become a widely utilized index of auditory perceptual dysfunction. In the visual system, deficits are especially observed in functions related to the magnocellular visual pathway, which projects preferentially to the dorsal visual stream. Dorsal stream dysfunction, in turn, contributes to

D. C. Javitt (✉)
Division of Experimental Therapeutics, Columbia University Medical Center, New York, NY, USA

Schizophrenia Research Division, Nathan Kline Institute, Orangeburg, NY, USA
e-mail: dcj2113@cumc.columbia.edu

Curr Topics Behav Neurosci (2025) 74: 1–40
https://doi.org/10.1007/7854_2025_582
Published Online: 02 April 2025

impairments in processes such as fragmented object recognition, face emotion recognition, and social cognition. Indices of visual perceptual dysfunction include visual P1 responses to magnocellular-biased stimuli, as well as fMRI activation levels within the dorsal and ventral streams. Perceptual level deficits are tied most directly to impaired function of N-methyl-D-aspartate glutamate receptors (NMDAR) and thus may serve both as pathophysiologically based nosological biomarkers and as targets for neuromodulatory and pharmacological treatment development.

Keywords Auditory · Magnocellular · Mismatch negativity (MMN) · Perception · Schizophrenia · Visual

1 Introduction

Schizophrenia is a severe mental disorder and a leading cause of functional disability. In addition to symptoms, schizophrenia is associated with severe cognitive impairments that appear to be primary determinants of the impaired functional outcome (Javitt 2023; Green et al. 2019). These deficits are increasingly being divided into "neurocognitive" deficits that reflect the ability to interact with inanimate objects and symbols and "social cognitive" deficits that reflect the ability to interact with conspecifics. The integrity of neuro- and social cognitive processes depends critically upon the integrity of perceptual-level processing within both the auditory and visual systems.

These processes are described in the Research Domain Criteria (RDoC) (Insel et al. 2010) matrix within the Auditory Perception and Visual Perception subconstructs of the Cognitive Systems domain and may be especially relevant to the domains of Reception of Facial and Non-facial Communication within the Social Processes Domain. Nevertheless, as opposed to other neuro- and social-cognitive constructs, perceptual-level measures have not yet been incorporated into standardized cognitive batteries for schizophrenia, such as the MATRICS consensus cognitive battery (MCCB) (Nuechterlein et al. 2008).

Perceptual level dysfunction thus remains a severely understudied process within both research and clinical settings in schizophrenia. Moreover, in many studies of schizophrenia, sensory-level cortices are viewed as passive transmitters of information to higher-level brain regions as originally proposed by Broadbent in the late 1950s (Broadbent 1957). In contrast, more contemporary models of brain function focus on the extensive local processing within subcortical and cortical brain regions that serve to isolate and prioritize information reaching subsequent processing stages, as well as the continuous feed-forward and feed-back communication between sensory and higher cortical regions (e.g. Pulvermuller et al. 2021).

In schizophrenia, severe perceptual deficits have been documented across sensory systems, with the greatest study of the auditory and visual perceptual systems. The pattern of dysfunction may indicate separate pathophysiological entities within the larger schizophrenia syndrome. Moreover, these perceptual-level deficits significantly contribute to other aspects of neuro- and social cognition. Finally, there is increasing evidence that interventions targeting basic perceptual deficits lead to improvements in higher-order processing. Even in the absence of remediation, compensatory strategies to overcome perceptual level dysfunction may be possible. Overall, therefore, perceptual level deficits in schizophrenia remain a critical contributor to the personal experience of individuals with schizophrenia (Javitt and Freedman 2015) as well as an important target for future interventions (Javitt 2023).

While beyond the scope of this chapter, other forms of perceptual disturbances, such as impaired proprioceptive (Javitt et al. 1999), introceptive (Ardizzi et al. 2016; Critchley et al. 2023; Torregrossa et al. 2022), and somatic (Melvin et al. 2021) processing, may also contribute to abnormal perceptual experience in schizophrenia. Furthermore, the neural mechanisms underlying introception in both health and disease remain a focus of active investigation (Berntson and Khalsa 2021; Khalsa et al. 2018; Quadt et al. 2018). Future studies, therefore, may permit direct comparison between the effects of extroceptive and introceptive processing abnormalities in schizophrenia.

2 Auditory Perceptual Disturbances in Schizophrenia

The human central auditory system begins in the cochlea and projects to the cortex via the superior olivary complex, inferior colliculus, and medial geniculate nucleus of the thalamus (Javitt and Sweet 2015). Schizophrenic individuals show minimal deficits in standard audiometric testing, which evaluates whether or not an individual can detect a pure sound embedded in noise. Such tests, however, depend primarily on peripheral cochlear function and give limited insight into the integrity of information processing at the level of the auditory sensory cortex. For example, performance is relatively intact even in individuals with large auditory cortex lesion (rev. in Javitt and Sweet (2015)). In individuals with auditory cortical lesions, the simplest process that is affected is the ability to compare the physical properties (e.g., pitch, duration) of two consecutive stimuli (Dykstra et al. 2012; Iversen and Mishkin 1973). Furthermore, in the absence of distractor stimuli, no further deficits are observed from lesions elsewhere in the brain (Chao and Knight 1995), suggesting that tone-matching ability may serve as an index of the initial stages of cortical auditory processing.

The earliest retrospective evidence for deficits in tone-matching ability in schizophrenia comes from two studies conducted in the 1960s/70s. In the first, the ability to identify common sounds (e.g., dentist's drill, train) was used as a control for the ability to detect threat based on sound. Contrary to expectation, schizophrenia was associated with significant impairments even in the "sound effect test" control condition, along with deficits in the ability to interpret intonation (Billingberg and Jonsson 1965). A follow-up test used an even simpler sensory control task, in which individuals were presented with stimuli that were either identical or differed by a specific physical feature – duration, loudness, or pitch. Contrary to expectation at the time, schizophrenia was associated with impairments in all 3 of these simple discriminations. Moreover, the reduction in discrimination ability correlated significantly with the inability to detect the intonation of neutral words that were read in a friendly, threatening, or neutral tone of voice, leading to the conclusion that even elementary features of auditory processing are impaired in schizophrenia (Jonsson and Sjostedt 1973).

"Echoic Memory" Dysfunction More systematic investigation of early auditory processing began in the early 1990s, based on the initial descriptions of the auditory echoic memory system as elaborated by Cowan (1984, 1988). Specifically, it was proposed that the echoic memory system consists of two separate perceptual-level stores: an initial short-term store with an apparent duration of up to 300 ms that permitted stimulus recognition and a subsequent longer-term "echoic" store that retained representation of the acoustic features of stimuli for up to ~20 s. This longer-term store converged with the first demonstrations of the mismatch negativity (MMN) auditory component to changes in simple auditory features such as pitch (Cowan et al. 1993; Hari et al. 1984), duration (Kaukoranta et al. 1989), intensity (Naatanen et al. 1987, 1989) or location (Paavilainen et al. 1989). As with the longer-term auditory store, MMN appeared to depend upon a memory system lasting tens of seconds.

An initial study (Strous et al. 1995) evaluated the ability of individuals to match tones following a brief (300 ms) delay that was chosen to be outside the interval for sound fusion but within the interval of auditory sensory ("echoic") memory. As expected, schizophrenia was associated with a marked reduction in tone-matching accuracy, with parallel changes in performance with increasing task difficulty. A subsequent series of studies evaluated the degree to which sensory-level deficits could be distinguished from potential consequences of attentional impairments.

Thus, an initial follow-up study evaluated the time course of retention of auditory information by matching performance across groups at 1 s, and then following the decline in performance over a subsequent 20 s. In signal detection analyses, the sensitivity and bias measures were also found to be indistinguishable when groups were matched for overall performance. Finally, both groups showed reduced performance, especially at long ISI, in a distraction condition in which they read digits out loud on a computer screen. Nevertheless, susceptibility to distraction was also similar across groups (Javitt et al. 1997). Individuals with schizophrenia were also no more susceptible to auditory backward masking than controls (March et al. 1999), again pointing to a reduction in processing precision within the early auditory store rather than impaired storage ability.

A final study (Rabinowicz et al. 2000) evaluated the sensitivity of individuals with schizophrenia to same-modality distraction, which is known to depend highly on the integrity of top-down attentional control mechanisms. In this paradigm, a series of alternating frequency tones was interspersed between the reference and test tones. In this study as well, schizophrenia was associated with significant tone

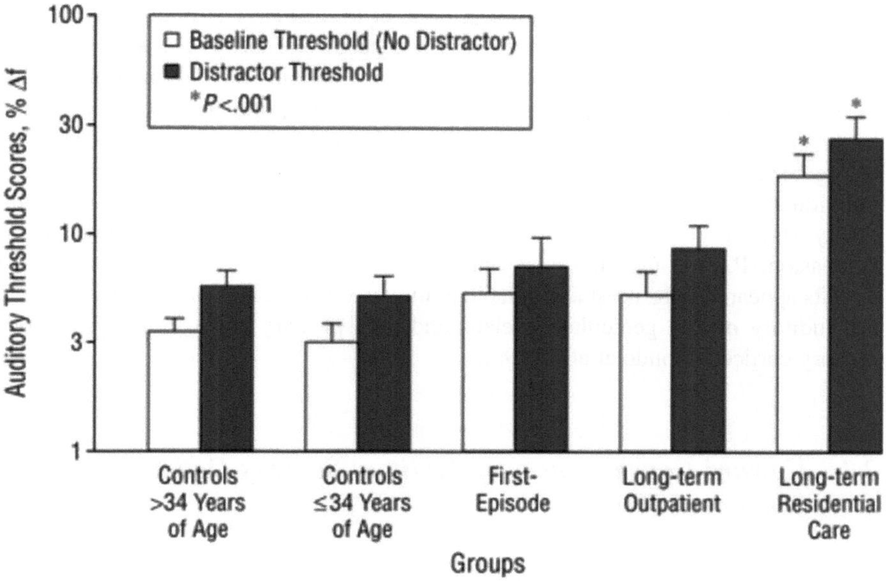

Fig. 1 Tone matching performance in schizophrenia individuals drawn from different clinical settings and similar-age controls, in the absence and presence of distractors. *$p < 0.001$ vs. controls. (From Rabinowicz et al. (2000). Reproduced with permission)

matching deficits relative to similar-age controls (Fig. 1). The deficits, moreover, were present in both first episode and long-term patients. All groups showed worse performance, as expected, in the presence vs. absence of the auditory distractor. Nevertheless, susceptibility to distraction was similar across groups, again supporting localization of the deficit to auditory sensory cortex rather than prefrontal control regions.

This study also provided the first observation that auditory tone matching measures might distinguish between pathophysiologically distinct subgroups of individuals with schizophrenia. Thus, individuals with long-term schizophrenia were recruited from two distinct recruitment sites: an outpatient facility that served individuals living in the community and a supervised long-term residential care setting. Despite similar length of illness, medication levels, and overall symptom severity, the two groups differed dramatically on tone-matching threshold. In a more detailed assessment, the groups showed similar positive symptom ratings, but the group with impaired tone matching showed greater severity of negative symptoms (Rabinowicz et al. 2000). While less studied, deficits in auditory spatial localization ability also appear to be impaired in schizophrenia and may be especially related to the pathophysiology of auditory verbal hallucinations (Perrin et al. 2010).

Similar deficits in auditory tone matching were observed across a range of research groups (Holcomb et al. 1995; Jahshan et al. 2013). In a meta-analysis conducted in 2017, a mean difference of ~1.2 sd was observed in schizophrenia relative to the normative population (Donde et al. 2017). The bimodality in tone matching scores was also confirmed in a retrospective analysis across >300 individuals (Donde et al. 2019a). Tone matching thus remains the only neurocognitive measure to show a bimodal distribution across schizophrenia individuals. In the larger analysis, the groups were also distinguished by the severity of conceptual disorganization and auditory-dependent neurocognitive processing. Interestingly, individuals with impaired early auditory processing, as reflected in tone-matching ability showed reduced rates of co-morbid illness (personality disorder, major depression, PTSD, and substance abuse). In imaging studies, the tone-matching deficits appeared to be most attributable to impaired functional connectivity between the auditory medial geniculate nucleus and both primary and secondary auditory sensory cortices (Donde et al. 2019a).

2.1 Upward Consequences of Auditory Sensory Deficits: Emotional Prosody

Given the consistent observation of impaired early auditory processing ability, a second series of investigations evaluated pathways by which early auditory processing impairments in schizophrenia might lead to impairments in higher-order cognition. In Western languages, one of the main uses of tonal modulation over time is to convey emotion. Thus, for example, when individuals are happy, they show characteristic changes in voice quality such that they tend to speak relatively loudly and with greater modulation than during more neutral periods. In contrast, when individuals are sad or fearful, they tend to speak more quietly and with lesser modulations. Fine gradations in pitch and rate characteristics convey fine gradations of emotion. Nevertheless, different individuals may "map" the physical features of speech to emotion in different ways, leading to different sensitivity to voice emotion across individuals (van Rijn and Larrouy-Maestri 2023).

The study of the interrelationship between specific perceptual abilities and affective discrimination is impeded somewhat by the absence of widely validated stimulus sets with well-defined physical features. Nevertheless, one set of studies was performed using a stimulus set published by Juslin and Laukka (2001) in which each of 8 professional actors provided 4 examples of each of 5 emotions – happiness, sadness, anger, disgust, and fear – along with a neutral expression. The four examples consisted of weak and strong portrayals, presented as either statements (declaratives) or questions (interrogatives). Each utterance contained the same neutral expression, "It is (or 'Is it') eleven o'clock").

The physical properties of each utterance were published. Key physical features included both the mean (F0M) and standard deviation (F0SD) of the fundamental frequency of each utterance; the mean (VoIntM) and standard deviation (VoIntSD) of voice intensity; and the presence or absence of high-frequency noise (HF500). Features with the strongest effect on emotion recognition were F0M, F0SD, VoIntM,

VoIntSD, HF500, and speech rate. Features with the strongest contribution to emotional intensity discrimination were F0M, VoIntM, VoIntSD, and speech rates.

Sensory-Specific vs. Supramodal Emotion Discrimination An initial study with this paradigm (Leitman et al. 2005) evaluated the relative degree to which emotion recognition deficits in schizophrenia, which are documented within both the auditory and visual domains, are most attributable to domain-general impairments in emotion recognition ability or domain-specific impairments within each domain. For this assessment, individuals participated in test of basic tone-matching ability, a "distorted tunes" task that assesses the ability to detect common melodies, and tests of both auditory (voice) and visual (face) emotion discrimination and identification. In the discrimination task, individuals indicated whether two sequentially presented utterances or faces portrayed the same or different emotion without naming the emotion. In the identification tasks, they were asked to explicitly name the emotion.

As expected, schizophrenia was associated with large effect-size deficits in both voice and face emotion discrimination and identification (both $d > 1.5$ sd). However,

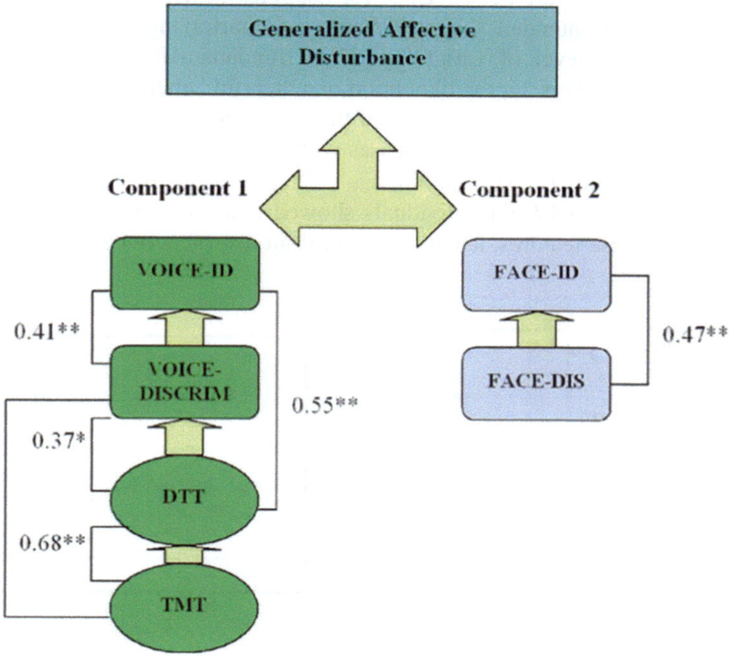

Fig. 2 Schematic diagram of interrelationship between sensory and affective measures. Values shown represent Spearman correlation coefficients between indicated measures.$*p < 0.05$, $**p < 0.01$. TMT Tone Matching Task, DTT Distorted Tunes Task, VOICE-DISCRIM Voice Emotion Discrimination Test, VOICE-ID Voice Emotion Identification Test, FACE-DISCRIM Face Emotion Discrimination Test, FACE-ID Face Emotion Identification Test. (From Leitman et al. (2005). Reproduced with permission)

in a principal components analysis within the schizophrenia group (Fig. 2), the sets of deficits segregated into two separate components, supporting the concept of domain-specific rather than domain-general deficits. Moreover, as expected, auditory sensory measures correlated with auditory, but not visual, emotion recognition impairments. Both voice and face emotion recognition scores correlated significantly with the severity of clinical conceptual disorganization scores. Voice emotion identification scores also correlated significantly with the severity of negative symptoms and with a proxy measure of global outcome. In a subsequent correlative imaging study, deficits in voice emotion identification correlated significantly with the integrity of radiations from the medial geniculate nucleus to primary auditory cortex, as well as higher-order auditory radiations (Leitman et al. 2007).

Effects by Cue A subsequent fine-grained analysis of auditory emotion recognition was conducted in which the ability of individuals to utilize specific cues was investigated (Leitman et al. 2010). These analyses took advantage of the natural variation in cue usage across utterances. Thus, in all cases, the actors creating the utterances were attempting to create well-recognizable examples of each emotion. Nevertheless, some attempts were more successful than others, which was in part attributable to the measured cues. Thus, for happiness, the likelihood that controls would recognize an intended "happy" utterance as portraying happiness increased significantly as the level of variability of the fundamental frequency increased (F0SD, Fig. 3a). Conversely, the likelihood of a "fearful" utterance being perceived as fearful increased as the average pitch of the base tone decreased. For disgust, the likelihood of correction detection increases with reduction in the mean pitch of the fundamental frequency (F0M, Fig. 3b).

In all cases, schizophrenia individuals showed reduced differentiation of utterances based on these features, leading to significant divergences between groups for

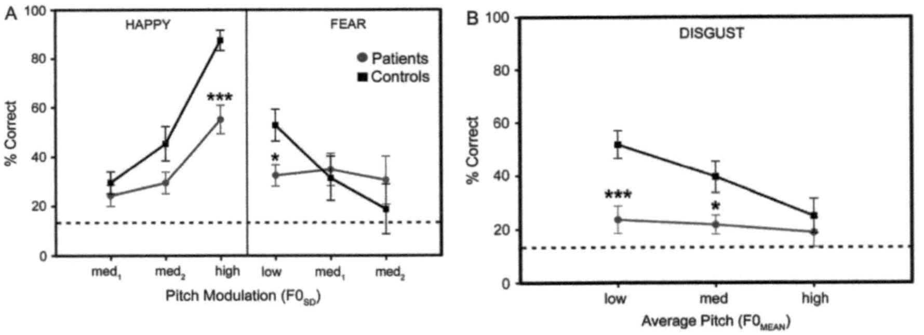

Fig. 3 Utilization of pitch cues in schizophrenia and control individuals. (**a**) Percent correct identification for intended happy (left) and fearful (right) stimuli as a function of degree of pitch modulation (F0SD). (**b**) Percent correct identification for intended disgust stimuli as a function of average pitch of the stimulus fundamental frequency (F0mean). (From Leitman et al. (2010). Reproduced with permission)

stimuli that were most recognizable for healthy volunteers. Two additional results of this analysis showed important nuances of the emotion recognition experience of individuals with schizophrenia. First, the results showed a critical difference between perception of "hot" vs. "cold" anger that should have been predicted a priori. Second, the analyses showed that individuals with schizophrenia, as expected, showed reduced ability to differentiate between weak and strong portrayals of emotions. Unexpectedly, however, the confusion resulted in relatively weak portrayals of emotion being perceived as strong, rather than strong portrayals being perceived as weak.

Hot vs. Cold Anger The concept of "hot" vs. "cold" anger was not a designed feature of the Juslin and Laukka battery but nevertheless was inherent in the structure. Thus, in typical English language usage, there are two vocal strategies to the portrayal of anger. The more typical approach termed "hot" anger is communicated by a large increase in both the intensity of voice and the degree of intensity modulation (i.e., "shouting"). The mechanics of shouting also increase the degree of high-frequency noise (HF500, "raspiness") that accompanies the utterance. In contrast, for weaker portrayals of anger, individuals often portray "cold" anger, in which voice qualities are "overcontrolled," leading to a drop in both F0M and F0SD.

In the Juslin & Laukka stimulus set, most "anger" utterances across actors and, especially, portrayals of strong anger used the "hot" anger strategy, so that anger portrayals differed from those of other emotions by overall voice intensity. When viewed as a function of increases in these voice features (Fig. 4), schizophrenia patients were as accurate as controls in using intensity cues to differentiate strong

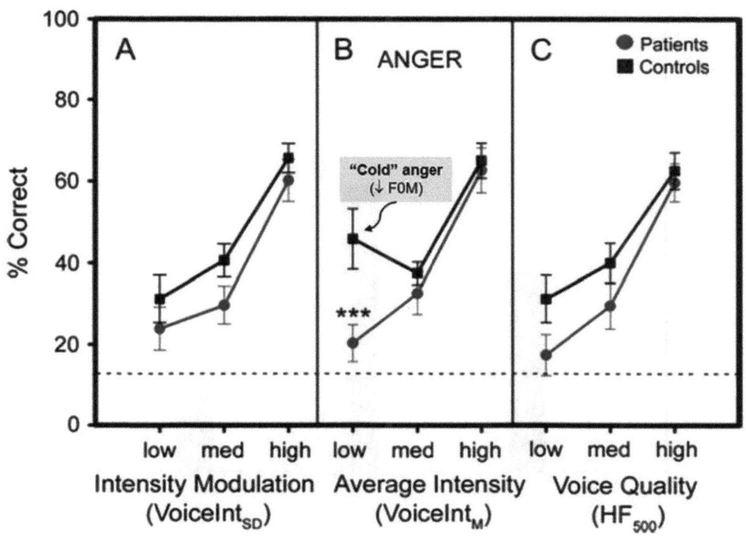

Fig. 4 Percent correct dentification for intended angry stimuli as a function of indicated properties. (From Leitman et al. (2010). Reproduced with permission)

and weak anger portrayals, leading to the equivalence in anger detection ability between groups. In contrast, in the "cold anger" stimuli, which were characterized by low voice intensity, schizophrenia individuals showed markedly impaired performance. Similarly, individuals with schizophrenia showed relatively intact ability to detect "elation," which is characterized by increased voice intensity, but not in "happiness," which is characterized by alterations in pitch. Thus, impairments in pitch discrimination in schizophrenia may force over-reliance on intensity cues but under-reliance on pitch cues, leading to misinterpretation of intended emotions during speech.

Intensity Perception by Group Because the actors were asked to provide both weak and strong portrayals of emotion, it was possible to evaluate the degree to which participants tracked the actors' intentions. Emotion intensity discrimination, like emotion identity discrimination, depends upon the ability to process tonal cues. Thus, it was expected that patients would show reduced ability to distinguish intended stimulus strength, which was confirmed in the analyses. Unexpectedly, however, individuals with schizophrenia tended to rate weak portrayals of emotion as strong, rather than strong as weak (Fig. 5). A potential explanation of this finding is that voice is often used to portray calmness and soothing properties, so failure to process these features may lead to failure to understand lack of emotional intensity. The deficits were subsequently confirmed across a larger sample and shown to correlate with impaired tone matching across large sample sizes (Gold et al. 2012).

Emotion by Abstract Tones A final line of evidence tying auditory emotion deficits to the physical features of the stimuli comes from a study in which frequency

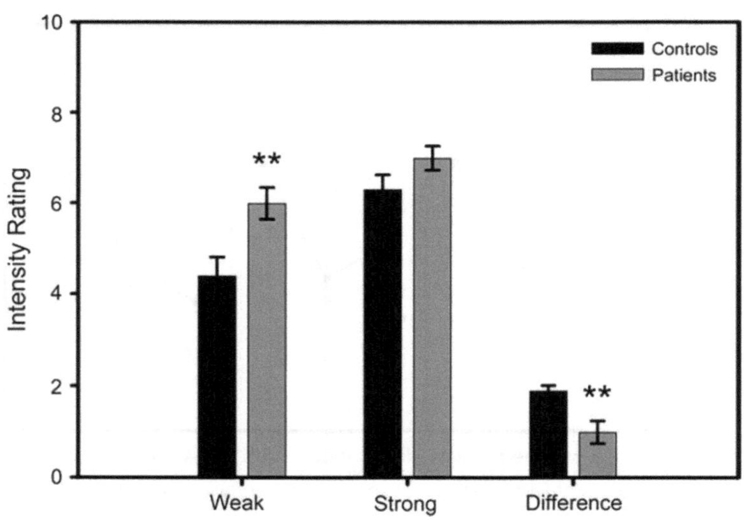

Fig. 5 Emotion intensity ratings by group. Intensity ratings for intended weak and strong portrayals of emotion were collapsed across emotion for both controls and patients. $^{**}p < 0.01$. (From Leitman et al. (2010). Reproduced with permission)

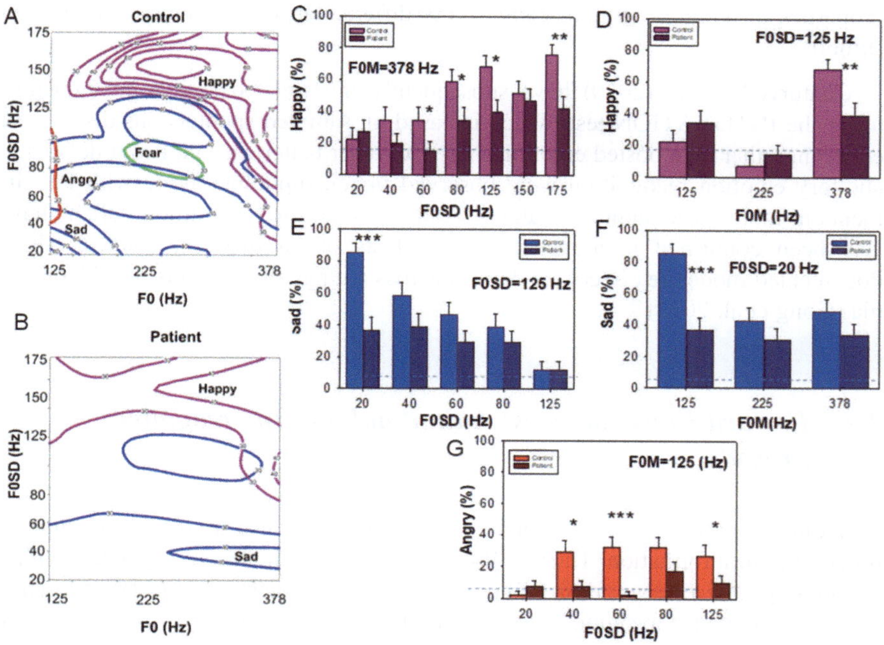

Fig. 6 (**a, b**): Contour maps showing the pattern of responses for emotions across changes in mean fundamental frequency (F0M) (X-axis) and variability (F0SD) (Y-axis) for controls (**a**) and patients (**b**) for frequency-modulated tone stimuli (without HF500). For illustration purposes, a cutoff of 30% is used to delineate regions relative to the chance performance level (20%). The numbers on the colored lines indicate the percent response associated with that area. (**c–g**): Bar graphs of the percent of subjects rating a stimulus a particular emotion across a given change in mean fundamental frequency (F0M) and variability (F0SD), and dashed line passing through chance level performance (20%). Happy 5 pink, sad 5 blue, angry 5 red, fear 5 green, and neutral 5 gray. Controls are on left in lighter shade. (From Kantrowitz et al. (2013). Reproduced with permission)

modulated (FM) tones were created that had physical characteristics resembling those of affective utterances (Kantrowitz et al. 2013). Thus, stimuli with high base pitch and pitch modulation were perceived as happy by control individuals, whereas those with low base pitch and variability were perceived as sad. Increases in both intensity and high-frequency noise increase the likelihood that tones would be perceived as angry.

Although the overall "weathermaps" of locations showed the same basic shape for schizophrenia as for controls, the degree of differentiation was substantially less in schizophrenia (Fig. 6). Thus, whereas controls showed a linear increase in the likelihood of describing an FM modulated tone as happy as the F0SD increased and sad as the F0SD decreased, this tendency was much reduced in schizophrenia. Although there are no correct or incorrect identifications for abstract tones, these findings nevertheless suggest that the perceptual experience of individuals with schizophrenia is one in which the reduced ability to discriminate based on perceptual properties of auditory stimuli leads to less differentiation of emotionally relevant stimuli.

Chaturvedi et al. (2020) have subsequently investigated similar relationships using the INTONATION test, which is based on sampled spontaneous displays of emotion, rather than posted emotions. As with other batteries, significant deficits in auditory emotion recognition were observed in schizophrenia along with impairments in more basic auditory features. Auditory emotion recognition deficits have also been confirmed in a recent meta-analysis of 18 different studies, which documented moderate effect-size deficits across a range of emotions in schizophrenia (Gong et al. 2021).

2.2 Upward Consequences: Attitudinal and Interrogative Prosody

In addition to affect, prosodic features of language are also used to communicate other types of information. Thus, in Western languages, the location of stress within a word (e.g., "desert" vs. "dessert") or sentence (e.g., subject vs. object) can change the overall sentence meaning. Intonation at the end of a sentence is used to convey whether the sentence is a statement (declarative) or question (interrogative). As with affective prosody, these processes are also significantly impaired in schizophrenia and correlated with more basic disturbances of early auditory processing (Leitman et al. 2007).

Prosody can also be used to convey attitude/counter-factuality through the use of sarcasm. Thus, sarcastic utterances contain the same words as sincere utterances. However, subtle changes in pitch and stress are used to convey that the utterance means the opposite of what might be inferred from the actual words. An initial study of sarcasm in schizophrenia demonstrated both a highly significant reduction in sarcasm detection ability ($d = 2.2$) and a significant correlation with more basic auditory perceptual abilities as well as voice emotion identification (Leitman et al. 2006). This finding was subsequently confirmed and linked to functional connectivity deficits involving auditory sensory cortex (Kantrowitz et al. 2014), as well as the ability to detect sarcasm in more naturalistic settings (Patel et al. 2020). Although the role of sarcasm detection ability has been less studied in schizophrenia than emotion discrimination, nevertheless it may play a more prominent role in predicting social outcome (Patel et al. 2020).

Tonal Language In tonal languages, such as Chinese, tonality may specifically affect word meaning. For example, in Mandarin, characters are represented in English by the Pinyin system, which uses the English alphabet to represent 417 separate syllables (e.g., "ya"). Different meanings are attributed to syllables by adding one of four tones, which are represented by adding a number or symbol. For example, (mao1), (mao2), (mao3), and (mao4) would have the same syllabic basis

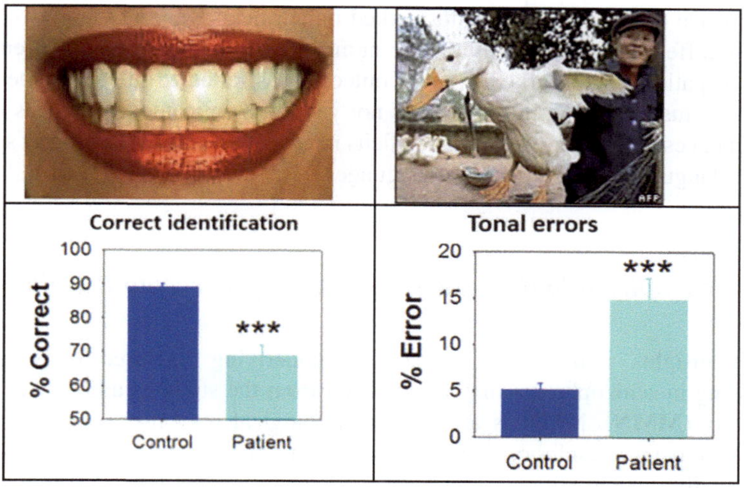

Fig. 7 Schematic illustration of the Mandarin auditory word recognition task. Individuals were shown pictures of objects that were represented by a syllable that differed only in tonality such as ya2 (tooth) versus ya1 (duck) along with distractor stimuli that were unrelated. Participants were then asked to point to the image corresponding to a recording of a word. Mandarin speaking participants showed reduced identification ability that was specific to objects that were represented by the same syllable but different tonality. ***$p < 0.001$. (From Yang et al. (2012). Reproduced with permission)

but would be pronounced with a different tone, thereby conveying different meanings, with (mao1) meaning "cat" and (mao4) meaning "hat." Thus, the inability to detect tonality in such languages could lead to alterations in word meaning.

In one study in which this was tested, Mandarin-speaking schizophrenia patients were asked to perform two different tasks. First, they heard the same syllable repeated twice with either the same or different tonality across the two repetitions. They were asked to state whether the two sets of syllables were the same or different. Second, they were played a syllable with a specific tonality and asked to choose between 4 pictures, one of which corresponded to the correct meaning of the syllable as played, a second of which corresponded to the meaning of the same syllable pronounced with a different tonality, and 2 pictures that were semantically unrelated to either word (Fig. 7).

As expected, Mandarin-speaking schizophrenia patients showed similar tone-matching deficits to those of previously studied English-speaking patients. In addition, the patients showed highly significant reductions both in the ability to discriminate across words and in the ability to identify word meaning by pointing to the correct picture. The differential error rate, moreover, was only for the semantically related words, whereas there was no significant increase in selection of nonrelated pictures. Reductions in both word discrimination and identification correlated highly with tone matching ability. Finally, based upon performance, the schizophrenia group ($n = 22$) could be divided into a group with relatively intact auditory performance ($n = 11$) vs. one with marked impairments ($n = 11$). The two groups did not differ in length of illness or number of hospitalizations. Nevertheless, impaired patients had markedly worse outcomes as reflected in employment level and job status. Although this study has not yet been replicated, it suggests that the consequences of auditory perceptual deficits may be even more disabling in speakers of tonal languages than in Western languages.

2.3 Neurophysiological Basis of Auditory Perceptual Deficits

Further insights into neural mechanisms underlying impaired early auditory processing in schizophrenia may be obtained from the study of auditory mismatch negativity (**MMN**). MMN is an electrophysiological response generated when a series of repetitive auditory stimuli is interrupted infrequently by a physically or conceptually deviant stimulus. In these circumstances, the deviant stimulus elicits activity within supragranular layers of primary and secondary auditory cortical regions that reflects detection of the deviant activity by local cortical ensembles (Javitt 2023; Javitt et al. 1996; Lakatos et al. 2020; Donde et al. 2023).

Deficits in MMN generation to pitch and duration deviant stimuli were first reported over 30 years ago and have been extensively replicated since that time (reviewed in Avissar et al. (2018)). Moreover, deficits in MMN generation have been shown to correlate with impaired tone-matching ability (Javitt et al. 2000) and to contribute directionally to impairments in higher-order auditory function (Abplanalp et al. 2024). In source localization studies, especially involving magnetoencephalography, primary generators for MMN are identified in auditory cortex, with significant differences between groups (Pekkonen et al. 2002). Because electrical activity reaching the scalp from subcortical structures in humans is weak, it is difficult to assess subcortical processes in clinical MMN studies. Nevertheless, in parallel fMRI studies, impaired deviant-related activity is observed in subcortical structures, including inferior colliculus and thalamus (Gaebler et al. 2020). In monkey studies, MMN-related activity is obtained from both medial geniculate and pulvinar nuclei of the thalamus (Lakatos et al. 2020).

MMN to Affective Contours Although MMN has been most extensively studied in schizophrenia to changes in simple stimulus features such as pitch, duration, or intensity, MMN has also been studied in response to more complex tonal alterations. Thus, with regards to prosody, MMN has been evaluated to FM-modulated tones that conveyed emotional content (Kantrowitz et al., 2015). As in the original study with these stimuli (Fig. 6), individuals with schizophrenia showed a less differentiated pattern of responses as a function of tonal features reflecting reduced use of characteristics such as F0M and F0SD. Schizophrenic individuals also showed the expected deficits in tone matching and auditory emotion recognition.

In the MMN study, schizophrenia individuals showed significantly reduced MMN generation to all deviant types (Fig. 8). Generators were localized to both

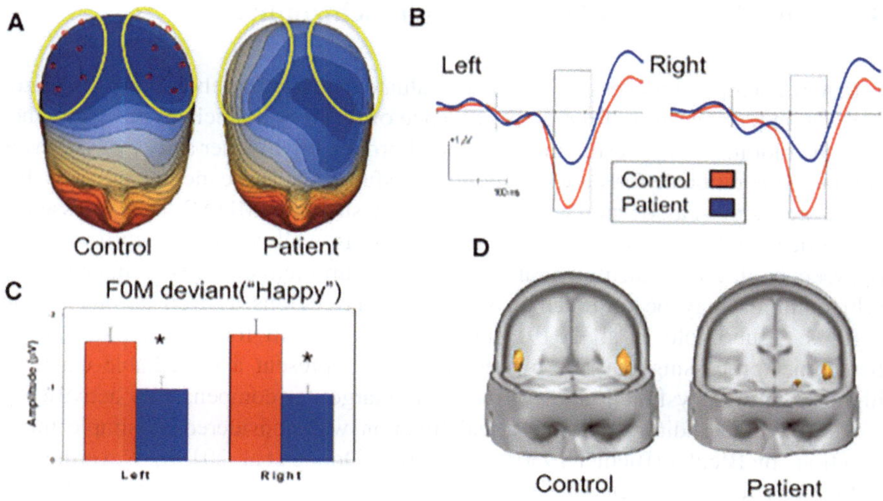

Fig. 8 MMN deficits to affective stimuli in schizophrenia. (**a**) Scalp distribution for control and schizophrenia participants. (**b**) Waveforms from indicated locations. (**c**) MMN amplitudes by group. (**d**) Source localization showing generators in auditory cortex and insula. (From Kantrowitz et al. (2015). Reproduced with permission)

the auditory cortex and insula using source localization approaches. When these were assessed independently, amplitude reductions were observed in both the auditory cortical and insula generators. Moreover, reduced MMN amplitudes correlated with impaired auditory emotion recognition. Reduced functional connectivity was also observed between auditory cortex and insula, as well as between insula and amygdala. Overall, these findings further support a model in which basic deficits in auditory perceptual processing contribute to social cognitive dysfunction, which, in turn, is highly related to functional outcome.

MMN to Other Prosodic Contours Although other contours have been studied to a lesser extent. Nevertheless, in one study, MMN was obtained from FM stimuli that approximated interrogative vs. declarative contours. As with MMN to simpler stimulus features, generators for interrogative MMN localized to primary auditory cortex, suggesting that MMN to linguistic features may also depend on lower-level auditory perceptual processing (Leitman et al. 2009). Nevertheless, no follow-up studies were conducted with this paradigm in schizophrenia. Finally, deficits in MMN generation have been shown to predict impaired ability in identifying common environmental sounds across a range of positive, negative, and neutral valences (e.g., crowds cheering, baby crying, lawnmower) (Joshi et al. 2018), demonstrating "real world" consequences to early auditory processing deficits.

3 Visual Perceptual Dysfunction in Schizophrenia

Auditory perceptual processing has been evaluated most comprehensively in schizophrenia, in part because of the preponderance of auditory hallucinations. Nevertheless, the rationale for investigating perceptual processing in other sensory systems is similar to the rationale for studying auditory dysfunction. These include the facts that (1) the genes most associated with schizophrenia, such as NMDAR, are expressed in subcortical and cortical sensory brain regions as well as higher cortical areas (Caspers et al. 2015; Zachlod et al. 2022), (2) postmortem changes associated with schizophrenia may be observed in visual sensory regions as well as other brain regions (Tsubomoto et al. 2019), and (3) if not tested for directly, deficits in perceptual processing may be missed and may represent a remediable cause of higher cognitive dysfunction, or, at the least, a target for compensatory activities.

As with the auditory system, visual function was considered a "simple intact function" by Bleuler (Bleuler 1950; Javitt 2009a; Donde et al. 2019b). Nevertheless, in literature review, deficits in early visual processing were first demonstrated in the early 1900s by Busch using the recently developed photographic shutter to deliver visual stimuli with precise durations (Donde et al. 2019b; Busch 1904). In the study, it was observed that individuals with schizophrenia need longer exposure times to detect a briefly presented stimulus than controls. While this finding was lost to the literature for many years, it was rediscovered in the early 1980s with the development of tachistoscopic backward masking approaches (Saccuzzo and Braff 1981). These studies not only replicated the initial findings of increased critical stimulus duration in schizophrenia but also demonstrated an increased susceptibility to backward masking that was especially related to the severity of negative symptoms (Green and Walker 1986; Green et al. 1994, 1997).

Over the past several decades, increasing literature has accumulated regarding both the causes and consequences of early visual processing disturbances in schizophrenia. An important finding has been that the disturbances primarily affect the functioning of the magnocellular component of the visual system that subserves "automatic" functions of which individuals are not consciously aware. Thus, unless specific tests are employed, early visual processing deficits are not readily detectable and may be attributed to higher cortical impairments.

Moreover, in several cases, the visual perceptual deficits combine with the auditory impairments to produce synergistic impairments in processes needed for successful social and occupational functioning. Thus, disturbances of both early auditory and early visual processing may be observed in disorders other than schizophrenia. However, in such disorders, intact function in one modality may compensate for impaired function in the other. For several processes, such as social cognitive appraisal and reading, this is not the case, potentially leading to the high rates of social and occupational dysfunction observed in the disorder.

3.1 Subcortical Visual Systems and Magnocellular Dysfunction in Schizophrenia

Current concepts of early visual processing derive largely from the dorsal/ventral pathway dichotomy first proposed by Ungerleider and Mishkin in the early 1980s (Ungerleider and Mishkin 1982) and subsequently refined by Milner and Goodale (1993). This model emphasizes the existence of two parallel visual systems that act in parallel to allow efficient use of complex visual information. In particular, although human visual fields encompass ~180° of visual space, high-resolution information can be extracted only from the foveal representation, which encompasses only ~6% of visual space. Thus, automatic systems are needed to detect objects outside of our attentional focus and rapidly deploy attention to regions of visual space requiring detailed evaluation.

In addition, even within the foveal representation, a number of objects may be present, of which only a subset may require detailed analysis. Systems are thus needed to parse complex visual information into discrete objects and, within the visual scene, select objects for detailed assessment. These systems must act in parallel with the high-resolution systems that primates in general, and humans in particular, have evolved for recognizing and manipulating objects and symbols. According to present conceptualizations, these processes are instantiated through the parallel actions of the dorsal and ventral stream visual pathways.

Dorsal and Ventral Visual Pathways The dorsal visual stream includes relays in deep layers of primary cortex (V1) along with secondary visual (e.g., V3A) and regions in posterior parietal cortex (e.g., lateral intraparietal region, LIP). It receives rapid, low-resolution visual input primarily from the subcortical magnocellular visual system, which furthermore is sensitive even to very low luminance contrast (e.g., <8% contrast) but saturates under high-contrast conditions. These specializations permit the dorsal stream to subserve processes such as the rapid allocation of visual attention to salient features of the environment, as further described below. Moreover, the low-resolution information transmitted through the dorsal stream to frontal brain regions may be processed prior to the slower but more detailed processing within the ventral visual stream. It has therefore been termed the "where" or the "vision for action" system (Milner and Goodale 2008). Finally, the magnocellular system preferentially decodes motion information for further processing (Kaas et al. 2022).

In contrast, the ventral stream includes regions primarily located in ventral visual and inferotemporal cortex. This system receives slower but higher-resolution input primarily from the parvocellular system, which integrates cone cell input from the fovea region of retina. The parvocellular system, moreover, has relatively low sensitivity to low-contrast stimuli but does not saturate, allowing it to provide greater sensitivity to "shades of gray." Finally, the parvocellular system is preferentially sensitive to chromatic contrast and thus subserves color vision. The parvocellular input into the ventral visual stream thus contains information needed for detailed

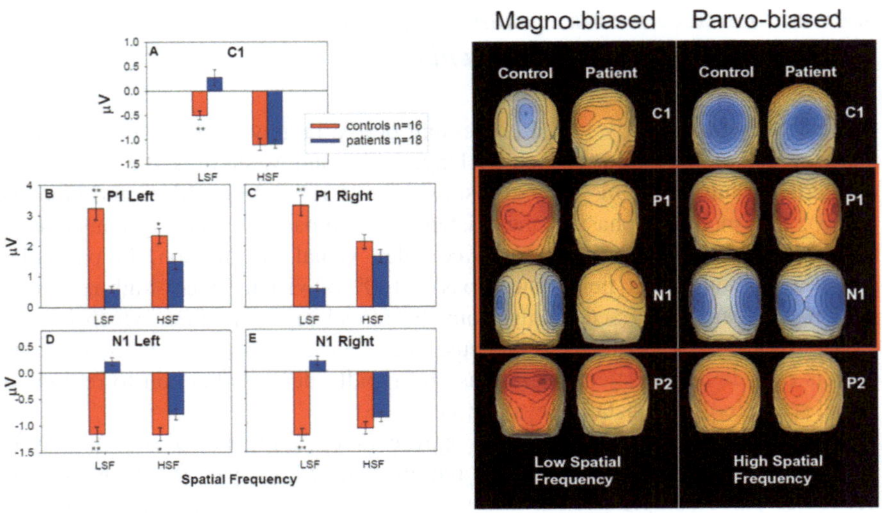

Fig. 9 Visual ERP responses to magno- vs. parvocellular-biased stimuli in control and schizophrenia individuals, showing reduced activation to magno-biased, low spatial frequency stimuli, but intact responses to parvo-biased, high spatial frequency stimuli. (From Butler et al. (2007). Reproduced with permission)

object recognition. The ventral stream thus has been termed the "what" or the "vision for identification" system (Milner and Goodale 2008).

The two systems can be interrogated separately using physiological measures such as event-related potentials (ERP) or fMRI responses to stimuli biased toward either the magnocellular or parvocellular pathways. A highly consistent finding across such studies (Butler et al. 2001, 2005, 2007, 2012; Javitt 2009b, 2015; Remy et al. 2023) is that individuals with schizophrenia show severe reductions in early visual responses to magnocellular vs. parvocellular biased stimuli. One of the most sensitive measures is the early P1 response to simple visual stimuli such as sinusoidally modulated "Gabor" patches that can be varied in contrast and spatial frequency.

In a representative study (Butler et al. 2007), ERPs were obtained in response to both low and high spatial frequency stimuli, which are biased toward the magnocellular and parvocelluar systems, respectively. Highly significant deficits were observed in the initial visual C1, P1, and N1 responses to low spatial-frequency stimuli, whereas responses to high spatial frequencies were intact (Fig. 9). Similarly, in imaging studies, reduced activation of early visual regions is observed in low-, but not high-, spatial frequency stimuli (Martinez et al. 2008, 2012a; Calderone et al. 2013). These findings complement behavioral results such as backward masking (Schechter et al. 2003; Slaghuis 2004), contrast sensitivity (Slaghuis 2004; O'Donnell et al. 2006), motion detection (Chen et al. 1999; Kim et al. 2006; Chen 2011), and global/local processing (Coleman et al. 2009; Johnson et al. 2005), which also document disturbances of early visual perceptual processing linked to magnocellular system dysfunction in schizophrenia.

3.2 *Fragmented Object Recognition*

Given the dependence of the dorsal visual stream on subcortical magnocellular, a predicted consequence of magnocellular system function is impaired function of the dorsal visual stream. One of the main roles of the dorsal stream visual system in reading is rapid "framing" of visual space to permit efficient use of processing resources. Such framing functions become increasingly important in the common situation in which a visual object is viewed in a fragmentary or ambiguous context, and non-visual brain regions must be recruited to help disambiguate the object. In such circumstances, the early visual system functions in a "frame and fill" mode in which rapid input to prefrontal brain regions via the magnocellular system activates stored templates of potential objects (the "frame") to which the actual sensory representation of the object is compared (the "fill"). Because of the differential timing of the magnocellular and parvocellular pathways, feedback to ventral stream object recognition from higher-order cortical regions coincides with the "feed forward sweep" of parvocellular information into ventral stream object recognition pathways, facilitating object recognition (Javitt 2009b; Kveraga et al. 2007; Sehatpour et al. 2008; Lamme and Roelfsema 2000; Tapia and Breitmeyer 2011).

An informative series of studies investigated the integrity of different aspects of this system in schizophrenia using a combination of fragmented images in which the identity of an object must be deduced based on partial information, as compared to responses to simpler stimuli termed Kanizsa figures in which illusory contours are formed by inducers located at the vertices of imaginary shapes (Fig. 10). In both cases, multimodal imaging studies were performed to evaluate neural-level processes underlying both intact and impaired function. The research focused on three separate ERP components: the early P1, which is thought to reflect initial activation within the magnocellular/dorsal stream pathway to visual stimuli; the early N1, which is thought to reflect the initial activation of the parvocellular/ventral visual

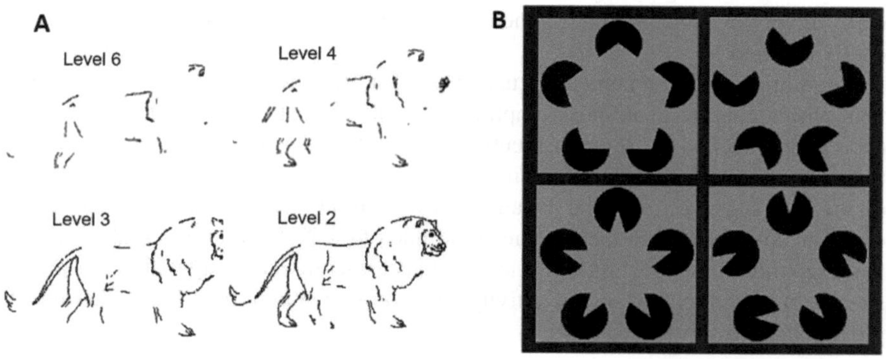

Fig. 10 Examples of (**a**) Fragmented objects at different levels of fragmentation. (From Doniger et al. (2001a)), (**b**) Kanizsa and control figures. (From Murray et al. (2002). Reproduced with permission)

stream; and the subsequent closure negativity (Ncl) component, which represents the joint activity within ventral stream cortex of magnocellular-driven recurrent information from higher brain regions combined with feedforward information into inferotemporal object recognition regions via the parvocellular visual stream.

An initial study evaluated the integrity of the closure process (Doniger et al. 2001a). In this study, patients were shown a series of increasingly defragmented stimuli, and the level at which they could identify the figures was recorded. On average, schizophrenia individuals required a nearly complete image in order to make the identification, which was significantly more complete than those required by controls. Nevertheless, individuals with schizophrenia were able to take advantage of two perceptual and cognitive manipulations. First, when a stimulus is repeated, individuals show much improved ability to identify it the second time even if they were not told to remember the image, reflecting automatic visual learning (Fig. 11a).

Second, if individuals are presented with a word (e.g., "lion") that potentially describes the object to be identified, when it does match the object (valid prompt), it greatly facilitates object recognition. By contrast, when it does not match the object (invalid prompt) or if a non-informative cue is used (e.g., a strong of X's), the identification level remains unchanged (Fig. 11b). In both instances, individuals with schizophrenia benefited from the manipulation equivalently to controls, suggesting that other elements of the object recognition circuit were unimpaired once allowance was made for the initial perceptual-level disturbance.

Closure Negativity As with auditory perceptual impairments, ERPs were employed to evaluate underlying neural mechanisms. For closure-related stimuli, two sets of ERP measures were of particular importance. First, as with simple visual stimuli (e.g., Gabors, Fig. 9), fragmented object stimuli elicited a P1 component that included both dorsal and ventral stream generators and an N1 component (~180 ms) that localized to the ventral stream (Fig. 12). These responses were not significantly affected by the level of fragmentation. As with simple visual stimuli, the P1 showed a highly significant reduction in amplitude in schizophrenia, especially involving dorsal stream generators, whereas N1 amplitude was not significantly affected.

In addition, the fragmented pictures elicited a "closure negativity" (Ncl) component with an approximate peak amplitude of 300 ms that increased in amplitude as objects approached the level of identification, with a discontinuous jump in amplitude at the level of identification. The scalp distribution of the Ncl was similar to that of the N1, suggesting similar localization to ventral stream regions of the lateral occipital complex. Nevertheless, the amplitude of the Ncl was substantially reduced in schizophrenia, suggesting dysfunction of late, re-entrant activity within the ventral stream, but relatively preserved activity elicited by the initial feed-forward sweep of parvocellular information.

Illusory Contour Processing Further insights into the integrity of the early visual system come from a parallel set of studies investigating illusory contour processing using a simpler type of stimuli termed "Kanizsa" figures. In this approach, inducers

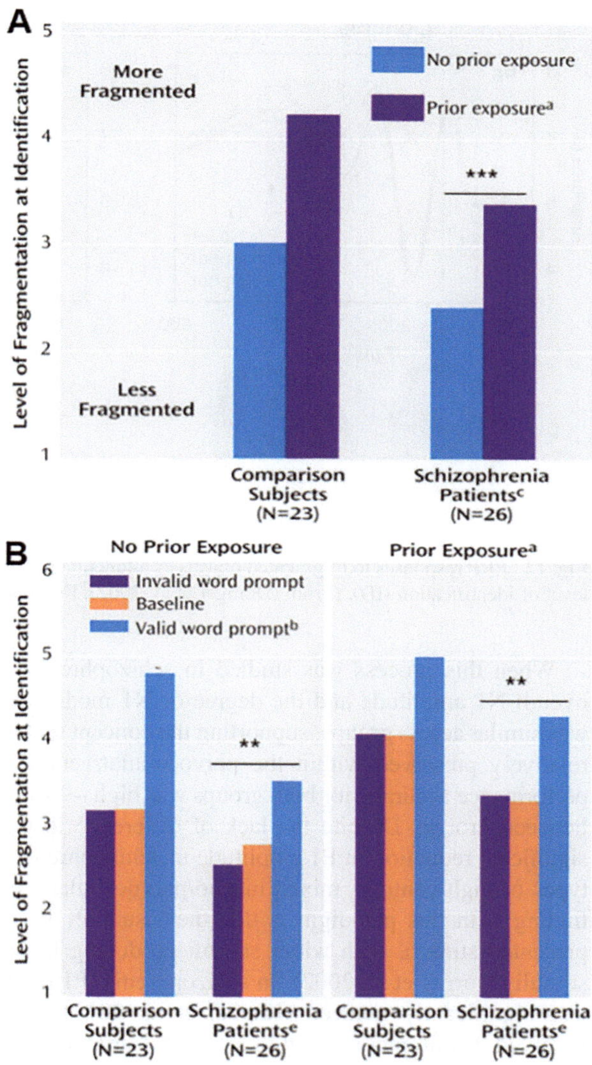

Fig. 11 Fragmented object identification in schizophrenia and control individuals. (**a**) Identification thresholds for first and repeat presentation. (**b**) Identification thresholds for objects preceded by valid vs. invalid prompts. $**p < 0.005$ schizophrenia vs. control across condition. $***p < 0.001$. (From Doniger et al. (2001a). Reproduced with permission)

are placed at the vertices of an imaginary shape, providing the strong illusion that a shape is present. As a control, the exact same inducers can be placed but in a rotated position, in which case no visual illusion is created. In order to obtain a behavioral read-out from the task, different illusory shapes are created by different inducer sets, and participants are asked to identify the shape, if present. In ERP studies, the presence vs. absence of the illusory shape produces a modulation of the visual N1 response, suggesting that the underlying process is localized to the ventral visual stream and depends only on the initial feedforward sweep.

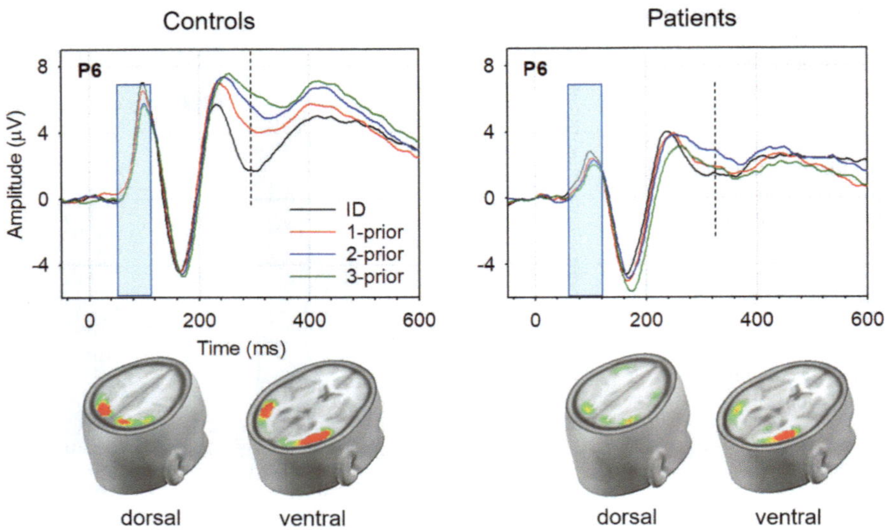

Fig. 12 ERP responses to fragmented objects as a function of level of fragmentation relative to the level of identification (ID). (From (Doniger et al. 2002). Reproduced with permission)

When this process was studied in schizophrenia (Foxe et al. 2005), both the overall N1 amplitude and the degree of N1 modulation by contour present/absent was similar across groups, supporting the concept that initial stages of processing are relatively preserved within the parvocellular/ventral stream system. In addition, performance accuracy for both groups was high ~90%, with no significant difference between groups. Despite the lack of difference in the N1 amplitude there was a significant reduction in P1 amplitude in schizophrenia similar to that seen for other types of high-contrast, mixed magno/parvocellular stimuli. Moreover, a secondary finding with this paradigm is that the visual P1 is modulated by the size of the presented stimuli, with wider stimuli producing larger P1 responses than narrow stimuli (Murray et al. 2002). In schizophrenia, P1 modulation by stimulus size was also significantly reduced (Knebel et al. 2011). The P1 modulations primarily involved recruitment of additional parietal structures within the dorsal stream visual system,

These findings were further refined by a modification of the illusory contour task in which the inducers are adjusted to create illusionary fat vs. thin shapes, which require additional processing beyond simply presence vs. absence of illusory shapes. Thus, the fat/thin distinction is indexed by activity within the Ncl rather than N1 latency range (Murray et al. 2004). When tested in these two conditions, schizophrenia patients showed intact illusory contour detection but impaired sensitivity to other stimulus features (Keane et al. 2014, 2019).

Intracranial Recording and Imaging Final information detailing the basis for perceptual closure comes from combined intracranial recording and neuroimaging studies. In the intracranial study (Sehatpour et al. 2008), ERP recordings were made from fragmented and unfragmented objects from 3 individuals with implanted electrodes for pre-surgical mapping of epileptogenic foci. Consistent with the surface activity, these recordings confirmed the localization of both the N1 and Ncl components to the ventral visual stream. In addition, the study revealed major nodes in lateral prefrontal cortex and hippocampus that relied on long-range synchronous beta (14–30 Hz) activity. These findings thus support a distributed network model in which rapid, magnocellularly driven projections to lateral prefrontal cortex activate stored templates in both prefrontal cortex and hippocampus to which the parvocellularly driven representation in lateral occipital cortex is compared.

If a “match” in configuration is observed, it permits object identification to be made at a relatively fragmented level (Sehatpour et al. 2008). Breakdown of this process in schizophrenia could thus lead to the observed deficits in perceptual closure. Although no word prompts were included in this study, a likely mechanism by which word prompts facilitate perceptual closure is by activating hippocampal representations of objects via a linguistic route, rather than through rapid forward projection of the low-resolution image through the dorsal visual stream.

In an initial fMRI study of the closure process, significant reductions were observed across all nodes of the network to fragmented but closeable images relative to scrambled versions that were non-closeable. In a path analysis, influences of the disease process were observed for the dorsal stream and frontal nodes only, whereas activation deficits within hippocampus and ventral visual stream were secondary to these reductions (Sehatpour et al. 2020). In a subsequent multimodal imaging study (Sehatpour et al. 2020), ERP and fMRI were obtained for both initial and repeated stimuli across levels of fragmentation. In contrast, for Kanizsa figures, the presence or absence of illusory contours modulates the initial N1 response, supporting the concept that the two processes differ in their dependence on feed-forward- vs. recurrent processing within the ventral visual stream (Doniger et al. 2001b). When ERPs were conducted on initial vs. repeat stimuli, a similar shift was observed in timing, suggesting that identification of repeat stimuli may also depend only on feedforward information within ventral cortical regions (Doniger et al. 2001b).

In a second ERP/fMRI study (Sehatpour et al. 2020), responses were assessed to both initial and repeat stimuli. In the ERP component, individuals with schizophrenia again showed reduced ability to identify fragmented objects as reflected in an increased threshold for identification. Patient showed impaired Ncl responses to initial stimuli but intact N1 modulation to repeat stimuli, supporting the concept that perceptual-level functions that depend only on the feedforward sweep within the ventral visual stream are intact, but those that depend upon additional recurrent processing driven via the dorsal visual stream are impaired.

In the fMRI component, significant reductions in activity were observed to first-time stimuli in both the dorsal and ventral visual stream nodes as well as the lateral prefrontal cortex, suggesting impaired activation of the distributed network. During repeat stimulation, activation over dorsal and ventral stream regions was normalized in schizophrenia, and prefrontal cortical activity was reduced for both groups, although the schizophrenia vs. control difference remained significant. Hippocampal

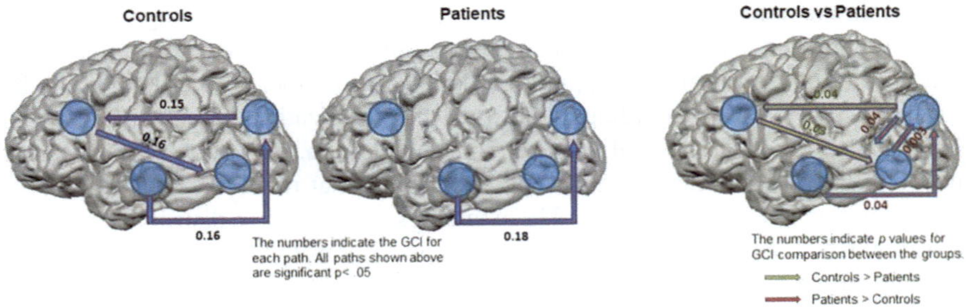

Fig. 13 Granger causality analysis of fragmented object closure in schizophrenia. (From Sehatpour et al. (2020). Reproduced with permission)

activity was similar in amplitude in both groups during the initial presentation, but substantially increased in schizophrenia on repeat stimulation, suggesting that individuals with schizophrenia may over-rely on the hippocampus to compensate for deficits elsewhere in the circuit (Fig. 13).

In a Granger causality analysis, significant causality was observed in controls in a pathway leading from dorsal stream visual cortex to lateral prefrontal cortex and then to ventral stream cortex. An additional path was observed from hippocampus to the dorsal visual stream. In schizophrenia, only the hippocampal-dorsal stream causality path showed significant activity. When differences were mapped across groups, a significant deficit was observed in schizophrenia in recruitment of the dorsal stream-frontal-ventral stream pathway. In contrast, schizophrenia was associated with over-recruitment of the hippocampal-dorsal stream pathway as well as bidirectional interaction between dorsal and ventral regions, indicating both deficits and compensations within the early perceptual system in schizophrenia. In healthy volunteers, transcranial magnetic stimulation (TMS) over the dorsal visual stream produces schizophrenia-like disruption of perceptual closure (Amiaz et al. 2016), further supporting dorsal stream dysfunction as a basis for perceptual closure impairments in schizophrenia.

Clinical Correlations Deficits in closure ability correlate consistently with the severity of negative symptoms across studies (Doniger et al. 2001a, 2002; Sehatpour et al. 2020). In addition, reductions in P1 amplitude and dorsal stream activation have been found to correlate with deficits in processing speed and working memory (Sehatpour et al. 2020). In addition, it has recently been reported that during shape completion tasks, individuals with schizophrenia overrecruit the dorsal attention network, which mediates goal-directed ("top down") selection for stimuli and selection of responses (Corbetta and Shulman 2002), while underrecruiting the secondary visual areas (Keane et al. 2023). The degree of over-recruitment, moreover, correlates with the severity of cognitive disorganization (Keane et al. 2023).

Overall, individuals' ability to identify objects in the real world depends upon activity within the ventral stream that can be indexed with ERP or fMRI measures. However, these processes depend upon recurrent circuits in the brain to make use of stored memories to assist in object identification and characterization. Those recurrent processes, in turn, depend upon rapid, low-resolution information transmitted to frontal regions via the dorsal visual stream. These processes thus highlight how breakdown of these networks, which operate outside of conscious awareness, may nevertheless contribute to the impoverished everyday experience of individuals with schizophrenia, as reflected in negative symptoms. Moreover, the need to overrecruit attentional systems to compensate for the failures in lower-level perceptual systems may contribute to the conceptual disorganization associated with the disorder.

3.3 Face Emotion Recognition and Third Visual Pathway Dysfunction

In addition to its role in various neurocognitive processes such as closure, visual perceptual information is also critical for multiple aspects of social cognition, especially face emotion recognition (FER), which is severely impaired in schizophrenia. FER is studied most commonly using tests such as the Penn Emotion Recognition – 40 faces (ER40) test, which uses static pictures of actors with posed facial expressions (Kohler et al. 2000). Schizophrenia is associated with reliable deficits in FER, which correlate with the severity of negative symptoms (Gao et al. 2021) and contribute to poor functional outcomes (Rassovsky et al. 2011; Abplanalp et al. 2023). When neural substrates of FER are assessed in schizophrenia using measures such as event-related potentials or fMRI, deficits are observed across a distributed network involving visual, limbic, and prefrontal regions (Li et al. 2010; Taylor et al. 2012; Delvecchio et al. 2013). Several lines of evidence, however, point to significant contributions of lower-level visual deficits to FER impairments, particularly in relation to magnocellular dysfunction.

First, FER deficits correlate with reduced contrast sensitivity to low-, but not high-, spatial frequency visual stimuli (Butler et al. 2009). Similarly, when faces are themselves filtered, FER deficits are differentially impaired to faces containing low, but not high, spatial-frequency information (Jahshan et al. 2017). In ERP studies, deficits are observed in the generation of the early visual P1 component to faces (Earls et al. 2016) along with the subsequent N170 and P250 components (Gao et al. 2021; McCleery et al. 2015). Furthermore, in an ERP study, when faces were manipulated to include low- vs. high-spatial frequency information, the P1 deficits were selective for the low spatial-frequency faces (Kim et al. 2015).

Second, in a study that used a contrast manipulation to bias activation toward the magnocellular vs. parvocellular visual pathway, deficits in FER as a function of contrast directly paralleled deficits in contrast sensitivity and impaired steady-state VEP elicited by magnocellular-biased stimuli (Butler et al. 2009). Thus, in this study, individuals showed reduced contrast sensitivity specifically to low spatial frequency stimuli (Fig. 14a), which correlated significantly with impaired FER performance.

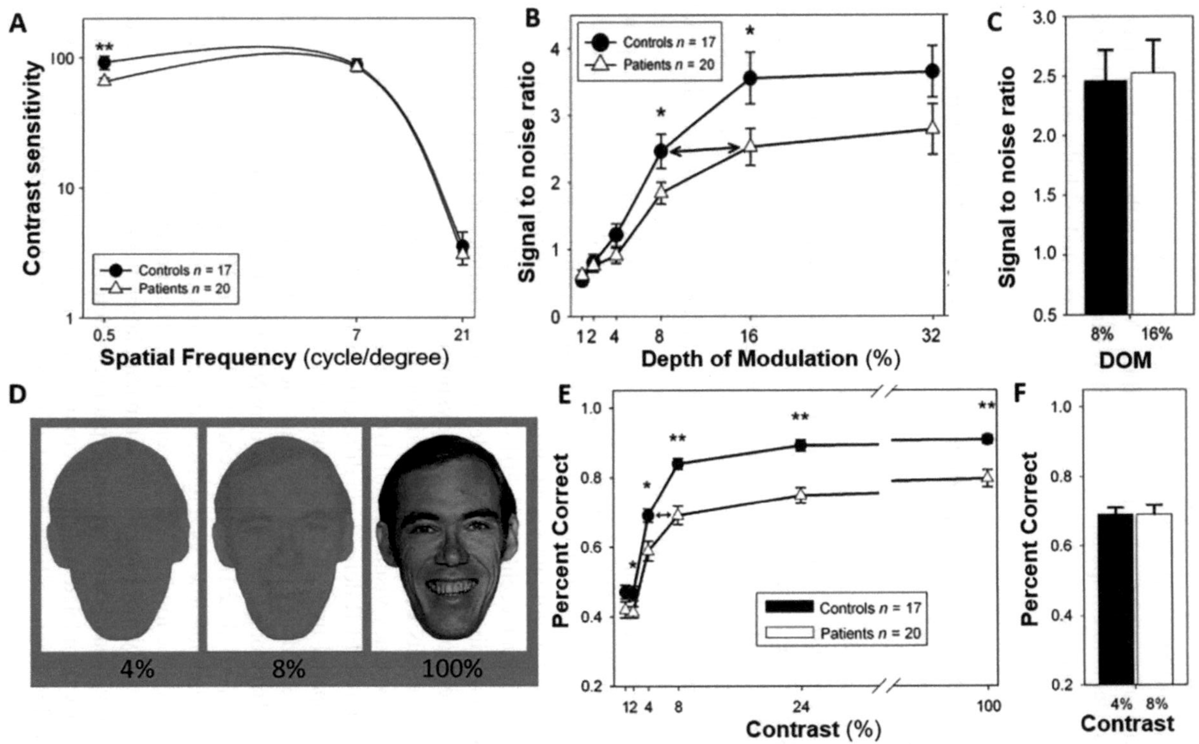

Fig. 14 FER deficits in schizophrenia as a function of luminance contrast. (**a**) Contrast sensitivity by groups, showing deficits to low spatial frequency stimuli. (**b**) ssVEP responses by depth of modulation (DOM). (**c**) DOM levels associated with equivalent ssVEP responses. (**d**) Contrast-modified emotional face. (**e**) FER performance by contrast level. (**f**) Contrast levels associated with equivalent performance. (From Butler et al. (2009). Reproduced with permission)

As had been observed previously (Butler et al. 2007), schizophrenia was also associated with a reduced ssVEP responses to magnocellular-biased stimuli, as reflected in reduced slope and plateau of the non-linear gain curves (Fig. 14b), as a result, the mean ssVEP response in schizophrenia to stimuli with 16% depth of modulation (i.e., ±8% contrast against a 0% contrast background) showed a response that was numerically equivalent to that of the control response to an 8% depth of modulation (Fig. 14c).

In order to investigate the interaction between these magnocellular ssVEP deficits and FER deficits, the emotional faces used for the FER study were then manipulated to have mean contrast levels of either 4%, 8%, 24%, or 100% (Fig. 14d). As expected, FER performance improved in both groups with increasing contrast. However, rather than showing continued improvement with increasing contrast as would be expected from a parvocellular-mediated function, the response curves showed a non-linear gain function characteristic of magnocellular function. Thus, in both groups, performance plateaued at 8% contrast with only marginal improvement thereafter. Moreover, as with the ssVEP function, individuals with schizophrenia showed a lower gain with increasing contrast and lower plateau level of performance (Fig. 14e).

Finally, as with the ssVEP function, schizophrenia individuals performed as well at 8% contrast as controls did with 4% contrast. These studies thus highlight the fact that even though many aspects of face processing (e.g., gender or age discrimination) benefit from high stimulus contrast levels that enable greater parvocellular engagement, the brain regions that process facial emotion do not appear to have access to this information. Thus, FER plateaus at low, magnocellular-selective contrast levels even as other aspects of face processing may continue to improve.

Third Visual Pathway Further insights into the physiological basis of these findings come from the recently elaborated "Third visual pathway" model of Pitcher and Ungerleider (Pitcher and Ungerleider 2021; Pitcher 2021). This model revises the original dorsal/ventral stream dichotomy to incorporate a third, primate-specific visual stream that is specialized for social cognitive processing, including FER. A unique feature of the third visual pathway is that it receives input preferentially through the magnocellular visual system and especially motion-sensitive visual regions within mid-temporal cortex (MT+ complex). Key nodes in this system include the posterior (pSTS), which is specifically involved in FER, and anterior (aSTS), which processes more complex social cognitive information (e.g., interactions among individuals). Thus, this system is preferentially sensitive to motion vs. static faces, as are encountered during normal human social interaction.

Deficits in pSTS dysfunction have been extensively documented in schizophrenia, especially in response to biological motion (Jimenez et al. 2018; Kim et al. 2011; Thakkar et al. 2014) and faces (Jimenez et al. 2018). In addition, recent fMRI studies provide direct evidence that failures of motion processing at the level of V1 and MT + contribute directly to impaired third visual pathway activation. Thus, in one study, motion processing thresholds and MT+ activation were assessed along with FER across individuals with schizophrenia and clinical high-risk individuals. Significant deficits in motion processing were observed across the two groups and correlated with impaired FER (Martinez et al. 2018). These results were further confirmed in a second study in which both motion sensitivity and motion-induced ERP measures were obtained across a sample of schizophrenia and autism spectrum disorder (ASD) individuals (Martinez et al. 2019).

In a third study (Martinez et al. 2021), fMRIs were obtained from static and dynamic emotional faces in schizophrenia. As predicted, schizophrenia was associated with reduced activation in both sensory regions (V1, MT+) and both anterior and posterior regions of pSTS. In a hierarchical regression analysis, deficits in sensory-level activation predicted impaired pSTS activation, which in turn predicted impaired FER. Deficits were also observed in the activation of subcortical nodes of the early visual system, especially pulvinar nucleus.

Most recently, fMRI studies with implied vs. real motion suggest that MT+ serves as the "gatekeeper" to the third visual pathway even to static faces (Martinez et al. 2024). Specifically, MT+ activation is observed even to static stimuli containing implied motion based upon the pose and configuration of objects in the picture. In such cases, MT+ activation appears to be driven "top down" via the mirror neuron system rather than "bottom up" through the magnocellular visual system and primary visual cortex. In schizophrenia, MT+ and pSTS activation deficits were observed in both implied and real motion, suggesting that impaired input to the third visual pathway via MT+ may be a primary deficit underlying FER and similar social cognitive impairments in schizophrenia.

4 Clinical Implications

At present, although perceptual deficits in schizophrenia are widely documented at the research level, they have not yet been integrated into the clinical management of the disorder. This "disconnect" between the clinical and research management stems, in part, from the traditional view that perceptual-level processes in schizophrenia are intact and in part because of the lack of standardized assessment batteries for perceptual-level assessment. This assessment issue is further exacerbated by the fact that much neurocognitive testing in schizophrenia continues to rely on "paper and pencil" type assessments, which cannot be used to assess perceptual-level processes. However, as testing increasingly migrates to tablet- and visor-based systems, the development of standardized perceptual-level assessment may be possible.

4.1 "Real World" Functions

Many tests of perceptual-level dysfunction, by necessity, use custom paradigms to isolate specific physiological processes. Thus, the use of the MMN oddball paradigm in the auditory system provides a readout of tone comparison processes, even though such sequences rarely occur in nature. Similarly, visual perceptual deficits can only be characterized using stimuli that isolate the differential properties of the magnocellular and parvocellular visual systems, although most "real world" stimuli contain a mixture of properties. Nevertheless, knowledge of these deficits is critical to understanding the drivers of impaired real-world function. Two such functions, reading and "mentalization," have been extensively studied and illustrate the critical clinical issues.

Reading The ability to read fluently and efficiently is an important contributor to occupational success in modern societies. Individuals with schizophrenia show relatively preserved premorbid reading ability, as reflected in their ability to correctly pronounce irregular "sight words" (e.g., knight, heresy). Nevertheless, when tested using batteries that assess their ability to process standard written texts, such as word passages, individuals with schizophrenia show a significant gap between their pre-morbid and current reading abilities, corresponding to a loss of 4 or more years of grade-equivalent function. Deficits in reading ability, moreover, correlate highly with global functional outcome (Donde et al. 2019c; Revheim et al. 2014; Carrion et al. 2015).

Given the extensive literature on neurocognitive impairments in schizophrenia, the default high-level neurocognitive hypothesis is that processes such as comprehension should be most affected, whereas more mechanistic aspects of reading should be relatively intact. However, when tested using detailed batteries, greatest deficits are observed in aspects such as rate and accuracy of word production, rather than higher-level functions such as comprehension (Donde et al. 2019c; Revheim et al. 2014). These deficits, in turn, are linked to disturbances in low-level auditory functions, such as the "sound out" of individual speech sounds ("phonemic awareness"), and visual processes, such as the ability to tightly control eye movements ("oculomotor programming") to land in the correct position on words (Dias et al. 2021).

Neural mechanisms underlying phonemic and oculomotor deficits, in turn, can be interrogated using physiological measures such as ERP or fMRI. In the case of phonemic processing, deficits correlate highly with impaired tone matching ability and MMN generation either to simple stimuli (e.g., pitch, duration deviants) (Donde et al. 2019c; Revheim et al. 2014) or to individual phonemes (Kawakubo et al. 2007; Fisher et al. 2019). In the case of oculomotor programming, deficits correlate with impaired magnocellular function and, specifically, the ability to use parafoveal information to "program" eye movements, leading to a markedly elevated rate of refixations during reading (Dias et al. 2021; Martinez et al. 2012b). At present, fluent (i.e., passage) reading ability is rarely tested in schizophrenia nor considered as a potential driver of impaired occupational outcome. If tested, remediation programs such as those used to treat developmental dyslexia may be effective.

Mentalization The ability to understand what other people are thinking and to tailor one's behavior accordingly is a critical component of daily life. This process is captured in the RDoC construct of "Understanding mental states" and has also been termed mentalization or “theory of mind.” Deficits in social cognition are strong drivers of impaired functional outcomes in schizophrenia. Although neurocognitive deficits have traditionally been viewed as primary drivers of impaired social cognition, recent Bayesian network analyses suggest dissociation between the two sets of processes (Abplanalp et al. 2023). Moreover, recent studies using The Awareness of Social Inference Test (TASIT), which is a video-based assessment of mentalizing ability, have demonstrated the importance of low-level deficits in motion detection, gaze control, and auditory sarcasm detection on overall function (Patel et al. 2020, 2021, 2023). These studies complement prior literature tying impaired mentalization ability to deficits in low-level auditory (Leitman et al. 2006; Kantrowitz et al. 2014) and visual (Rassovsky et al. 2011; Green et al. 2012; Sergi and Green 2003; Sergi et al. 2006), and emphasize the consequences of sensory-level dysfunction to overall functional outcome.

4.2 Sensory-Informed Cognitive Remediation and Clinical Intervention

Given the increasing evidence for contributions of sensory-level dysfunction to overall functional outcome, a critical issue is whether targeting these deficits can lead to improved function. Two series of studies emphasize the importance of assessment and remediation of sensory level dysfunction. The first used a computerized cognitive training program designed to address deficits in basic auditory processing. This training resulted in significant gains in both auditory perception and verbal learning, along with a reduction in hallucinations (Thomas et al. 2018), with durable effects over time (Loewy et al. 2022).

The second evaluated auditory abilities at baseline and then randomized individuals to receive auditory training that either did or did not include basic auditory training modules. Individuals with pre-existing auditory perceptual deficits showed cognitive gains only when randomized to receive sensory level. Moreover, sensory-level function predicted improvements in both verbal learning ability and functional outcome (Medalia et al. 2019). Overall, these findings highlight the need to consider perceptual-level dysfunction in designing cognitive remediation approaches for schizophrenia.

Sensory brain regions may also serve as important targets for neuromodulatory intervention using non-invasive brain stimulation approaches such as transcranial direct current stimulation (tDCS). Thus, several studies have demonstrated that tDCS targeted at left auditory cortex leads to significant improvement in persistent auditory hallucinations (Brunelin et al. 2012; Kantrowitz et al. 2019; Yang et al. 2019). Similarly, tDCS targeted at visual cortex may improve both motion detection ability (Pobric et al. 2018) and visuomotor processing (Sehatpour et al. 2023). Future studies are therefore required, combining both cognitive remediation and neuromodulatory approaches to plasticity enhancement.

4.3 Translational Biomarkers

Finally, NMDAR dysfunction or dysregulation remains the presumptive cause of both symptoms and neurocognitive deficits associated with schizophrenia (Javitt 2023; Javitt and Zukin 1991). Nevertheless, there are at present no approved treatments that target NMDAR, in part because of the lack of biomarkers that can be used for early-stage drug development. Perceptual level measures have proven highly translatable across species. Thus, for example, MMN can be implemented in both rodent (Lee et al. 2018) and primate models and has shown to be sensitive to NMDAR antagonist administration (Javitt et al. 1996; Lakatos et al. 2020; Lee et al. 2018; Gil-da-Costa et al. 2013). Visual measures have been last studied from a neurochemical perspective. Nevertheless, visual MMN has been successfully implemented in rodents and shown to depend on both NMDAR function and somatostatin interneuron function (Ross and Hamm 2020; Van Derveer et al. 2021), mimicking the deficits in schizophrenia. Similarly, schizophrenia-like alterations in visual perceptual grouping may be induced in rodents by ketamine administration, supporting underlying NMDAR involvement (Kurylo et al. 2007; Kurylo and Gazes 2008).

In general, perceptual level function is preserved across rodents, primates, and humans more strongly than higher-order processes and, moreover, can be evaluated even in untrained animals. Oscillatory structure is also well preserved across species despite marked differences in brain size (Buzsaki et al. 2013). As a result, perceptual level measures such as MMN are increasingly being used as translational biomarkers during preclinical and early-stage treatment development (O'Donnell et al. 2023; Sehatpour et al. 2022). Thus, consideration of perceptual-level functioning may aid in both the characterization and future treatment of schizophrenia.

5 Conclusion

Perceptual-level processes were once viewed as simple, intact functions in schizophrenia but increasingly are viewed as a key component of the disorder. Deficits in perceptual processing have been extensively documented in both the auditory and visual systems in schizophrenia and contribute to both neuro- and social cognitive dysfunction, as well as overall functional outcome. The specific pattern of perceptual-level deficit, moreover, may help differentiate schizophrenia from other cognitive disorders and may help identify pathophysiologically distinct subtypes with the overall schizophrenia syndrome. Tools for sensory-level assessment when available are considered by clinicians to provide useful information (Medalia et al. 2023). The development of readily accessible tools for both auditory and visual function must therefore be viewed as a research priority, which can help guide both personalized remediation and clinical intervention approaches. Neuromodulatory approaches targeted at perceptual level dysfunction may be of clinical benefit even in advance of improved pharmacological intervention approaches.

Acknowledgments Supported in part by grants R01MH49334, R01MH121449, and R01MH134118 to DCJ.

References

Abplanalp SJ, Lee J, Horan WP, Kern RS, Penn DL, Green MF (2023) A Bayesian network approach to social and nonsocial cognition in schizophrenia: are some domains more fundamental than others? Schizophr Bull 49(4):997–1006

Abplanalp SJ, Braff DL, Light GA, Joshi YB, Nuechterlein KH, Green MF et al (2024) Clarifying directional dependence among measures of early auditory processing and cognition in schizophrenia: leveraging Gaussian graphical models and Bayesian networks. Psychol Med 1–10

Amiaz R, Vainiger D, Gershon AA, Weiser M, Lavidor M, Javitt DC (2016) Applying Transcranial Magnetic Stimulation (TMS) over the dorsal visual pathway induces schizophrenia-like disruption of perceptual closure. Brain Topogr 29(4):552–560

Ardizzi M, Ambrosecchia M, Buratta L, Ferri F, Peciccia M, Donnari S et al (2016) Interoception and positive symptoms in schizophrenia. Front Hum Neurosci 10:379

Avissar M, Xie S, Vail B, Lopez-Calderon J, Wang Y, Javitt DC (2018) Meta-analysis of mismatch negativity to simple versus complex deviants in schizophrenia. Schizophr Res 191:25–34

Berntson GG, Khalsa SS (2021) Neural circuits of interoception. Trends Neurosci 44(1):17–28

Billingberg O, Jonsson CO (1965) The ability of schizophrenic patients to interpret intonation. Acta Psychiatr Scand 41(2):218–226

Bleuler E (1950) In: Zinkin J (ed) Dementia praecox of the group of the schizophrenias. International Universities Press, New York

Broadbent DE (1957) A mechanical model for human attention and immediate memory. Psychol Rev 64(3):205–215

Brunelin J, Mondino M, Gassab L, Haesebaert F, Gaha L, Suaud-Chagny MF et al (2012) Examining transcranial direct-current stimulation (tDCS) as a treatment for hallucinations in schizophrenia. Am J Psychiatry 169(7):719–724

Busch A (1904) The power of comprehension and observation in Dementia Praecox. Psychol Arbeiten

Butler PD, Schechter I, Zemon V, Schwartz SG, Greenstein VC, Gordon J et al (2001) Dysfunction of early-stage visual processing in schizophrenia. Am J Psychiatry 158(7):1126–1133

Butler PD, Zemon V, Schechter I, Saperstein AM, Hoptman MJ, Lim KO et al (2005) Early-stage visual processing and cortical amplification deficits in schizophrenia. Arch Gen Psychiatry 62(5):495–504

Butler PD, Martinez A, Foxe JJ, Kim D, Zemon V, Silipo G et al (2007) Subcortical visual dysfunction in schizophrenia drives secondary cortical impairments. Brain 130(Pt 2):417–430

Butler PD, Abeles IY, Weiskopf NG, Tambini A, Jalbrzikowski M, Legatt ME et al (2009) Sensory contributions to impaired emotion processing in schizophrenia. Schizophr Bull 35(6): 1095–1107

Butler PD, Chen Y, Ford JM, Geyer MA, Silverstein SM, Green MF (2012) Perceptual measurement in schizophrenia: promising electrophysiology and neuroimaging paradigms from CNTRICS. Schizophr Bull 38(1):81–91

Buzsaki G, Logothetis N, Singer W (2013) Scaling brain size, keeping timing: evolutionary preservation of brain rhythms. Neuron 80(3):751–764

Calderone DJ, Martinez A, Zemon V, Hoptman MJ, Hu G, Watkins JE et al (2013) Comparison of psychophysical, electrophysiological, and fMRI assessment of visual contrast responses in patients with schizophrenia. NeuroImage 67:153–162

Carrion RE, Cornblatt BA, McLaughlin D, Chang J, Auther AM, Olsen RH et al (2015) Contributions of early cortical processing and reading ability to functional status in individuals at clinical high risk for psychosis. Schizophr Res 164(1–3):1–7
Caspers J, Palomero-Gallagher N, Caspers S, Schleicher A, Amunts K, Zilles K (2015) Receptor architecture of visual areas in the face and word-form recognition region of the posterior fusiform gyrus. Brain Struct Funct 220(1):205–219
Chao LL, Knight RT (1995) Human prefrontal lesions increase distractability to irrelevant sensory inputs. Neuroreport 6:1605–1610
Chaturvedi R, Kraus M, Keefe RSE (2020) A new measure of authentic auditory emotion recognition: application to patients with schizophrenia. Schizophr Res 222:450–454
Chen Y (2011) Abnormal visual motion processing in schizophrenia: a review of research progress. Schizophr Bull 37(4):709–715
Chen Y, Nakayama K, Levy DL, Matthysse S, Holzman PS (1999) Psychophysical isolation of a motion-processing deficit in schizophrenics and their relatives and its association with impaired smooth pursuit. Proc Natl Acad Sci USA 96(8):4724–4729
Coleman MJ, Cestnick L, Krastoshevsky O, Krause V, Huang Z, Mendell NR et al (2009) Schizophrenia patients show deficits in shifts of attention to different levels of global-local stimuli: evidence for magnocellular dysfunction. Schizophr Bull 35(6):1108–1116
Corbetta M, Shulman GL (2002) Control of goal-directed and stimulus-driven attention in the brain. Nat Rev Neurosci 3(3):201–215
Cowan N (1984) On short and long auditory stores. Psychol Bull 96(2):341–370
Cowan N (1988) Evolving conceptions of memory storage, selective attention, and their mutual constraints within the human information-processing system. Psychol Bull 104:163–191
Cowan N, Winkler I, Teder W, Naatanen R (1993) Memory prerequisites of mismatch negativity in the auditory event- related potential (ERP). J Exp Psychol 19(4):909–921
Critchley HD, Sherrill SP, Ewing DL, van Praag CG, Habash-Bailey H, Quadt L et al (2023) Cardiac interoception in patients accessing secondary mental health services: a transdiagnostic study. Auton Neurosci 245:103072
Delvecchio G, Sugranyes G, Frangou S (2013) Evidence of diagnostic specificity in the neural correlates of facial affect processing in bipolar disorder and schizophrenia: a meta-analysis of functional imaging studies. Psychol Med 43(3):553–569
Dias EC, Sheridan H, Martinez A, Sehatpour P, Silipo G, Rohrig S et al (2021) Neurophysiological, oculomotor, and computational modeling of impaired reading ability in schizophrenia. Schizophr Bull 47(1):97–107
Donde C, Luck D, Grot S, Leitman DI, Brunelin J, Haesebaert F (2017) Tone-matching ability in patients with schizophrenia: a systematic review and meta-analysis. Schizophr Res 181:94–99
Donde C, Martinez A, Kantrowitz JT, Silipo G, Dias EC, Patel GH et al (2019a) Bimodal distribution of tone-matching deficits indicates discrete pathophysiological entities within the syndrome of schizophrenia. Transl Psychiatry 9(1):221
Donde C, Avissar M, Weber MM, Javitt DC (2019b) A century of sensory processing dysfunction in schizophrenia. Eur Psychiatry 59:77–79
Donde C, Martinez A, Sehatpour P, Patel GH, Kraut R, Kantrowitz JT et al (2019c) Neural and functional correlates of impaired reading ability in schizophrenia. Sci Rep 9(1):16022
Donde C, Kantrowitz JT, Medalia A, Saperstein AM, Balla A, Sehatpour P et al (2023) Early auditory processing dysfunction in schizophrenia: mechanisms and implications. Neurosci Biobehav Rev 148:105098
Doniger GM, Silipo G, Rabinowicz EF, Snodgrass JG, Javitt DC (2001a) Impaired sensory processing as a basis for object-recognition deficits in schizophrenia. Am J Psychiatry 158(11):1818–1826
Doniger GM, Foxe JJ, Schroeder CE, Murray MM, Higgins BA, Javitt DC (2001b) Visual perceptual learning in human object recognition areas: a repetition priming study using high-density electrical mapping. NeuroImage 13(2):305–313

Doniger GM, Foxe JJ, Murray MM, Higgins BA, Javitt DC (2002) Impaired visual object recognition and dorsal/ventral stream interaction in schizophrenia. Arch Gen Psychiatry 59(11):1011–1020

Dykstra AR, Koh CK, Braida LD, Tramo MJ (2012) Dissociation of detection and discrimination of pure tones following bilateral lesions of auditory cortex. PLoS One 7(9):e44602

Earls HA, Curran T, Mittal V (2016) Deficits in early stages of face processing in schizophrenia: a systematic review of the P100 component. Schizophr Bull 42(2):519–527

Fisher DJ, Rudolph ED, Ells EML, Knott VJ, Labelle A, Tibbo PG (2019) Mismatch negativity-indexed auditory change detection of speech sounds in early and chronic schizophrenia. Psychiatry Res Neuroimaging 287:1–9

Foxe JJ, Murray MM, Javitt DC (2005) Filling-in in schizophrenia: a high-density electrical mapping and source-analysis investigation of illusory contour processing. Cereb Cortex 15(12):1914–1927

Gaebler AJ, Zweerings J, Koten JW, Konig AA, Turetsky BI, Zvyagintsev M et al (2020) Impaired Subcortical Detection of Auditory Changes in Schizophrenia but Not in Major Depression. Schizophr Bull 46(1):193–201

Gao Z, Zhao W, Liu S, Liu Z, Yang C, Xu Y (2021) Facial emotion recognition in Schizophrenia. Front Psychiatry 12:633717

Gil-da-Costa R, Stoner GR, Fung R, Albright TD (2013) Nonhuman primate model of schizophrenia using a noninvasive EEG method. Proc Natl Acad Sci USA 110(38):15425–15430

Gold R, Butler P, Revheim N, Leitman DI, Hansen JA, Gur RC et al (2012) Auditory emotion recognition impairments in schizophrenia: relationship to acoustic features and cognition. Am J Psychiatry 169(4):424–432

Gong B, Li Q, Zhao Y, Wu C (2021) Auditory emotion recognition deficits in schizophrenia: a systematic review and meta-analysis. Asian J Psychiatr 65:102820

Green M, Walker E (1986) Symptom correlates of vulnerability to backward masking in schizophrenia. Am J Psychiatry 143(2):181–186

Green MF, Nuechterlein KH, Mintz J (1994) Backward masking in schizophrenia and mania. I. Specifying a mechanism. Arch Gen Psychiatry 51(12):939–944

Green MF, Nuechterlein KH, Breitmeyer B (1997) Backward masking performance in unaffected siblings of schizophrenic patients. Evidence for a vulnerability indicator. Arch Gen Psychiatry 54(5):465–472

Green MF, Hellemann G, Horan WP, Lee J, Wynn JK (2012) From perception to functional outcome in schizophrenia: modeling the role of ability and motivation. Arch Gen Psychiatry 69(12):1216–1224

Green MF, Horan WP, Lee J (2019) Nonsocial and social cognition in schizophrenia: current evidence and future directions. World Psychiatry 18(2):146–161

Hari R, Hamalainen M, Ilmoniemi R, Kaukoranta E, Reinikainen K, Salminen J et al (1984) Responses of the primary auditory cortex to pitch changes in a sequence of tone pips: neuromagnetic recordings in man. Neurosci Lett 50(1–3):127–132

Holcomb HH, Ritzl EK, Medoff DR, Nevitt J, Gordon B, Tamminga CA (1995) Tone discrimination performance in schizophrenic patients and normal volunteers: impact of stimulus presentation levels and frequency differences. Psychiatry Res 57(1):75–82

Insel T, Cuthbert B, Garvey M, Heinssen R, Pine DS, Quinn K et al (2010) Research domain criteria (RDoC): toward a new classification framework for research on mental disorders. Am J Psychiatry 167(7):748–751

Iversen SD, Mishkin M (1973) Comparison of superior temporal and inferior prefrontal lesions on auditory and non-auditory tasks in rhesus monkeys. Brain Res 55:355–367

Jahshan C, Wynn JK, Green MF (2013) Relationship between auditory processing and affective prosody in schizophrenia. Schizophr Res 143(2–3):348–353

Jahshan C, Wolf M, Karbi Y, Shamir E, Rassovsky Y (2017) Probing the magnocellular and parvocellular visual pathways in facial emotion perception in schizophrenia. Psychiatry Res 253:38–42

Javitt DC (2009a) Sensory processing in schizophrenia: neither simple nor intact. Schizophr Bull 35(6):1059–1064
Javitt DC (2009b) When doors of perception close: bottom-up models of disrupted cognition in schizophrenia. Annu Rev Clin Psychol 5:249–275
Javitt DC (2015) Meeting overview: sensory perception and schizophrenia, Lausanne, Switzerland June 31–July 1, 2014. Schizophr Res Cogn 2(2):42–45
Javitt DC (2023) Cognitive impairment associated with schizophrenia: from pathophysiology to treatment. Annu Rev Pharmacol Toxicol 63:119–141
Javitt DC, Freedman R (2015) Sensory processing dysfunction in the personal experience and neuronal machinery of schizophrenia. Am J Psychiatry 172(1):17–31
Javitt DC, Sweet RA (2015) Auditory dysfunction in schizophrenia: integrating clinical and basic features. Nat Rev Neurosci 16(9):535–550
Javitt DC, Zukin SR (1991) Recent advances in the phencyclidine model of schizophrenia. Am J Psychiatry 148(10):1301–1308
Javitt DC, Steinschneider M, Schroeder CE, Arezzo JC (1996) Role of cortical N-methyl-D-aspartate receptors in auditory sensory memory and mismatch negativity generation: implications for schizophrenia. Proc Natl Acad Sci USA 93(21):11962–11967
Javitt DC, Strous RD, Grochowski S, Ritter W, Cowan N (1997) Impaired precision, but normal retention, of auditory sensory ("echoic") memory information in schizophrenia. J Abnorm Psychol 106(2):315–324
Javitt DC, Liederman E, Cienfuegos A, Shelley AM (1999) Panmodal processing imprecision as a basis for dysfunction of transient memory storage systems in schizophrenia. Schizophr Bull 25(4):763–775
Javitt DC, Shelley A, Ritter W (2000) Associated deficits in mismatch negativity generation and tone matching in schizophrenia. Clin Neurophysiol 111(10):1733–1737
Jimenez AM, Lee J, Reavis EA, Wynn JK, Green MF (2018) Aberrant patterns of neural activity when perceiving emotion from biological motion in schizophrenia. NeuroImage Clin 20:380–387
Johnson SC, Lowery N, Kohler C, Turetsky BI (2005) Global-local visual processing in schizophrenia: evidence for an early visual processing deficit. Biol Psychiatry 58(12):937–946
Jonsson CO, Sjostedt A (1973) Auditory perception in schizophrenia: a second study of the Intonation test. Acta Psychiatr Scand 49(5):588–600
Joshi YB, Breitenstein B, Tarasenko M, Thomas ML, Chang WL, Sprock J et al (2018) Mismatch negativity impairment is associated with deficits in identifying real-world environmental sounds in schizophrenia. Schizophr Res 191:5–9
Juslin PN, Laukka P (2001) Impact of intended emotion intensity on cue utilization and decoding accuracy in vocal expression of emotion. Emotion 1(4):381–412
Kaas JH, Qi HX, Stepniewska I (2022) Escaping the nocturnal bottleneck, and the evolution of the dorsal and ventral streams of visual processing in primates. Philos Trans R Soc Lond Ser B Biol Sci 377(1844):20210293
Kantrowitz JT, Leitman DI, Lehrfeld JM, Laukka P, Juslin PN, Butler PD et al (2013) Reduction in tonal discriminations predicts receptive emotion processing deficits in schizophrenia and schizoaffective disorder. Schizophr Bull 39(1):86–93
Kantrowitz JT, Hoptman MJ, Leitman DI, Silipo G, Javitt DC (2014) The 5% difference: early sensory processing predicts sarcasm perception in schizophrenia and schizo-affective disorder. Psychol Med 44(1):25–36
Kantrowitz JT, Hoptman MJ, Leitman DI, Moreno-Ortega M, Lehrfeld JM, Dias E et al (2015) Neural substrates of auditory emotion recognition deficits in schizophrenia. J Neurosci 35(44): 14909–14921
Kantrowitz JT, Sehatpour P, Avissar M, Horga G, Gwak A, Hoptman MJ et al (2019) Significant improvement in treatment resistant auditory verbal hallucinations after 5 days of double-blind, randomized, sham controlled, fronto-temporal, transcranial direct current stimulation (tDCS): a replication/extension study. Brain Stimul 12(4):981–991

Kaukoranta E, Sams M, Hari R, Hamalainen M, Naatanen R (1989) Reactions of human auditory cortex to a change in tone duration. Hear Res 41(1):15–21

Kawakubo Y, Kamio S, Nose T, Iwanami A, Nakagome K, Fukuda M et al (2007) Phonetic mismatch negativity predicts social skills acquisition in schizophrenia. Psychiatry Res 152(2–3):261–265

Keane BP, Joseph J, Silverstein SM (2014) Late, not early, stages of Kanizsa shape perception are compromised in schizophrenia. Neuropsychologia 56:302–311

Keane BP, Paterno D, Kastner S, Krekelberg B, Silverstein SM (2019) Intact illusory contour formation but equivalently impaired visual shape completion in first- and later-episode schizophrenia. J Abnorm Psychol 128(1):57–68

Keane BP, Krekelberg B, Mill RD, Silverstein SM, Thompson JL, Serody MR et al (2023) Dorsal attention network activity during perceptual organization is distinct in schizophrenia and predictive of cognitive disorganization. Eur J Neurosci 57(3):458–478

Khalsa SS, Adolphs R, Cameron OG, Critchley HD, Davenport PW, Feinstein JS et al (2018) Interoception and mental health: a roadmap. Biol Psychiatry Cogn Neurosci Neuroimaging 3(6): 501–513

Kim D, Wylie G, Pasternak R, Butler PD, Javitt DC (2006) Magnocellular contributions to impaired motion processing in schizophrenia. Schizophr Res 82(1):1–8

Kim J, Park S, Blake R (2011) Perception of biological motion in schizophrenia and healthy individuals: a behavioral and FMRI study. PLoS One 6(5):e19971

Kim DW, Shim M, Song MJ, Im CH, Lee SH (2015) Early visual processing deficits in patients with schizophrenia during spatial frequency-dependent facial affect processing. Schizophr Res 161(2–3):314–321

Knebel JF, Javitt DC, Murray MM (2011) Impaired early visual response modulations to spatial information in chronic schizophrenia. Psychiatry Res 193(3):168–176

Kohler CG, Bilker W, Hagendoorn M, Gur RE, Gur RC (2000) Emotion recognition deficit in schizophrenia: association with symptomatology and cognition. Biol Psychiatry 48(2):127–136

Kurylo DD, Gazes Y (2008) Effects of Ketamine on perceptual grouping in rats. Physiol Behav 95: 152–156

Kurylo DD, Pasternak R, Silipo G, Javitt DC, Butler PD (2007) Perceptual organization by proximity and similarity in schizophrenia. Schizophr Res 95(1–3):205–214

Kveraga K, Boshyan J, Bar M (2007) Magnocellular projections as the trigger of top-down facilitation in recognition. J Neurosci 27(48):13232–13240

Lakatos P, O'Connell MN, Barczak A, McGinnis T, Neymotin S, Schroeder CE et al (2020) The thalamocortical circuit of auditory mismatch negativity. Biol Psychiatry 87(8):770–780

Lamme VAF, Roelfsema PR (2000) The distinct modes of vision offered by feedforward and recurrent processing. Trends Neurosci 23:571–579

Lee M, Balla A, Sershen H, Sehatpour P, Lakatos P, Javitt DC (2018) Rodent mismatch negativity/theta neuro-oscillatory response as a translational neurophysiological biomarker for N-methyl-D-aspartate receptor-based new treatment development in schizophrenia. Neuropsychopharmacology 43(3):571–582

Leitman DI, Foxe JJ, Butler PD, Saperstein A, Revheim N, Javitt DC (2005) Sensory contributions to impaired prosodic processing in schizophrenia. Biol Psychiatry 58(1):56–61

Leitman DI, Ziwich R, Pasternak R, Javitt DC (2006) Theory of Mind (ToM) and counterfactuality deficits in schizophrenia: misperception or misinterpretation? Psychol Med 36(8):1075–1083

Leitman DI, Hoptman MJ, Foxe JJ, Saccente E, Wylie GR, Nierenberg J et al (2007) The neural substrates of impaired prosodic detection in schizophrenia and its sensorial antecedents. Am J Psychiatry 164(3):474–482

Leitman DI, Sehatpour P, Shpaner M, Foxe JJ, Javitt DC (2009) Mismatch negativity to tonal contours suggests preattentive perception of prosodic content. Brain Imaging Behav 3(3): 284–291

Leitman DI, Laukka P, Juslin PN, Saccente E, Butler P, Javitt DC (2010) Getting the cue: sensory contributions to auditory emotion recognition impairments in schizophrenia. Schizophr Bull 36(3):545–556

Li H, Chan RC, McAlonan GM, Gong QY (2010) Facial emotion processing in schizophrenia: a meta-analysis of functional neuroimaging data. Schizophr Bull 36(5):1029–1039

Loewy R, Fisher M, Ma S, Carter C, Ragland JD, Niendam TA et al (2022) Durable cognitive gains and symptom improvement are observed in individuals with recent-onset schizophrenia 6 months after a randomized trial of auditory training completed remotely. Schizophr Bull 48(1):262–272

March L, Cienfuegos A, Goldbloom L, Ritter W, Cowan N, Javitt DC (1999) Normal time course of auditory recognition in schizophrenia, despite impaired precision of the auditory sensory ("echoic") memory code. J Abnorm Psychol 108(1):69–75

Martinez A, Hillyard SA, Dias EC, Hagler DJ Jr, Butler PD, Guilfoyle DN et al (2008) Magnocellular pathway impairment in schizophrenia: evidence from functional magnetic resonance imaging. J Neurosci 28(30):7492–7500

Martinez A, Hillyard SA, Bickel S, Dias EC, Butler PD, Javitt DC (2012a) Consequences of magnocellular dysfunction on processing attended information in schizophrenia. Cereb Cortex 22(6):1282–1293

Martinez A, Revheim N, Butler PD, Guilfoyle DN, Dias EC, Javitt DC (2012b) Impaired magnocellular/dorsal stream activation predicts impaired reading ability in schizophrenia. NeuroImage Clin 2:8–16

Martinez A, Gaspar PA, Hillyard SA, Andersen SK, Lopez-Calderon J, Corcoran CM et al (2018) Impaired motion processing in schizophrenia and the attenuated psychosis syndrome: etiological and clinical implications. Am J Psychiatry 175(12):1243–1254

Martinez A, Tobe R, Dias EC, Ardekani BA, Veenstra-VanderWeele J, Patel G et al (2019) Differential patterns of visual sensory alteration underlying face emotion recognition impairment and motion perception deficits in schizophrenia and autism spectrum disorder. Biol Psychiatry 86(7):557–567

Martinez A, Tobe RH, Gaspar PA, Malinsky D, Dias EC, Sehatpour P et al (2021) Disease-specific contribution of pulvinar dysfunction to impaired emotion recognition in schizophrenia. Front Behav Neurosci 15:787383

Martinez A, Gaspar PA, Bermudez DH, Belen Aburto-Ponce M, Beggel O, Javitt DC (2024) Disrupted third visual pathway function in schizophrenia: evidence from real and implied motion processing. NeuroImage Clin 41:103570

McCleery A, Lee J, Joshi A, Wynn JK, Hellemann GS, Green MF (2015) Meta-analysis of face processing event-related potentials in schizophrenia. Biol Psychiatry 77(2):116–126

Medalia A, Saperstein AM, Qian M, Javitt DC (2019) Impact of baseline early auditory processing on response to cognitive remediation for schizophrenia. Schizophr Res 208:397–405

Medalia A, Saperstein A, Javitt DC, Qian M, Meyler S, Styke S (2023) Feasibility and clinical utility of using the tone matching test for assessment of early auditory processing in schizophrenia. Psychiatry Res 323:115152

Melvin K, Crossley J, Cromby J (2021) The feeling, embodiment and emotion of hallucinations in first episode psychosis: a prospective phenomenological visual-ecological study using novel multimodal unusual sensory experience (MUSE) maps. EClinicalMedicine 41:101153

Milner AD, Goodale MA (1993) Visual pathways to perception and action. Prog Brain Res 95:317–337

Milner AD, Goodale MA (2008) Two visual systems re-viewed. Neuropsychologia 46(3):774–785

Murray MM, Wylie GR, Higgins BA, Javitt DC, Schroeder CE, Foxe JJ (2002) The spatiotemporal dynamics of illusory contour processing: combined high-density electrical mapping, source analysis, and functional magnetic resonance imaging. J Neurosci 22(12):5055–5073

Murray MM, Foxe DM, Javitt DC, Foxe JJ (2004) Setting boundaries: brain dynamics of modal and amodal illusory shape completion in humans. J Neurosci 24(31):6898–6903

Naatanen R, Paavilainen P, Alho K, Reinikainen K, Sams M (1987) The mismatch negativity to intensity changes in an auditory stimulus sequence. Electroencephalogr Clin Neurophysiol Suppl 40:125–131

Naatanen R, Paavilainen P, Alho K, Reinikainen K, Sams M (1989) Do event-related potentials reveal the mechanism of the auditory sensory memory in the human brain? Neurosci Lett 98(2): 217–221

Nuechterlein KH, Green MF, Kern RS, Baade LE, Barch DM, Cohen JD et al (2008) The MATRICS Consensus Cognitive Battery, part 1: test selection, reliability, and validity. Am J Psychiatry 165(2):203–213

O'Donnell BF, Bismark A, Hetrick WP, Bodkins M, Vohs JL, Shekhar A (2006) Early stage vision in schizophrenia and schizotypal personality disorder. Schizophr Res 86(1–3):89–98

O'Donnell P, Dong C, Murthy V, Asgharnejad M, Du X, Summerfelt A et al (2023) The D-amino acid oxidase inhibitor luvadaxistat improves mismatch negativity in patients with schizophrenia in a randomized trial. Neuropsychopharmacology 48(7):1052–1059

Paavilainen P, Karlsson ML, Reinikainen K, Naatanen R (1989) Mismatch negativity to change in spatial location of an auditory stimulus. Electroencephalogr Clin Neurophysiol 73(2):129–141

Patel GH, Arkin SC, Ruiz-Betancourt DR, DeBaun HM, Strauss NE, Bartel LP et al (2020) What you see is what you get: visual scanning failures of naturalistic social scenes in schizophrenia. Psychol Med 1–10

Patel GH, Arkin SC, Ruiz-Betancourt DR, Plaza FI, Mirza SA, Vieira DJ et al (2021) Failure to engage the temporoparietal junction/posterior superior temporal sulcus predicts impaired naturalistic social cognition in schizophrenia. Brain 144(6):1898–1910

Patel GH, Gruskin DC, Arkin SC, Jamerson EC, Ruiz-Betancourt DR, Klim CC et al (2023) The road not taken: disconnection of a human-unique cortical pathway underlying naturalistic social perception in schizophrenia. Biol Psychiatry Glob Open Sci 3(3):398–408

Pekkonen E, Katila H, Ahveninen J, Karhu J, Huotilainen M, Tiihonen J (2002) Impaired temporal lobe processing of preattentive auditory discrimination in schizophrenia. Schizophr Bull 28(3): 467–474

Perrin MA, Butler PD, DiCostanzo J, Forchelli G, Silipo G, Javitt DC (2010) Spatial localization deficits and auditory cortical dysfunction in schizophrenia. Schizophr Res 124(1–3):161–168

Pitcher D (2021) Characterizing the Third Visual Pathway for Social Perception. Trends Cogn Sci 25(7):550–551

Pitcher D, Ungerleider LG (2021) Evidence for a Third Visual Pathway Specialized for Social Perception. Trends Cogn Sci 25(2):100–110

Pobric G, Hulleman J, Lavidor M, Silipo G, Rohrig S, Dias E et al (2018) Seeing the world as it is: mimicking veridical motion perception in schizophrenia using non-invasive brain stimulation in healthy participants. Brain Topogr 31(5):827–837

Pulvermuller F, Tomasello R, Henningsen-Schomers MR, Wennekers T (2021) Biological constraints on neural network models of cognitive function. Nat Rev Neurosci 22(8):488–502

Quadt L, Critchley HD, Garfinkel SN (2018) The neurobiology of interoception in health and disease. Ann N Y Acad Sci 1428(1):112–128

Rabinowicz EF, Silipo G, Goldman R, Javitt DC (2000) Auditory sensory dysfunction in schizophrenia: imprecision or distractibility? Arch Gen Psychiatry 57(12):1149–1155

Rassovsky Y, Horan WP, Lee J, Sergi MJ, Green MF (2011) Pathways between early visual processing and functional outcome in schizophrenia. Psychol Med 41(3):487–497

Remy I, Bernardin F, Ligier F, Krieg J, Maillard L, Schwan R et al (2023) Association between retinal and cortical visual electrophysiological impairments in schizophrenia. J Psychiatry Neurosci 48(3):E171–E1E8

Revheim N, Corcoran CM, Dias E, Hellmann E, Martinez A, Butler PD et al (2014) Reading deficits in schizophrenia and individuals at high clinical risk: relationship to sensory function, course of illness, and psychosocial outcome. Am J Psychiatry 171(9):949–959

Ross JM, Hamm JP (2020) Cortical microcircuit mechanisms of mismatch negativity and its underlying subcomponents. Front Neural Circuits 14:13

Saccuzzo DP, Braff DL (1981) Early information processing deficit in schizophrenia. New findings using schizophrenic subgroups and manic control subjects. Arch Gen Psychiatry 38(2):175–179

Schechter I, Butler PD, Silipo G, Zemon V, Javitt DC (2003) Magnocellular and parvocellular contributions to backward masking dysfunction in schizophrenia. Schizophr Res 64(2–3): 91–101

Sehatpour P, Molholm S, Schwartz TH, Mahoney JR, Mehta AD, Javitt DC et al (2008) A human intracranial study of long-range oscillatory coherence across a frontal-occipital-hippocampal brain network during visual object processing. Proc Natl Acad Sci USA 105(11):4399–4404

Sehatpour P, Bassir Nia A, Adair D, Wang Z, DeBaun HM, Silipo G et al (2020) Multimodal computational modeling of visual object recognition deficits but intact repetition priming in schizophrenia. Front Psychiatry 11:547189

Sehatpour P, Javitt DC, De Baun HM, Carlson M, Beloborodova A, Margolin DH et al (2022) Mismatch negativity as an index of target engagement for excitation/inhibition-based treatment development: a double-blind, placebo-controlled, randomized, single-dose cross-over study of the serotonin type-3 receptor antagonist CVN058. Neuropsychopharmacology 47(3):711–718

Sehatpour P, Kreither J, Lopez-Calderon J, Shastry AM, De Baun HM, Martinez A et al (2023) Network-level mechanisms underlying effects of transcranial direct current stimulation (tDCS) on visuomotor learning in schizophrenia. Transl Psychiatry 13(1):360

Sergi MJ, Green MF (2003) Social perception and early visual processing in schizophrenia. Schizophr Res 59(2–3):233–241

Sergi MJ, Rassovsky Y, Nuechterlein KH, Green MF (2006) Social perception as a mediator of the influence of early visual processing on functional status in schizophrenia. Am J Psychiatry 163(3):448–454

Slaghuis WL (2004) Spatio-temporal luminance contrast sensitivity and visual backward masking in schizophrenia. Exp Brain Res 156(2):196–211

Strous RD, Cowan N, Ritter W, Javitt DC (1995) Auditory sensory ("echoic") memory dysfunction in schizophrenia. Am J Psychiatry 152(10):1517–1519

Tapia E, Breitmeyer BG (2011) Visual consciousness revisited: magnocellular and parvocellular contributions to conscious and nonconscious vision. Psychol Sci 22(7):934–942

Taylor SF, Kang J, Brege IS, Tso IF, Hosanagar A, Johnson TD (2012) Meta-analysis of functional neuroimaging studies of emotion perception and experience in schizophrenia. Biol Psychiatry 71(2):136–145

Thakkar KN, Peterman JS, Park S (2014) Altered brain activation during action imitation and observation in schizophrenia: a translational approach to investigating social dysfunction in schizophrenia. Am J Psychiatry 171(5):539–548

Thomas ML, Bismark AW, Joshi YB, Tarasenko M, Treichler EBH, Hochberger WC et al (2018) Targeted cognitive training improves auditory and verbal outcomes among treatment refractory schizophrenia patients mandated to residential care. Schizophr Res 202:378–384

Torregrossa LJ, Amedy A, Roig J, Prada A, Park S (2022) Interoceptive functioning in schizophrenia and schizotypy. Schizophr Res 239:151–159

Tsubomoto M, Kawabata R, Zhu X, Minabe Y, Chen K, Lewis DA et al (2019) Expression of transcripts selective for GABA neuron subpopulations across the cortical visuospatial working memory network in the healthy state and schizophrenia. Cereb Cortex 29(8):3540–3550

Ungerleider LG, Mishkin M (1982) In: Ingle DJ, Goodale MA, Mansfield RJW (eds) Two cortical visual systems. The MIT Press, Cambridge, M.A

Van Derveer AB, Bastos G, Ferrell AD, Gallimore CG, Greene ML, Holmes JT et al (2021) A role for somatostatin-positive interneurons in neuro-oscillatory and information processing deficits in schizophrenia. Schizophr Bull 47(5):1385–1398

van Rijn P, Larrouy-Maestri P (2023) Modelling individual and cross-cultural variation in the mapping of emotions to speech prosody. Nat Hum Behav 7(3):386–396

Yang L, Chen S, Chen CM, Khan F, Forchelli G, Javitt DC (2012) Schizophrenia, culture and neuropsychology: sensory deficits, language impairments and social functioning in Chinese-speaking schizophrenia patients. Psychol Med 42(7):1485–1494

Yang F, Fang X, Tang W, Hui L, Chen Y, Zhang C et al (2019) Effects and potential mechanisms of transcranial direct current stimulation (tDCS) on auditory hallucinations: a meta-analysis. Psychiatry Res 273:343–349

Zachlod D, Bludau S, Cichon S, Palomero-Gallagher N, Amunts K (2022) Combined analysis of cytoarchitectonic, molecular and transcriptomic patterns reveal differences in brain organization across human functional brain systems. NeuroImage 257:119286

How Layer-Specific fMRI Can Contribute to Understanding Perceptual Disturbances Across Psychiatric Disorders

Joost Haarsma and Peter Kok

Contents

Abstract Perceptual disturbances occur across various sensory domains and contribute to significant suffering in numerous psychiatric and neurological conditions. Despite decades of research into the neural mechanisms underlying these distressing experiences, progress has been relatively limited. Here we explore the potential of layer-specific fMRI to enhance our understanding of these phenomena. We posit that perceptual disturbances can stem from alterations in the neural integration of internally generated signals–such as memory, imagination, prediction, and expectations–with sensory evidence being used to optimize inferences about the world. Emerging evidence suggests that these key computations are distributed across different cortical layers, highlighting the utility of layer-specific imaging in identifying the mechanisms driving such disruptions. We review recent findings that underscore the promise of layer-specific fMRI in elucidating these neural processes and discuss how pharmacological layer-specific fMRI could further advance this understanding. Finally, we address the current limitations of layer-specific fMRI and the progress made toward overcoming these challenges.

J. Haarsma (✉) and P. Kok
Department of Imaging Neuroscience, UCL Queen Square Institute of Neurology, University College London, London, UK
e-mail: j.haarsma@ucl.ac.uk

Curr Topics Behav Neurosci (2025) 74: 41–72
https://doi.org/10.1007/7854_2025_595
Published Online: 12 July 2025

Keywords Hallucinations · Layer-fMRI · Precision-psychiatry · Predictive coding

1 Introduction

Perceptual disturbances play an important role in various psychiatric and neurological disorders, including psychosis, autism, anxiety disorders, eating disorders, bipolar disorder, Parkinson's disease and dementia (Berntson and Khalsa 2021; Dakin and Frith 2005; Khalsa et al. 2018; O'Brien et al. 2020; Sterzer et al. 2018). Yet despite the prevalence of these symptoms, their centrality to the illness, and their relation to poor disease outcomes, little is understood about the nature and mechanistic underpinnings of these experiences.

Perceptual disturbances occur across sensory domains, including audition, vision, somatosensation, olfaction, interoception, and proprioception (Eversfield and Orton 2019; Fenelon et al. 2002; Lewandowski et al. 2009; Lim et al. 2016). Within modalities, the types of experiences that can be considered to fall under the term perceptual disturbances are varied as well. Considering just the visual domain, across different disorders, patients might report changes in the appearance of objects, where they might appear unusually bright or intense, or find that objects seem warped, flat, or perceived to look like cardboard cutouts. We might refer to such experiences, where objects are perceived altered, as "illusory." In contrast, full blown hallucinations can also emerge in psychosis or Parkinson's disease, where de novo percepts arise, without a corresponding external stimulus. These hallucinatory experiences range from geometrical patterns, all the way to the experience of fully formed individuals situated in space (Bunney et al. 1999; Dudley et al. 2019). Much theoretical work has attempted to explain these phenomena as deficits in various cognitive or perceptual systems. Some have argued that these disturbances are the result of deficits in attention, reality monitoring, corollary discharge, or the result of an excessive influence of expectations on perception (Ford and Mathalon 2005; McGHIE and Chapman 1961; Powers et al. 2016; Simons et al. 2017; Sterzer et al. 2018).

Changes in interoception–the nervous system's ability to sense internal bodily states–are increasingly implicated in anxiety disorders, eating disorders, and disturbances in the sense of self (Allen et al. 2022; Khalsa et al. 2018; Seth and Tsakiris 2018). In anxiety disorders, heightened interoceptive sensitivity or misinterpretation of bodily signals, such as an elevated heart rate, can amplify fear responses and perpetuate a cycle of hypervigilance (Paulus and Stein 2010). Similarly, in eating disorders, altered interoception can manifest as a disconnect between physiological hunger and satiety cues, contributing to disordered eating behaviors and distorted body image (Khalsa et al. 2015). Moreover, disturbances in interoception may undermine the sense of self, as the ability to integrate bodily sensations with emotional and cognitive processes is central to maintaining a coherent self-concept (Woelk and Garfinkel 2024). These experiences can be construed as perceptual disturbances in and among themselves, where a failure to make proper inferences about one's own bodily states can lead to maladaptive behaviors (Berntson and Khalsa 2021).

As will become clear, there is a good reason to assume that many of these forms of perceptual disturbances can arise from altered information signaling on the level of the cortical microcircuit. Advances in the field of high-field neuroimaging has made it possible to study these microcircuits in a noninvasive way, using a method that has become known as layer-specific fMRI. In this chapter, we introduce the method of layer-specific fMRI and how it can be used to study the neural mechanisms that underlie perceptual disturbances. In order to achieve this, we start by describing a theoretical framework of how laminar mechanisms might underlie perceptual inference. Having such a framework to guide our thinking will be useful, as conceptualizing how perception might go awry requires some understanding of the mechanisms underlying normative perception. We will not attempt to give a comprehensive account of these theories, as insights in how perception is implemented through these mechanisms will likely change over time and are expected to coevolve with how we think of perceptual disturbances. However, we provide enough detail to allow the reader to start thinking about how different layers might give rise to perceptual disturbances. Once this has been covered, we move on to discuss how layer-specific fMRI has been applied in recent years to tackle novel questions in the field of perception and cognitive neuroscience more generally. We consider how comparative studies between clinical and pharmacological groups can start to highlight some of the key mechanisms that underlie these experiences. The field of layer-specific fMRI is new, and it should be noted that at the present moment of writing this review, studies using layer-specific fMRI to study psychiatric or neurological disorders are still sparse. Therefore, our aim is to describe what its potential might be in exploring the neural mechanisms of perceptual disturbance, as well as some of the limitations that still need to be addressed. Central to the discussion here will be its ability to study the diversity of nuanced mechanisms that possibly underlie different psychiatric disorders, in contrast to attempting to reconciling all forms of perceptual disturbances under a single theoretical umbrella.

2 Frameworks to Understand Perceptual Inference

Before comprehending how perception might go awry, we first need a model of normative perceptual inference. Traditionally, perception has been construed as directly reflecting the external world (Gibson 1950). This approach is reflected in neural theories of perception as predominantly a feedforward process, building progressively and sequentially from simple low-level representations (edges and colors) to higher-order cognition like object recognition (Serre et al. 2007). This construes perception as a data-driven, mostly passive process of accumulation of sensory data (see, for example, Marr's computational theory of vision (Marr 2010). In contrast, the constructive approach construes perception as context dependent and

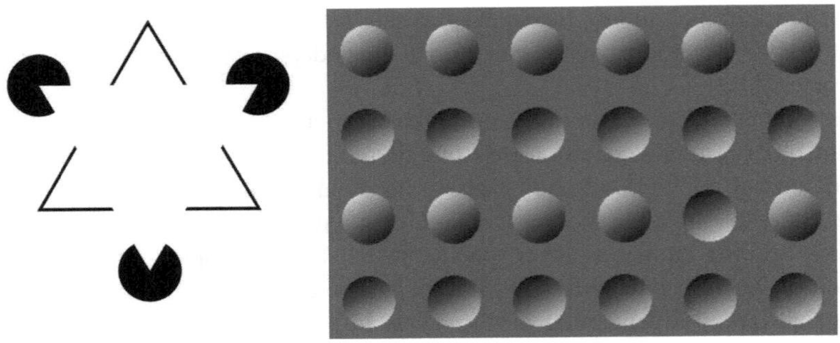

Fig. 1 The constructionist approach suggests that perception is an active process. Examples highlighting the constructionist nature are illusions like the Kanizsa triangle and the convex-concave illusion. Here contextual information leads to perceptual changes, resulting in the perception of illusory contours, or perceiving the same stimulus as concave or convex depending on how lighting is interpreted

active. This view puts a particular importance on top-down influences on perception, most clearly exemplified by visual illusions (Fig. 1) (Gregory 1997; Gregory et al. 1980).

One constructionist view that has become popular is the view that the brain forms a generative model of the world (Clark 2013). Given this assumption, much of cognitive neuroscience has been dedicated to understanding how the brain might form such an internal model in computational terms (Spratling 2017), as well as how such a model might be implemented on a neural level (Bastos et al. 2012, 2020; Friston 2009). Although an extensive discussion of the different theories that are currently popular in the field of cognitive neuroscience is beyond the scope of the current chapter, it is worth highlighting one of these theories, as it helps us understand how perception might go awry in different ways, as well as how layer-specific fMRI might be used to test predictions resulting from them.

Here we use the predictive coding theory to guide the discussion around how layer-specific fMRI can be applied to understanding perceptual disturbances. The benefit of using this theory is that the hypothesized mechanisms that underly perception are believed to be segregated and implemented in different cortical layers (Bastos et al. 2012; Clark 2013; Friston 2009; Spratling 2017). While there are various iterations of this theory that make slightly different claims about the neural mechanisms through which the brain forms a model, there is a common theme among them (see Spratling 2017 for an overview). The general claim in predictive coding theories is that the brain models its environment by making iterative predictions about sensory inputs. These predictions need to somehow be updated by virtue of new sensory evidence to approximate an optimal estimate of the outside world. Most predictive coding theories suggest that this is achieved through the use of a precision-weighted prediction errors, signaling the difference between sensory input and prediction (Bastos et al. 2012). This notion of integrating prior knowledge with new sensory input in a weighted fashion resonates with Bayesian theories of the brain, which suggest that optimal inference requires integrating precision weighted prior knowledge with precision weighted evidence (Friston 2012; Yon and Frith 2021).

How might the brain implement these computational mechanisms? Again, predictive coding theories answer this question in slightly different ways, but we can delineate some commonalities. Many of these theories start from the assumption that the brain is a hierarchically organized system, with feedforward and feedback connections interwoven to form a dense neural network of interconnected systems. Indeed, anatomical work has revealed that in the primate visual cortex, there are many feedforward and feedback connections that seem to make up a hierarchical system (Angelucci et al. 2002; Felleman and Van Essen 1991). Interestingly, they observed a preference for feedback neurons originating from the deep layers (Layers V and VI) of higher-order regions to terminate in the deep and superficial (Layers I, II, and III) layers of the cortex. In contrast, feedforward neurons preferentially originate from the superficial layers and preferentially terminate in the middle layers (Layer IV) of higher-order regions (Fig. 2b). As such, different brain regions form a hierarchical network through layer-specific connections between regions. Notably, the posterior granular insular cortex displays similar laminar differentiation, although the anterior insular cortex lacks a distinguished layer IV (Flynn 1999; Gogolla 2017). Given these anatomical findings, predictive coding theories have suggested there is a functional segregation on the laminar level in the sensory cortex, attributing different functions to different layers (Bastos et al. 2012, 2020; Friston 2018). That is, feedback signals from the deep layers are typically believed to convey predictive information, terminating in the superficial and deep layers of lower

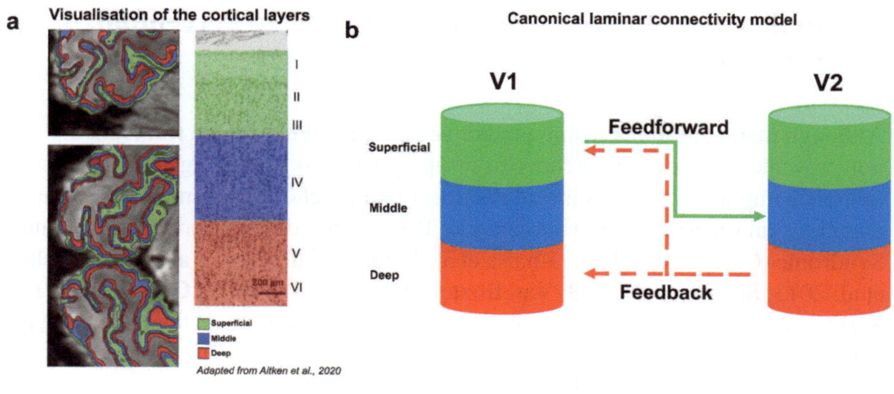

Fig. 2 (**a**) The superficial, middle, and deep layers can be visualized on a functional EPI scan by dividing the grey matter into three layers with equal volume. Histologically, in the primary visual cortex, the superficial layers capture layers I-III, the middle layer capture layer IV, while the deep layers capture layer V&VI (de Sousa et al. 2010). (**b**) The layers of the early visual cortex have a typical connectivity pattern, where feedforward signals are sent from the superficial layers of lower sensory regions to the middle layers of higher-order regions. In contrast, feedback signals are sent from the deep layers of higher sensory regions to the agranular (deep and superficial) layers of sensory regions

sensory regions, explaining away sensory inputs from lower regions. Feedforward sensory input instead arrives in the middle layers, from where it is sent to the nearby superficial layers, and compared to predictive information from higher-order regions. This is then suggested to result in the computation of a prediction error, capturing as yet unexplained sensory input, which is used to update predictions upstream (Bastos et al. 2012; de Lange et al. 2018; Lawrence et al. 2019a; Stephan et al. 2019). While some elements of predictive coding theory are well supported, other aspects await empirical support or refutation. One aspect that has been well supported is that the brain generates predictive signals, about both present and future states (see evidence discussed below). The brain also precision-weights prediction errors during learning (Haarsma et al. 2021). In the sensory domain it is well known that unexpected stimuli can generate stronger cortical responses, but whether these actually represent prediction errors remains unclear (Garrido et al. 2009). Finally, the hierarchical nature of sensory cortices lends them well to hierarchical inference as in predictive coding (see sections below). However, some postulates of predictive coding theory remain less well supported. In particular whether the above-mentioned prediction and error signals are computed by different neural units, as well as the minimization of prediction error through reciprocal exchange between different hierarchical levels, remain less strongly supported (Walsh et al. 2020). However, although the empirical fate of predictive coding theory remains to be decided, it serves as a useful heuristic for the purpose of this chapter in allowing us to frame how layer-specific imaging can be used to further our understanding of how perception can go awry in clinical and neurological disorders.

3 Application of Predictive Coding Theory to Perceptual Disturbances

Although many questions within predictive coding still need to be resolved, this has not prevented clinical researchers from applying it to theorizing how perceptual inference might go awry in clinical disorders like psychosis, autism, anxiety, eating disorders, and neurological disorders like Parkinson's disease and Charles Bonnet Syndrome (Corlett et al. 2019; Khalsa et al. 2018; O'Callaghan et al. 2017; Reichert et al. 2013; Sterzer et al. 2018; Van Boxtel and Lu 2013; Van De Cruys et al. 2014; Zarkali et al. 2019). This is not necessarily a problem. Progress on understanding the neural mechanisms of various pathologies does not require the science on the relevant cognitive mechanisms to be fully settled. One might picture here two rock climbers securing each other's ascent to the top of a mountain. The higher-up climber is in the position to scout the rockface and secure pins, allowing the climber lower-down to follow on more solid footing. The point of this analogy is that a safe ascent of the second climber does not require the top of the mountain to have been reached. Incremental advancement of both climbers is sufficient for successful progress. In concordance with the rise in popularity of normative predictive coding theories of perception and inference, theorists and researchers have applied it to understanding aberrant perceptual inference too. The first of such theories focused on psychosis and emerged almost two decades ago, which suggested that visual and auditory hallucinations could be conceived to be the result of overly strong prior expectations, potentially due to aberrant modulatory neurotransmission, shifting perceptual inference toward priors (Stephan et al. 2006). Indeed, there has been increasing evidence, mostly behavioral, that individuals who report experiencing hallucinations show increased reliance on prior expectations in perception, in the visual and auditory domains (Cassidy et al. 2018; Haarsma et al. 2020; Powers et al. 2017; Schmack et al. 2013; Stuke et al. 2021; Teufel et al. 2015; Zarkali et al. 2019). However, a second strain of studies have complicated this picture somewhat, demonstrating that those who experience psychosis also often display perceptual processing abnormalities that are more in line with a weakened influence on perceptual processing. Findings relating to misinterpreted inner speech (Crapse and Sommer 2008; Feinberg 1978), weaker susceptibility to illusions (Dima et al. 2009, 2010; Haarsma et al. 2020; Pearl et al. 2009; White et al. 2014), and weaker attenuation of sensory consequences in action (Blakemore et al. 2000; Ford and Mathalon 2005; Shergill et al. 2005) might be better understood as instances of weaker priors. Theorists have attempted to resolve this potential contradiction by arguing that the balance between top-down predictive signals and sensory input is shifted in different ways for different symptoms. That is, whereas disturbances in agency might result from diminished predictive signals (Griffin and Fletcher 2017), hallucinations might be the result of overly strong predictive signals (Corlett et al. 2019). Others have suggested that there are subtle differences depending on disease stage or the presence of psychosis in addition to hallucinations (Corlett et al. 2011; Haarsma et al. 2020; Powers et al. 2017; Schmack et al. 2013). Finally, some have suggested that the relative weakening of priors is not specifically correlated with the presence of hallucinations, whereas overly stronger priors are (Corlett et al. 2019; Sterzer et al. 2018).

These contradictions and inconsistencies point to a broader issue, which is that predictive coding theories and Bayesian theories of perception more generally have been difficult to properly constrain, as they tend to be overparameterized. That is, almost any finding can be accounted for by changing some parameter in the theory, and the same finding can often be explained in a multitude of ways, making it difficult to falsify (Haarsma et al. 2022). For example, evidence demonstrating heightened reliance on priors could also be explained by weakened reliance on sensory evidence. These different hypotheses matter, as they might suggest different disease etiologies (see below for a further discussion on this). Layer-specific fMRI could potentially provide some solutions by interrogating more precisely what the role of different feedforward and feedback signals is in perception (Haarsma et al. 2022; Lawrence et al. 2019a; Stephan et al. 2019).

4 Layer-Specific fMRI as a Tool to Study the Computational Architecture of Perception

Much of the development of layer-specific fMRI is a direct result of the wider use of high-field MRI scanners with magnetic field strengths of 7 Tesla and beyond. High-field MRI scanners offer one particular advantage over lower field strengths for layer-specific fMRI studies: the higher magnetic field strengths provide enhanced signal-to-noise ratio (SNR) and therefore allow greater spatial resolution. This allows researchers to distinguish finer structures within the cortex, such as cortical layers and columns, allowing the study of layer-specific activity patterns (Dumoulin et al., 2018; Lawrence et al. 2019b) (Fig. 2a). With the increase in SNR, existing gradient-echo and spin-echo sequences can be optimized to exploit the laminar signal, thereby providing more detailed information about the cortical layers (Han et al. 2019; Norris 2012).

With the advent of high-resolution imaging, the layer-specific fMRI field entered a stage of validation-oriented studies. The primary goal was to compare layer-specific signals to a known gold standard, to build confidence in its utility to reliably measure laminar signals. These studies were therefore not so much focused on revealing novel mechanisms within the human brain but rather on finding concordance between ground truth and results obtained by layer-specific fMRI. Many of the earliest studies were encouraged by finding sensory input primarily modulating the middle layers, both in early monkey work as well as in humans, suggesting that it is possible in principle to differentiate activity in different layers (Chen et al. 2013; Goense and Logothetis 2006; Koopmans et al. 2010, 2011; Polimeni et al. 2010) (Fig. 3a). Subsequent experiments explored the laminar mechanisms underlying more complex processes, e.g., by comparing scrambled and unscrambled figures and stimuli differentially targeting parvo- and magnocellular pathways, revealing differential laminar profiles across these conditions (Olman et al. 2012).

After establishing that layer-specific fMRI can pick up laminar specific signals in the visual cortex, others started to apply it to study more complex perceptual processes. These studies often added to the confidence in the reliability of layer-specific fMRI, as there was considerable overlap with laminar findings from nonhuman animals (Self et al. 2019). Naturally, many studies using layer-specific fMRI are interested in the interplay between feedforward and feedback signaling, and thus many experimental manipulations contrast bottom-up signals and top-down processes such as working memory, imagery, illusory perception, and expectations. Most of these studies have focused on studying the primary visual cortex, as the laminar organization of this area is best understood, and the methods used in the earliest laminar studies were optimized for this region. We therefore discuss work in the visual cortex first.

One of the first studies to explore more complex perceptual processes studied contextual feedback processing in visual scene construction. To test whether specific layers contribute to filling in contextual information, the researchers occluded a quadrant of a visual scene that was previously perceived as a whole (Muckli et al.

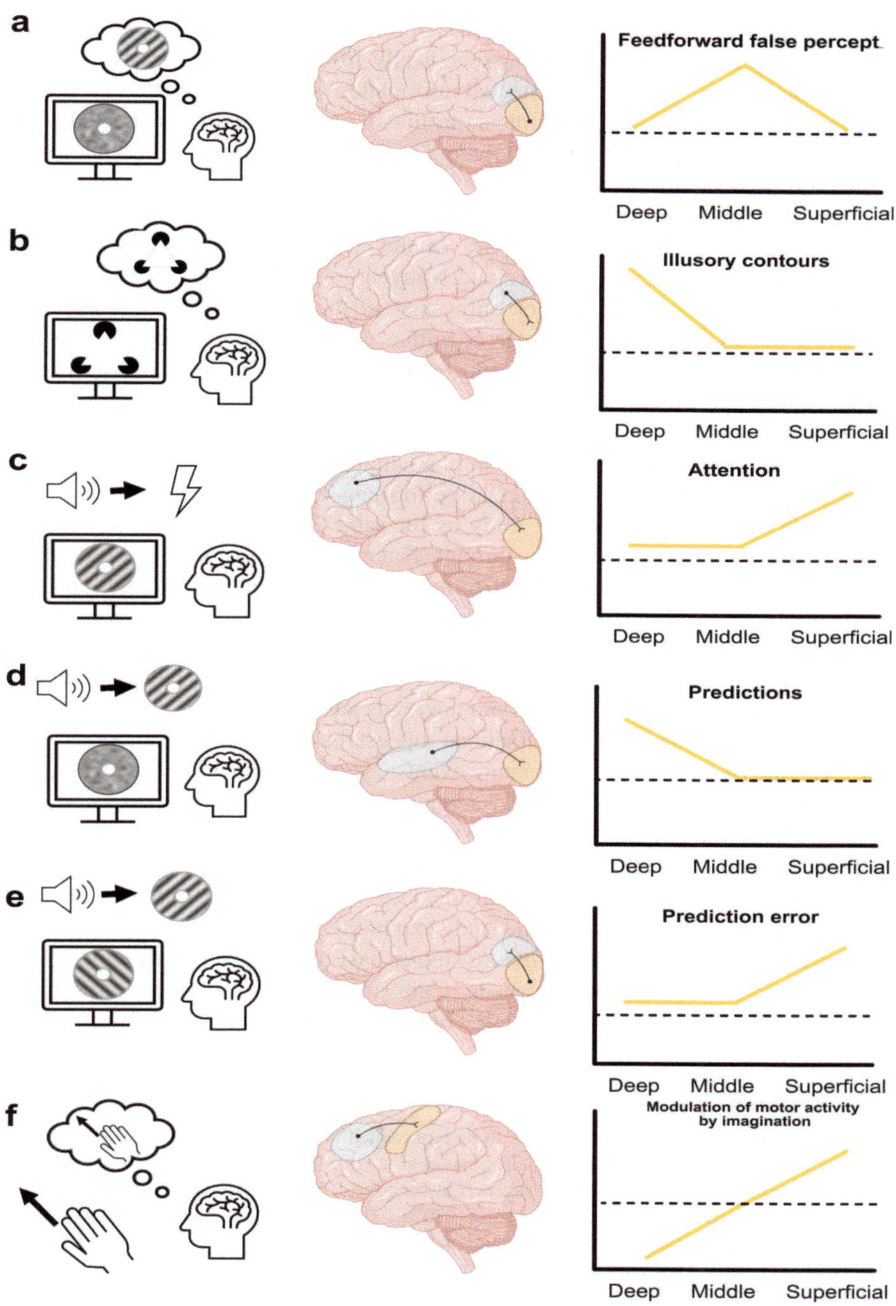

Fig. 3 Layer-specific profiles of different perceptual mechanisms, with a focus on the visual system. Blue regions represent higher-order regions, whereas orange regions represent lower order regions. Connectivity patterns are based on separate studies and were typically not measured in the respective layer fMRI experiments themselves. The laminar profiles reflect the region in which the depicted neural connections terminate. (**a**) The orientation of expected gratings has

2015). Using multivariate techniques, they found that they could decode the scene context from the occluded quadrant specifically from the superficial layers of the visual cortex, suggesting that contextual information is fed back into these layers. A second study focused on the laminar profile of illusory figures induced by the Kanizsa triangle illusion. Using retinotopic mapping, an area of V1 was identified that corresponded to the area in the visual field where the illusory figure was perceived. They found that specifically in conditions where the illusory figure was present, there was heightened activity in the deep layers of V1, in line with nonhuman physiological work (Lee and Nguyen 2001; Pak et al. 2020) (Fig. 3b). Notably, feedback signals do not only exist in the context of low-level visual illusions but extend to higher-order cognitive domains like working memory as well. For example, one study demonstrated that merely maintaining a specific orientated grating in mind was enough to evoke orientation-specific activity in the deep and superficial layers of the primary visual cortex (Lawrence et al. 2018). Again, these findings align nicely with nonhuman animal work, demonstrating similar agranular activity in monkeys performing a working memory task (van Kerkoerle et al. 2017). A second study by the same group reported separate layer-specific modulations of feedback, through feature-based attention, and bottom-up input, with attention most strongly affecting superficial layers, and bottom-up input particularly modulating the middle layers (Lawrence et al. 2019b) (Fig. 3a, c).

The previously discussed studies provide strong evidence that the agranular layers play an important role in a range of top-down perceptual processes, such as contextual processing and working memory. Because much experimental work on the mechanisms underlying hallucinations suggest that they are the result of strongly overweighted expectations about future events, a key question will be whether these expected events are represented in the agranular layers of the early sensory cortex as well. If this is found to be the case, such a mechanism is ideally situated to modulate sensory processing, potentially driving abnormal perceptual experience. Indeed, a series of studies have identified considerable support for the representation of expected events in these layers. One study explored the layer-specific representation of expected sensory templates. Previous studies have shown that merely expecting to see an oriented grating induces a sensory template of that stimulus in the early visual cortex (Kok et al. 2014, 2017). In line with predictive coding theories, subsequent

Fig. 3 (continued) repeatedly been found to be represented in the deep layers of the visual cortex (Aitken et al. 2020; Haarsma et al. 2023; Thomas et al. 2024). Working memory seems to rely on a combination of the deep and superficial layers (Lawrence et al. 2018). (**b**) False percepts can emerge from activity in the middle layers (Haarsma et al. 2023). The middle layers are also often modulated by bottom-up input (Chen et al. 2013; Goense and Logothetis 2006; Koopmans et al. 2010, 2011; Polimeni et al. 2010; Lawrence et al. 2019b; Yu et al. 2019). (**c**) Low-level perceptual illusions rely on activity in the deep layers (Kok et al. 2017). (**d**) Attention has been shown to modulate the superficial layers (De Martino et al. 2015; Gau et al. 2020; Lawrence et al. 2019b, but see also Klein et al. 2018). (**e**) Prediction errors are represented in the superficial layers (Thomas et al. 2024). (**f**) Imagined movement modulates deep and superficial layers of the motor cortex in opposite directions (Persichetti et al. 2020)

work showed that merely expecting a grating on the basis of a visual cue was enough to induce a sensory template specifically in the deep layers of the visual cortex (Aitken et al. 2020) (Fig. 3d). This finding was replicated in a second study where participants had to make judgments about the orientation of a grating embedded in noise, while an auditory cue predicted the most likely orientation. Interestingly, here the deep layers once more represented the expected grating, despite most participants not being aware that the cues predicted the most likely orientation. This suggests that even implicitly the brain can generate sensory templates on the basis of previously conditioned stimuli (Haarsma et al. 2023). In a third study, actions were predictive of upcoming stimuli, with different levels of validity. Once more, only in the valid conditions did the deep layers represent the expected orientation, in line with an important role for these layers in signaling expected stimuli (Thomas et al. 2024). Together, these studies suggest there is converging evidence, at least with regard to simple orientation-like stimuli, that the deep visual cortical layers indeed represent expected stimuli.

Another important claim of predictive coding theories is that there should be different neural populations representing errors and predictions across the cortical hierarchy. This component of predictive coding is one of the less substantiated claims (Walsh et al. 2020). Some indirect evidence in line with this comes from histological work suggesting that feedforward and feedback connections originate from different cell populations (Markov et al. 2014). Electrophysiological work suggests that feedforward and feedback signals are transmitted at different frequencies (Arnal and Giraud 2012; Bastos et al. 2020). More direct evidence comes from studies in mice and monkeys showing that different cell populations in V1 report mismatch and prediction signals (Attinger et al. 2017; Bell et al. 2016; Fiser et al. 2016; Keller et al. 2012; Keller and Mrsic-Flogel 2018; Leinweber et al. 2017), but see also (Muzzu and Saleem 2021) for an alternative view. However, evidence that prediction errors originate from feedforward projecting neurons in the superficial layers has remained sparse (Walsh et al. 2020). A recent layer-specific fMRI study aimed to test this prediction directly (Thomas et al. 2024) (Fig. 3e). Here, the orientation of a grating could either be expected or unexpected following a particular action. When orientations were expected, all layers represented the orientation that was presented. However, unexpected orientations, or prediction errors, were specifically represented in the superficial layers, in line with previous findings from animal work (Bastos et al. 2020).

Layer-specific fMRI has also been used in other sensory modalities, such as the somatosensory domain. For example, one study explored how predictable and unpredictable sensory input preferentially activates different layers in the early somatosensory cortex (S1) and found that like findings in the visual cortex described above, thalamic sensory input preferentially activated the middle layers, whereas predictability mostly modulated the agranular superficial and deep layers (Yu et al. 2019). Another study explored laminar specific signals underlying action. Here, participants were required to tap their fingers while layer-specific activity in the superficial and deep layers in the early motor cortex (M1) was measured. Finger tapping evoked activity in the superficial and deep layers, whereas imagined movements only activated the superficial layers. Further, imagined movement repressed signals in superficial layers but enhanced signals in the deep layers (Persichetti et al. 2020) (Fig. 3f). This further underscores the importance of layer-specific fMRI to distinguish between suppression and sharpening theories of perception (Thomas et al. 2024).

The auditory domain has received considerable attention in layer-specific fMRI studies as well. One study explored frequency preference across layers, as well as its modulation by attention. They found that attention particularly modulated the superficial layers of the primary auditory cortex (De Martino et al. 2015). More recently, the same researchers studied how stimulus complexity is represented across the primary auditory cortex and found that the superficial layers in particular represented more complex features of sounds (Moerel et al. 2019). Studying cross-modal processing and attention, one study found that visual-audio stimuli modulated activity in the deep layers specifically, whereas attention modulated the superficial layers in both primary sensory regions (Gau et al. 2020). Taken together, there is support that just like in the visual cortex, the agranular layers of auditory cortex are primarily involved in feedback processing. Speculatively, content-based feedback signals seem to be represented more often in the deep layers, while attention tends to modulate superficial layers (although note that some have found content in the superficial (Muckli et al. 2015) and attention in deep layers (Klein et al. 2018)). The exact functional roles of the superficial and deep layers remain to be resolved, but their shared role in conveying feedback signals is well established.

Fewer studies have targeted regions outside of the primary sensory regions, creating knowledge gaps waiting to be filled. The insula, with its clear histological differentiation (Flynn 1999; Gogolla 2017), will be a prime future target for layer-specific studies targeted to understand the microcircuits underpinning aberrant interoceptive inference. Further, laminar imaging of higher-order regions like the association cortex will be important in gaining a full understanding of how perceptual disturbances arise, due to the importance of these areas in predictive coding theories. Some have argued that this is now in the realm of possibilities (Finn et al. 2021). One study explored layer-specific responses in the prefrontal cortex during a working memory task, showing heightened activity in the superficial layers during maintenance and deep layers during responses (Finn et al. 2019). Notably, this is in line with work in nonhuman primates (Bastos et al. 2018). A study in word reading demonstrated a particular involvement of the deep layers of the dorsolateral cortex in interpreting words versus pseudowords (Sharoh et al. 2019). Finally, the entorhinal cortex, a crucial interface of the hippocampal complex, has a laminar differentiation in connectivity where superficial and deep layers provide input to and receive output from the hippocampus, respectively (Lavenex and Amaral 2000). Recent laminar studies have demonstrated that it is possible to image the entorhinal layers in a meaningful way (Koster et al. 2018; Maass et al. 2014). This is crucial as the hippocampus is believed to play a critical role in signaling contextual information like perceptual expectations and can play an important role in testing predictive coding theories of hallucinations (Aitken and Kok 2022; Hindy et al. 2016; Kok and Turk-Browne 2018; Schapiro et al. 2012; Stachenfeld et al. 2017; Whittington et al. 2020).

5 Potential Future Avenues in the Study of Perceptual Disturbances

As perception is believed to rely on the interplay of laminar signals integrating across different levels of the cortical hierarchy, perceptual disturbances as seen in various psychiatric and neurological disorders likely have a natural root in these processes. While there have been very few studies that have applied layer-specific fMRI to studying these perceptual disturbances, we can start to speculate how layer-specific fMRI can contribute to understanding them. We will start out with discussing the potential laminar mechanisms that might underlie false inferences in psychosis for two reasons. First, theories considering hallucinations in psychosis have been developed extensively over recent years. Second, the primary modalities affected in psychosis, vision and auditory, have been studied the most using layer fMRI and are therefore the best understood.

As described above, hallucinations as seen in psychosis have been theorized to arise due to increased influences of perceptual expectation signals (Corlett et al. 2019; Powers et al. 2016; Sterzer et al. 2018), which recent studies have reliably linked to signals in the deep layers of the visual cortex (Aitken et al. 2020; Haarsma et al. 2023; Thomas et al. 2024). However, expectations alone are not sufficient to generate perceptual experiences. So how might these expectations give rise to perceptual experiences?

One hypothesis is that it is simply a matter of signal magnitude. If the modulatory drive in the deep layers is strong enough, it might ultimately lead to a concurrent conscious experience. Therefore, what might differentiate these signals from ordinary working memory and expectation signals is that they are simply stronger. Alternatively, it might not so much be the strength of the signal, but rather the level of sensory detail that these feedback signals contain. Indeed, it has been suggested that the brain's ability to separate imagination from reality relies on an inference on this sensory detail. If an experience is vivid and detailed, it is more likely to reflect reality, whereas if an experience is fuzzy and undetailed, it likely reflects imagination, or a different top-down process like expectation or working memory (Dijkstra et al. 2022). This hypothesis could in principle be tested using multivariate techniques in combination with layer fMRI. If this hypothesis is correct, the critical factor in turning a top-down perceptual expectation signal into a hallucinatory perceptual experience is an increased sharpening of the representation of the expected stimulus in the feedback layers of the visual cortex (Abdelhack and Kamitani 2018).

Finally, a third hypothesis may be that hallucinatory experiences arise through modulation of feedforward activity. For example, an expectation of a stimulus might prime pyramidal neurons in the deep cortical layers through receptors on their apical dendrites (Larkum 2013; Spruston 2008). Targeting the apical dendrites would not be expected to drive these pyramidal neurons directly but allow them to function as coincidence detectors (Larkum 2013). In turn, deep layer neurons can modulate incoming sensory input through their projections onto the middle layers (Binzegger

2004; Kim et al. 2014), allowing sensory input concurrent with expectations to be processed more quickly, giving them a head start in signal processing (Antic et al. 2010; Kim et al. 2014; Major et al. 2013). This suggests that a combination of expectations and noisy sensory input is critical in forming hallucinations. What follows from this view is that hallucinations in the absence of such noisy input might be rare. Indeed, this view, where feedback connections modulate sensory inputs rather than having the ability to directly drive activity in the input layers (Kim et al. 2014), may explain why many studies use concurrent noisy input to induce hallucinations (Haarsma et al. 2020; Kafadar et al. 2020; Powers et al. 2017; Schmack et al. 2013, 2021; Stuke et al. 2021; Teufel et al. 2015). That is, if there is no sensory noise to modulate, the feedback signals will not be in the position to give rise to false inferences. In practice, sensory noise could be environmental (e.g., in the dark at night or in the rain) or due to a loss of sensory precision in the early sensory cortex. Indeed, studies have consistently reported decreased sensitivity to sensory input in people with psychosis across various paradigms. For example, schizophrenia patients often show impaired detection of sensory stimuli across different sensory domains (Dondé et al. 2019). This could potentially reflect a separate compounding factor, allowing the hallucinations to manifest. Such a phenomenon is clearly seen in neurological disorders such as Charles Bonnet syndrome, where a loss of sensory input is key in driving the abnormal perceptual experiences (Burke 2002; Ffytche et al. 1998). Recently, behavioral studies have started to show support for this hypothesis in normative nonclinical samples, demonstrating reduced sensitivity to sensory input secondary to heightened hallucinatory reports on stimulus detection tasks (Haarsma et al. 2023; Benrimoh et al. 2024). Layer-specific fMRI could be used to simultaneously study the representation of sensory input and expectation signals and test how the two interact to give rise to hallucinatory experiences.

The hypotheses outlined thus far have pertained mostly to laminar signals in the early sensory cortex, as this is where most of the normative research has been conducted. Ultimately, in order to deepen our understanding of how expectations might shape perception in clinical populations, studies will need to move beyond the sensory cortices. Critical brain regions that are likely key in modulating these effects, and for some of which layer-specific fMRI has been demonstrated to contribute to novel insights, are the hippocampus and medial temporal lobe (Koster et al. 2018; Maass et al. 2014), prefrontal and association cortices (Finn et al. 2019, 2021), cross-modal sensory regions (Gau et al. 2020), and the insula in the case of interoception (Haufler et al. 2022; Nord et al. 2021). All of these regions have been demonstrated to play an integral role in integrating top-down contextual information with sensory signals and might therefore contribute to perceptual disturbances in different ways across clinical conditions (Bar 2007).

While enhanced sensory expectations might underlie some forms of hallucinations, there is significant behavioral evidence that demonstrates a reduced influence of prior expectations in psychosis, typically measured with various auditory and visual illusions, as discussed above. Once more layer-specific fMRI can contribute to our understanding of what underlies these changes in perception. Studies in normative samples that investigated visual and auditory illusions have found a role for agranular layers in mediating these auditory and visual illusions (Gau et al. 2020; Kok et al. 2016). These paradigms could be applied to study the mechanisms that result in the breaking of these illusions in psychosis, i.e., either a weakened influence of feedback signals or an increased reliance on sensory input, thereby outweighing sensory feedback signals.

Predictive coding accounts of hallucinations are not the only theories that could benefit from layer-specific fMRI. A different framework which has been considerably influential in recent years is reality monitoring. This framework rests on the proposal that in order for the mind to keep track of internally and externally generated signals like inner speech and external voices, the brain needs to somehow keep track of which signals are generated internally (Barnes et al. 2003; Bentall et al. 1991; Griffin and Fletcher 2017; Mondino et al. 2019). Central to many of these accounts is the idea that reality monitoring is implemented through comparing sensory experiences to so-called efference copies. These can be conceptualized as predictions of self-generated perceptual experiences, which can modulate sensory processing to give rise to the feeling of agency. Therefore, disturbances in the feeling of agency, as in delusions of control, or attributing inner speech to an external source, could involve aberrant signaling of efference copies or their integration with sensory signals (Griffin and Fletcher 2017). These models could be tested using layer-specific fMRI. Consider recent work that demonstrated laminar specific modulation of somatosensory responses and action signals by predictability and imagined action (Persichetti et al. 2020; Yu et al. 2019). An interesting future avenue of research would be to test whether this extends to the auditory domain and explore whether self-generated speech modulates the agranular layers and whether this is perturbed in individuals with auditory hallucinations. Indeed, layer-specific imaging of speech-sensitive regions has been shown to be feasible (Sharoh et al. 2019). Delusions of agency are also common in psychosis and have been thought to arise from a failure to modulate self-generated sensory signals (Frith 2012). Therefore, another direction of future research could be to explore how agency modulates somatosensory and movement-related signals in S1 and M1, respectively, in psychosis. One might expect that the top-down modulation found in the above-mentioned studies would be diminished in psychosis, providing strong evidence that disturbances in agency arise through diminished top-down modulation of self-generated sensory signals. Further, the posterior insula, a key hub for integrating bodily signals and therefore critical to interoception, has distinct cortical layers (Flynn 1999; Gogolla 2017). The exact function of the different layers remains to be clarified. However, it seems plausible that given the key role that the insula plays in interoception, disturbances in interoception could arise from alterations in how information is integrated on the level of the microcircuit. Indeed, recent studies in schizophrenia demonstrated aberrant cardiac interoception (Ardizzi et al. 2016; Jeganathan et al. 2024; Koreki et al. 2021; Torregrossa et al. 2022), while metacognitive awareness of these disturbances seems intact (Torregrossa et al. 2022). Interestingly, these disturbances seem to correlate with positive symptoms (Ardizzi et al. 2016; Koreki et al. 2021, although not in Jeganathan et al. 2024).

Given the importance of the insula in interoception (Fermin et al. 2022; Hassanpour et al. 2018; Nord et al. 2021; Simmons et al. 2013; Vicario et al. 2020), it is likely that such disturbances in psychotic disorders like schizophrenia find their origin in dysfunctional information processing on the level of the insular microcircuit.

Thus, the emergence of layer-specific fMRI provides a great opportunity to explore how perception might go awry across multiple sensory domains and clinical disorders, promising to shine a new light on unanswered questions.

6 Pharmacological Layer-Specific fMRI

Predictive coding theories of perception attribute important roles to various neurotransmitter systems in mediating the balance between feedback and feedforward signaling (Sterzer et al. 2018). If changes occur in these neuromodulatory systems, this balance can become perturbed, leading to false inferences in a myriad of ways. Imaging the laminar mechanisms that underlie perception while participants undergo different drug challenges could therefore provide invaluable insights into the different ways perception might go awry in clinical disorders. By characterizing how different signals can become perturbed and comparing these changes to what is observed in a disorder, we can start to see the similarities and differences between the clinical and pharmacological theories that are aimed at capturing the same phenomena. For instance, such studies might show that while psychosis patients and individuals undergoing a dopamine challenge might both increase their rates of false alarms on a detection task, the laminar mechanisms through which these behavioral reports manifest might be different. Thus, comparative pharmacological and clinical layer-specific imaging studies will be key in ultimately understanding these disorders. Here we review several possible novel directions of research.

One neurotransmitter system that has long been linked to the etiology of schizophrenia is the NMDA-receptor. The NMDA-receptor theory of schizophrenia goes back a number of decades (Olney et al. 1999) but has gained increasing support through various lines of genetic, molecular, and immunological evidence that suggest an important role for the glutamate system and the NMDA-receptor specifically in modulating some of the symptoms of schizophrenia (Lennox et al. 2017; Merritt and Egerton 2017; Schizophrenia Working Group of the Psychiatric Genomics Consortium 2014). Neuroimaging studies have demonstrated that NMDA-receptor agonists like ketamine perturb error signaling in associative learning paradigms (Corlett et al. 2006, 2016), as well as in mismatch negativity paradigms (Weber et al. 2020). Given the prevalence of NMDA-receptors on pyramidal feedback cells (Fox et al. 1989; Rosier et al. 1993), a plausible hypothesis would be that these perturbations in learning and sensory processing arise due to interference with feedback signaling. Indeed, there is some work in humans that shows that ketamine interferes with feedback signaling (van Loon et al. 2016). Alternatively, the described experimental effects could be explained by enhanced feedforward signaling. Here, layer-specific fMRI could be used to distinguish between these two competing hypotheses by investigating whether NMDA-antagonists interfere specifically with signaling in the deep layers or whether they alternatively enhance feedforward signaling by modulating the middle layers.

While the NMDA-receptor has mostly been associated with feedback signals, the cholinergic system has often been associated with modulating feedforward precision (Moran et al. 2013; Stephan et al. 2006). The cholinergic system might play an important part in the etiology of schizophrenia. For example, there are studies suggesting a loss of sensory precision in schizophrenia due to reduced cholinergic tone (Higley and Picciotto 2014). Further, in neurological conditions like Alzheimer's disease, Lewy-body dementia, and Parkinson's disease, reduced cholinergic function has been consistently related to an increase in visual hallucinations (O'Brien et al. 2020). Experimental evidence using an acetylcholine agonist has been shown to induce hallucinations by sensory conditioning (Warburton et al. 1985), suggesting a mechanistic role for cholinergic disturbances in hallucinations. Again, layer-specific imaging could be used to study whether cholinergic modulation primarily affects feedforward signaling through modulations of the middle and perhaps superficial layers or whether it perturbs feedback signaling in the deep layers instead.

The serotonin system, particularly the 5HT2a receptor that is modulated by classical psychedelics, is another promising neurotransmitter system that is likely to play a role in perceptual disturbances. Although research linking hallucinations in schizophrenia to the serotonin system is somewhat limited, there is increasing evidence that in Parkinson's disease it is a key contributor to visual hallucinations. Previous research has linked 5HT2a receptors in the visual ventral system to the prevalence of visual hallucinations in Parkinson's disease (Ballanger et al. 2010; Huot et al. 2010). Moreover, Pimavanserin, a partial 5HT2a agonist, has shown promising results in treating visual hallucinations in Parkinson's disease (Cummings et al. 2014; O'Brien et al. 2020). The exact mechanisms through which the 5HT2a serotonin system perturbs perceptual processing remain unclear. Some have argued that 5HT2a agonists weaken the influence of prior beliefs (Carhart-Harris and Friston 2019). Indeed, there is some indirect evidence for this. For example, some sensory processing phenomena believed to rely on feedback processing are reduced under psychedelics, like mismatch negativity responses (Timmermann et al. 2018), and the experience of illusory contours (Kometer et al. 2011, 2013), and binocular rivalry is altered as well (Carter et al. 2005, 2007). Neurophysiological studies have demonstrated that these substances alter glutamatergic signaling in layer V neurons, which, given their role in feedback processing, gives credence to the altered feedback processing hypothesis (Aghajanian and Marek 1999). However, the complex hallucinations that are sometimes experienced under these substances seem less compatible with the weaker prior account. Here, laminar imaging can help elucidate this apparent contradiction and test directly whether deep layer signaling of visual illusions (Kok et al. 2016) is perturbed by administration of psychedelics, whereas perhaps more cognitive expectation signals in the deep layers (Aitken et al. 2020; Haarsma et al. 2023) are enhanced. The origin of the top-down drive in these two instances could be different, with the former originating from higher regions within the visual cortex (Pak et al. 2020), and the latter from the hippocampus (Aitken and Kok 2022).

Finally, dopamine has long been implicated in the etiology of psychosis, both in schizophrenia and Parkinson's disease (Howes and Kapur 2009; Meltzer and Stahl 1976). Given the lack of dopamine receptors in the sensory cortex, it is unlikely that dopamine directly affects early sensory processing. Instead, what dopamine might do is enhance the influence of higher-order beliefs on lower-level sensory processing. Indeed, recent evidence from animal studies has suggested that dopamine might do exactly this (Schmack et al. 2021). In a series of experiments, it was demonstrated that hallucination-like behavior was preceded and causally affected by dopamine levels in the caudate nucleus. One hypothesis might be that activity in the caudate enhances feedback signals in the deep layers of the early auditory cortex, which could be studied using laminar imaging. Such evidence would strongly substantiate a role for dopamine in enhancing sensory prediction signals.

Thus, the combination of pharmacological and clinical layer-specific fMRI studies could provide key insights into the various ways perception can go awry. Comparative approaches are particularly encouraged here, as the same behavioral phenomenon might arise through different mechanisms, which layer-specific fMRI could help uncover.

7 Relevance of Layer-Specific fMRI to Precision Medicine

In addition to layer-specific fMRI's potential to test different hypotheses about what underlies perceptual aberrancies in illnesses like psychosis or Parkinson's disease, one of the more exciting prospects is that it allows taking a personalized approach to understanding these mechanisms. There are a range of theoretical papers that have been rightfully influential in how we currently understand hallucinations. However, a potential drawback of such theoretical accounts is that there is often an underlying assumption that there is a single explanation that underlies the broad category of hallucinatory phenomena. While this may be true to some extent, perhaps in the sense that there is a final common pathway that underlies all conscious experience, there is a potential risk that this approach loses sight of the various ways that perception can go awry. Perhaps an alternative approach might be to start with a theoretical framework in which to understand perception and use this to specify the diversity of mechanisms through which perception can go awry. This type of approach, which we have taken in this chapter, can provide a roadmap toward understanding hallucinations that releases us from the obligation to find a single explanation that subsumes all hallucinatory phenomena. A major advantage is that it instead allows us to think about the various routes that may lead to hallucinations, something that may be valuable for explaining individual variations across and even within conditions. Taking such an approach, auditory hallucinations as seen in schizophrenia need not arise through the same mechanisms as hallucinations seen in Parkinson's disease. In psychosis, the types of experiences that are typically considered to be hallucinations can range from low-level perceptual disturbances, such as geometrical patterns, to high-level hallucinations such as seeing fully formed entities or hearing voices, referred to as minor and major phenomena respectively (Mocellin et al. 2006; Pagonabarraga et al. 2016). Given their differences in phenomenology, these might well arise from different mechanisms, and methods like layer-specific fMRI could allow for a more nuanced understanding of these mechanisms.

Embracing this view could allow for a more personalized approach in which future layer-specific imaging studies could play a role. For example, in a recent study, participants occasionally reported perceiving oriented gratings that were not actually presented, with high levels of confidence. In this instance, false percepts were not driven by cued expectations and were reflected in the middle input layers of the visual cortex, potentially reflecting a form of feedforward hallucinations (Haarsma et al. 2023). Expectation-driven hallucinations, on the other hand, are more likely to result from deep layer signals (Aitken et al. 2020; Haarsma et al. 2023; Thomas et al. 2024). Therefore, in future, layer-specific fMRI could be used as a tool to identify patients' unique mechanistic profiles, allowing us to potentially tailor treatment accordingly.

As discussed in the previous section, extending layer-specific fMRI to include pharmacology will be key in this endeavor, and can aid our understanding of the different ways perception can go wrong. For example, false inferences during perceptual discrimination tasks might be the result of excessive feedforward signaling in some patients, whereas they result from enhanced feedback signals in others. If neuromodulatory systems are primarily involved in feedback over feedforward mechanisms, and vice versa, these systems could then be future targets of intervention depending on whether a given patient presents a feedforward or feedback layer profile during false inferences. Thus, although it is still likely a long way off, in theory laminar imaging has the potential to identify personalized treatment targets.

8 Challenges

As layer-specific fMRI is still a relatively novel method, there are still several challenges that limit its application in clinical groups, and overcoming these challenges will likely increase its potential impact. One well-known issue is that large draining veins in the cortex cause a bias towards the superficial layers, as BOLD signal arising from neural activity in deeper layers flows upward to the pial surface (Uludağ and Blinder 2018). This phenomenon complicates interpretation of conventional BOLD-based methods (Kay et al. 2019). The earliest studies attempted to identify voxels containing veins, which was reasonably successful in removing some bias (Koopmans et al. 2010). Alternatively, using non-BOLD-based fMRI sequences could address the root cause of the problem (Huber et al. 2019). One such method is CBV-weighed VASO (Lu et al. 2003), which has been demonstrated to have more local specificity and a more equally weighted contrast across the different cortical depths. However, the overall sensitivity of the obtained signals is reduced compared to more conventional Gradient Echo (GE) EPI sequences (Huber et al. 2015).

Another issue relates to restricted brain coverage, which limits layer-specific fMRI's ability to study whole-brain laminar dynamics. Although SS-SI-VASO or MAGEC_VASO sequences seem to provide superior brain coverage over conventional methods while not relying on the BOLD response for their signals (Huber et al. 2019). Given the importance of achieving high spatial resolution, motion artefacts can be another confounding factor, which might be particularly pertinent for older populations that struggle to lie still for the long periods of time layer-specific imaging requires. Prospective motion correction can limit the amount of pre-processing required once the data has been acquired (Bause et al. 2020). Further, developments in hardware to make participants more comfortable help to reduce motion.

Geometric distortions commonly occur in echo planar imaging (EPI) fMRI sequences. Worryingly these issues are exacerbated at higher magnetic field strengths, as the inhomogeneities in the b0 field become exacerbated, limiting the spatial selectivity of the fMRI signal. This is a considerable problem for layer-specific fMRI, which naturally requires high spatial selectivity. These issues can be partially corrected by acquiring reversed encoding images, which allow the researcher to measure the degree of distortion and adjust for them. Indeed, recently this method has been shown to work well for 3D-epi sequences, where the distortions were corrected in the areas where they typically occur the strongest, i.e., the frontal and temporal areas of the brain. Given the amount of distortions in these areas at higher field strengths, deploying distortion correction will be particularly pertinent when layer-specific imaging will move from the early sensory regions toward higher-order regions like the association and frontal cortices (Malekian et al. 2023).

Currently, layer-specific imaging studies most commonly achieve a voxel-resolution of 0.8 mm isotropic, which is sufficient to separate activity into three separate layers within the visual cortex, as the typical thickness of the grey matter is 2.5–3 mm on average. This means that activity in various neighboring laminae, such as layers V and VI, will be averaged together, limiting the extent to which we can make inferences about the contribution of different layers to perception. Preliminary work has achieved higher resolution images (0.2 mm) by using a line-scanning method (Morgan et al. 2020). If this was to become the standard, further delineation of the contribution of different laminae might come within reach.

A well-known limitation of fMRI is that it suffers from poor temporal resolution, which, given the importance of the timing of various neural events to theories of perception, limits its potential. Some potentially promising, but preliminary, work in anesthetized mice has been reported to be able to overcome the temporal limitation of fMRI by adopting a novel sequence referred to as DIANA (short for Direct Imaging of Neuronal Activity) (Toi et al. 2021). Using this sequence, the researchers were able to obtain laminar-specific neural (rather than hemodynamic) signals from the mouse somatosensory cortex with very high temporal resolution. However, some recent replication attempts have failed so far; thus its application in humans is still far from certain (Choi et al. 2023; Hodono et al. 2023).

To complement layer-specific fMRI, other avenues of noninvasive imaging could be explored. In recent years laminar MEG has emerged as a potentially exciting tool to study perception and cognition. Due to its ability to measure fluctuations in the magnetic field that are a direct consequence of electrical neural signals, it reflects a more direct measure of neural activity and is thereby only constrained by the quality of the data and the models used to make inferences about it (Bonaiuto et al. 2018a, b; Liuzzi et al. 2017; Meyer et al. 2017; Troebinger et al. 2014a, b).

Finally, in order to properly interpret layer-specific fMRI results, one needs to make a reliable assessment of the distribution of the histological layers across cortical depth. While the distribution of layers in the early sensory regions is fairly stable, in higher-order regions like the association cortex, the distribution of layers across depths is more variable (Finn et al. 2021). However, progress can be made by using evidence from anatomical images to gain information about the laminar distribution in higher-order regions like the association cortex. These include diffusion MRI, T1-weighted imaging of the myeloarchitecture, and magnetic susceptibility imaging, as well as using information from whole-brain laminar atlases, which will allow for more accurate estimations of the distribution of layers across cortical depth (Callaghan et al. 2014; Dinse et al. 2015; Finn et al. 2021; Trampel et al. 2019; Wagstyl et al. 2020).

9 Conclusion

Layer-specific fMRI holds great potential for drilling down on some of the leading theories of how perceptual disturbances emerge across various clinical and neurological disorders. By combining it with pharmacological theories of perception, we can start to see the similarities and divergences in how perception can go awry. Ultimately, these methods could aid in developing personalized treatments, helping to alleviate the suffering resulting from perceptual disturbances more effectively.

References

Abdelhack M, Kamitani Y (2018) Sharpening of hierarchical visual feature representations of blurred images. Eneuro 5(3):ENEURO.0443-17.2018. https://doi.org/10.1523/ENEURO.0443-17.2018

Aghajanian GK, Marek GJ (1999) Serotonin, via 5-HT2A receptors, increases EPSCs in layer V pyramidal cells of prefrontal cortex by an asynchronous mode of glutamate release. Brain Res 825(1–2):161–171. https://doi.org/10.1016/S0006-8993(99)01224-X

Aitken F, Kok P (2022) Hippocampal representations switch from errors to predictions during acquisition of predictive associations. Nat Commun 13(1):3294. https://doi.org/10.1038/s41467-022-31040-w

Aitken F, Menelaou G, Warrington O, Koolschijn RS, Corbin N, Callaghan MF, Kok P (2020) Prior expectations evoke stimulus-specific activity in the deep layers of the primary visual cortex. PLoS Biol 18(12):e3001023. https://doi.org/10.1371/journal.pbio.3001023

Allen M, Levy A, Parr T, Friston KJ (2022) In the body's eye: the computational anatomy of interoceptive inference. PLoS Comput Biol 18(9):e1010490. https://doi.org/10.1371/journal.pcbi.1010490

Angelucci A, Levitt JB, Walton EJS, Hupé J-M, Bullier J, Lund JS (2002) Circuits for local and global signal integration in primary visual cortex. J Neurosci 22(19):8633–8646. https://doi.org/10.1523/JNEUROSCI.22-19-08633.2002

Antic SD, Zhou W-L, Moore AR, Short SM, Ikonomu KD (2010) The decade of the dendritic NMDA spike. J Neurosci Res 88(14):2991–3001. https://doi.org/10.1002/jnr.22444

Ardizzi M, Ambrosecchia M, Buratta L, Ferri F, Peciccia M, Donnari S, Mazzeschi C, Gallese V (2016) Interoception and positive symptoms in schizophrenia. Front Hum Neurosci 10. https://doi.org/10.3389/fnhum.2016.00379

Arnal LH, Giraud A-L (2012) Cortical oscillations and sensory predictions. Trends Cogn Sci 16(7):390–398. https://doi.org/10.1016/j.tics.2012.05.003

Attinger A, Wang B, Keller GB (2017) Visuomotor coupling shapes the functional development of mouse visual cortex. Cell 169(7):1291–1302.e14. https://doi.org/10.1016/j.cell.2017.05.023

Ballanger B, Strafella AP, Van Eimeren T, Zurowski M, Rusjan PM, Houle S, Fox SH (2010) Serotonin 2A receptors and visual hallucinations in Parkinson disease. Arch Neurol 67(4). https://doi.org/10.1001/archneurol.2010.35

Bar M (2007) The proactive brain: using analogies and associations to generate predictions. Trends Cogn Sci 11(7):280–289. https://doi.org/10.1016/j.tics.2007.05.005

Barnes J, Boubert L, Harris J, Lee A, David AS (2003) Reality monitoring and visual hallucinations in Parkinson's disease. Neuropsychologia 41(5):565–574. https://doi.org/10.1016/S0028-3932(02)00182-3

Bastos AM, Usrey WM, Adams RA, Mangun GR, Fries P, Friston KJ (2012) Canonical microcircuits for predictive coding. Neuron 76(4):695–711. https://doi.org/10.1016/j.neuron.2012.10.038

Bastos AM, Loonis R, Kornblith S, Lundqvist M, Miller EK (2018) Laminar recordings in frontal cortex suggest distinct layers for maintenance and control of working memory. Proc Natl Acad Sci 115(5):1117–1122. https://doi.org/10.1073/pnas.1710323115

Bastos AM, Lundqvist M, Waite AS, Kopell N, Miller EK (2020) Layer and rhythm specificity for predictive routing. Proc Natl Acad Sci 117(49):31459–31469. https://doi.org/10.1073/pnas.2014868117

Bause J, Polimeni JR, Stelzer J, In M-H, Ehses P, Kraemer-Fernandez P, Aghaeifar A, Lacosse E, Pohmann R, Scheffler K (2020) Impact of prospective motion correction, distortion correction methods and large vein bias on the spatial accuracy of cortical laminar fMRI at 9.4 Tesla. NeuroImage 208:116434. https://doi.org/10.1016/j.neuroimage.2019.116434

Bell AH, Summerfield C, Morin EL, Malecek NJ, Ungerleider LG (2016) Encoding of stimulus probability in macaque inferior temporal cortex. Curr Biol 26(17):2280–2290. https://doi.org/10.1016/j.cub.2016.07.007

Bentall RP, Baker GA, Havers S (1991) Reality monitoring and psychotic hallucinations. Br J Clin Psychol 30(3):213–222. https://doi.org/10.1111/j.2044-8260.1991.tb00939.x

Benrimoh D, Fisher VL, Seabury R, Sibarium E, Mourgues C, Chen D, Powers A (2024) Evidence for reduced sensory precision and increased reliance on priors in hallucination-prone individuals in a general population sample. Schizophrenia Bull 50(2):349–362

Berntson GG, Khalsa SS (2021) Neural circuits of interoception. Trends Neurosci 44(1):17–28. https://doi.org/10.1016/j.tins.2020.09.011

Binzegger T (2004) A quantitative map of the circuit of cat primary visual cortex. J Neurosci 24(39):8441–8453. https://doi.org/10.1523/JNEUROSCI.1400-04.2004

Blakemore S-J, Smith J, Steel R, Johnstone EC, Frith CD (2000) The perception of self-produced sensory stimuli in patients with auditory hallucinations and passivity experiences: evidence for a breakdown in self-monitoring. Psychol Med 30(5):1131–1139. https://doi.org/10.1017/S0033291799002676

Bonaiuto JJ, Meyer SS, Little S, Rossiter H, Callaghan MF, Dick F, Barnes GR, Bestmann S (2018a) Lamina-specific cortical dynamics in human visual and sensorimotor cortices. eLife 7: e33977. https://doi.org/10.7554/eLife.33977

Bonaiuto JJ, Rossiter HE, Meyer SS, Adams N, Little S, Callaghan MF, Dick F, Bestmann S, Barnes GR (2018b) Non-invasive laminar inference with MEG: comparison of methods and source inversion algorithms. NeuroImage 167:372–383. https://doi.org/10.1016/j.neuroimage.2017.11.068

Bunney WE, Hetrick WP, Bunney BG, Patterson JV, Jin Y, Potkin SG, Sandman CA (1999) Structured interview for assessing perceptual anomalies (SIAPA). Schizophr Bull 25(3): 577–592. https://doi.org/10.1093/oxfordjournals.schbul.a033402

Burke W (2002) The neural basis of Charles Bonnet hallucinations: A hypothesis. J Neurol Neurosurg Psychiatry 73(5):535–541. https://doi.org/10.1136/jnnp.73.5.535

Callaghan MF, Freund P, Draganski B, Anderson E, Cappelletti M, Chowdhury R, Diedrichsen J, FitzGerald THB, Smittenaar P, Helms G, Lutti A, Weiskopf N (2014) Widespread age-related differences in the human brain microstructure revealed by quantitative magnetic resonance imaging. Neurobiol Aging 35(8):1862–1872. https://doi.org/10.1016/j.neurobiolaging.2014.02.008

Carhart-Harris RL, Friston KJ (2019) REBUS and the anarchic brain: toward a unified model of the brain action of psychedelics. Pharmacol Rev 71(3):316–344. https://doi.org/10.1124/pr.118.017160

Carter OL, Pettigrew JD, Hasler F, Wallis GM, Liu GB, Hell D, Vollenweider FX (2005) Modulating the rate and rhythmicity of perceptual rivalry alternations with the mixed 5-HT2A and 5-HT1A agonist psilocybin. Neuropsychopharmacology 30(6):1154–1162. https://doi.org/10.1038/sj.npp.1300621

Carter OL, Hasler F, Pettigrew JD, Wallis GM, Liu GB, Vollenweider FX (2007) Psilocybin links binocular rivalry switch rate to attention and subjective arousal levels in humans. Psychopharmacology 195(3):415–424. https://doi.org/10.1007/s00213-007-0930-9

Cassidy CM, Balsam PD, Weinstein JJ, Rosengard RJ, Slifstein M, Daw ND, Abi-Dargham A, Horga G (2018) A perceptual inference mechanism for hallucinations linked to striatal dopamine. Curr Biol 28(4):503–514.e4. https://doi.org/10.1016/j.cub.2017.12.059

Chen G, Wang F, Gore JC, Roe AW (2013) Layer-specific BOLD activation in awake monkey V1 revealed by ultra-high spatial resolution functional magnetic resonance imaging. NeuroImage 64:147–155. https://doi.org/10.1016/j.neuroimage.2012.08.060

Choi S-H, Im GH, Choi S, Yu X, Bandettini PA, Menon RS, Kim S-G (2023) No replication of direct neuronal activity-related (DIANA) fMRI in anesthetized mice [preprint]. Neuroscience. https://doi.org/10.1101/2023.05.26.542419

Clark A (2013) Whatever next? Predictive brains, situated agents, and the future of cognitive science. Behav Brain Sci 36(3):181–204. https://doi.org/10.1017/S0140525X12000477

Corlett PR, Honey GD, Aitken MRF, Dickinson A, Shanks DR, Absalom AR, Lee M, Pomarol-Clotet E, Murray GK, McKenna PJ, Robbins TW, Bullmore ET, Fletcher PC (2006) Frontal responses during learning predict vulnerability to the psychotogenic effects of ketamine: linking cognition, brain activity, and psychosis. Arch Gen Psychiatry 63(6):611. https://doi.org/10.1001/archpsyc.63.6.611

Corlett PR, Honey GD, Krystal JH, Fletcher PC (2011) Glutamatergic model psychoses: prediction error, learning, and inference. Neuropsychopharmacology 36(1):294–315. https://doi.org/10.1038/npp.2010.163

Corlett PR, Honey GD, Fletcher PC (2016) Prediction error, ketamine and psychosis: an updated model. J Psychopharmacol 30(11):1145–1155. https://doi.org/10.1177/0269881116650087

Corlett PR, Horga G, Fletcher PC, Alderson-Day B, Schmack K, Powers AR (2019) Hallucinations and strong priors. Trends Cogn Sci 23(2):114–127. https://doi.org/10.1016/j.tics.2018.12.001

Crapse TB, Sommer MA (2008) Corollary discharge across the animal kingdom. Nat Rev Neurosci 9(8):587–600. https://doi.org/10.1038/nrn2457

Cummings J, Isaacson S, Mills R, Williams H, Chi-Burris K, Corbett A, Dhall R, Ballard C (2014) Pimavanserin for patients with Parkinson's disease psychosis: a randomised, placebo-controlled phase 3 trial. Lancet 383(9916):533–540. https://doi.org/10.1016/S0140-6736(13)62106-6
Dakin S, Frith U (2005) Vagaries of visual perception in autism. Neuron 48(3):497–507. https://doi.org/10.1016/j.neuron.2005.10.018
de Lange FP, Heilbron M, Kok P (2018) How do expectations shape perception? Trends Cogn Sci 22(9):764–779. https://doi.org/10.1016/j.tics.2018.06.002
De Martino F, Moerel M, Ugurbil K, Goebel R, Yacoub E, Formisano E (2015) Frequency preference and attention effects across cortical depths in the human primary auditory cortex. Proc Natl Acad Sci 112(52):16036–16041. https://doi.org/10.1073/pnas.1507552112
de Sousa AA, Sherwood CC, Schleicher A, Amunts K, MacLeod CE, Hof PR, Zilles K (2010) Comparative cytoarchitectural analyses of striate and extrastriate areas in hominoids. Cereb Cortex 20(4):966–981. https://doi.org/10.1093/cercor/bhp158
Dijkstra N, Kok P, Fleming SM (2022) Perceptual reality monitoring: neural mechanisms dissociating imagination from reality. Neurosci Biobehav Rev 135:104557. https://doi.org/10.1016/j.neubiorev.2022.104557
Dima D, Roiser JP, Dietrich DE, Bonnemann C, Lanfermann H, Emrich HM, Dillo W (2009) Understanding why patients with schizophrenia do not perceive the hollow-mask illusion using dynamic causal modelling. NeuroImage 46(4):1180–1186. https://doi.org/10.1016/j.neuroimage.2009.03.033
Dima D, Dietrich DE, Dillo W, Emrich HM (2010) Impaired top-down processes in schizophrenia: A DCM study of ERPs. NeuroImage 52(3):824–832. https://doi.org/10.1016/j.neuroimage.2009.12.086
Dinse J, Härtwich N, Waehnert MD, Tardif CL, Schäfer A, Geyer S, Preim B, Turner R, Bazin P-L (2015) A cytoarchitecture-driven myelin model reveals area-specific signatures in human primary and secondary areas using ultra-high resolution in-vivo brain MRI. NeuroImage 114: 71–87. https://doi.org/10.1016/j.neuroimage.2015.04.023
Dondé C, Avissar M, Weber MM, Javitt DC (2019) A century of sensory processing dysfunction in schizophrenia. Eur Psychiatry 59:77–79. https://doi.org/10.1016/j.eurpsy.2019.04.006
Dudley R, Aynsworth C, Mosimann U, Taylor J-P, Smailes D, Collerton D, McCarthy-Jones S, Urwyler P (2019) A comparison of visual hallucinations across disorders. Psychiatry Res 272: 86–92. https://doi.org/10.1016/j.psychres.2018.12.052
Dumoulin SO, Fracasso A, Van Der Zwaag W, Siero JCW, Petridou N (2018) Ultra-high field MRI: advancing systems neuroscience towards mesoscopic human brain function. NeuroImage 168: 345–357. https://doi.org/10.1016/j.neuroimage.2017.01.028
Eversfield CL, Orton LD (2019) Auditory and visual hallucination prevalence in Parkinson's disease and dementia with Lewy bodies: a systematic review and meta-analysis. Psychol Med 49(14):2342–2353. https://doi.org/10.1017/S0033291718003161
Feinberg I (1978) Efference copy and corollary discharge: implications for thinking and its disorders*. Schizophr Bull 4(4):636–640. https://doi.org/10.1093/schbul/4.4.636
Felleman DJ, Van Essen DC (1991) Distributed hierarchical processing in the primate cerebral cortex. Cereb Cortex 1(1):1–47. https://doi.org/10.1093/cercor/1.1.1
Fenelon G, Thobois S, Bonnet A-M, Broussolle E, Tison F (2002) Tactile hallucinations in Parkinson's disease. J Neurol 249(12):1699–1703. https://doi.org/10.1007/s00415-002-0908-9
Fermin ASR, Friston K, Yamawaki S (2022) An insula hierarchical network architecture for active interoceptive inference. R Soc Open Sci 9(6):220226. https://doi.org/10.1098/rsos.220226
Ffytche DH, Howard RJ, Brammer MJ, David A, Woodruff P, Williams S (1998) The anatomy of conscious vision: an fMRI study of visual hallucinations. Nat Neurosci 1(8):738–742. https://doi.org/10.1038/3738
Finn ES, Huber L, Jangraw DC, Molfese PJ, Bandettini PA (2019) Layer-dependent activity in human prefrontal cortex during working memory. Nat Neurosci 22(10):1687–1695. https://doi.org/10.1038/s41593-019-0487-z

Finn ES, Huber L, Bandettini PA (2021) Higher and deeper: bringing layer fMRI to association cortex. Prog Neurobiol 207:101930. https://doi.org/10.1016/j.pneurobio.2020.101930

Fiser A, Mahringer D, Oyibo HK, Petersen AV, Leinweber M, Keller GB (2016) Experience-dependent spatial expectations in mouse visual cortex. Nat Neurosci 19(12):1658–1664. https://doi.org/10.1038/nn.4385

Flynn FG (1999) Anatomy of the insula functional and clinical correlates. Aphasiology 13(1): 55–78. https://doi.org/10.1080/026870399402325

Ford JM, Mathalon DH (2005) Corollary discharge dysfunction in schizophrenia: can it explain auditory hallucinations? Int J Psychophysiol 58(2–3):179–189. https://doi.org/10.1016/j.ijpsycho.2005.01.014

Fox K, Sato H, Daw N (1989) The location and function of NMDA receptors in cat and kitten visual cortex. J Neurosci 9(7):2443–2454. https://doi.org/10.1523/JNEUROSCI.09-07-02443.1989

Friston K (2009) The free-energy principle: a rough guide to the brain? Trends Cogn Sci 13(7): 293–301. https://doi.org/10.1016/j.tics.2009.04.005

Friston K (2012) The history of the future of the Bayesian brain. NeuroImage 62(2):1230–1233. https://doi.org/10.1016/j.neuroimage.2011.10.004

Friston K (2018) Does predictive coding have a future? Nat Neurosci 21(8):1019–1021. https://doi.org/10.1038/s41593-018-0200-7

Frith C (2012) Explaining delusions of control: the comparator model 20 years on. Conscious Cogn 21(1):52–54. https://doi.org/10.1016/j.concog.2011.06.010

Garrido MI, Kilner JM, Stephan KE, Friston KJ (2009) The mismatch negativity: a review of underlying mechanisms. Clin Neurophysiol 120(3):453–463. https://doi.org/10.1016/j.clinph.2008.11.029

Gau R, Bazin P-L, Trampel R, Turner R, Noppeney U (2020) Resolving multisensory and attentional influences across cortical depth in sensory cortices. eLife 9:e46856. https://doi.org/10.7554/eLife.46856

Gibson JJ (1950) The perception of the visual world. Houghton Mifflin, p xii, 242

Goense JBM, Logothetis NK (2006) Laminar specificity in monkey V1 using high-resolution SE-fMRI. Magn Reson Imaging 24(4):381–392. https://doi.org/10.1016/j.mri.2005.12.032

Gogolla N (2017) The insular cortex. Curr Biol 27(12):R580–R586. https://doi.org/10.1016/j.cub.2017.05.010

Gregory RL (1997) Knowledge in perception and illusion. Philos Trans R Soc Lond B Biol Sci 352(1358):1121–1127. https://doi.org/10.1098/rstb.1997.0095

Gregory RL, Longuet-Higgins HC, Sutherland NS (1980) Perceptions as hypotheses. Philos Trans R Soc London B: Biol Sci 290(1038):181–197. https://doi.org/10.1098/rstb.1980.0090

Griffin JD, Fletcher PC (2017) Predictive processing, source monitoring, and psychosis. Annu Rev Clin Psychol 13(1):265–289. https://doi.org/10.1146/annurev-clinpsy-032816-045145

Haarsma J, Knolle F, Griffin JD, Taverne H, Mada M, Goodyer IM, The NSPN Consortium, Fletcher PC, Murray GK (2020) Influence of prior beliefs on perception in early psychosis: effects of illness stage and hierarchical level of belief. J Abnorm Psychol 129(6):581–598. https://doi.org/10.1037/abn0000494

Haarsma J, Fletcher PC, Griffin JD, Taverne HJ, Ziauddeen H, Spencer TJ, Miller C, Katthagen T, Goodyer I, Diederen KMJ, Murray GK (2021) Precision weighting of cortical unsigned prediction error signals benefits learning, is mediated by dopamine, and is impaired in psychosis. Mol Psychiatry 26(9):5320–5333. https://doi.org/10.1038/s41380-020-0803-8

Haarsma J, Kok P, Browning M (2022) The promise of layer-specific neuroimaging for testing predictive coding theories of psychosis. Schizophr Res 245:68–76. https://doi.org/10.1016/j.schres.2020.10.009

Haarsma J, Deveci N, Corbin N, Callaghan MF, Kok P (2023) Expectation cues and false percepts generate stimulus-specific activity in distinct layers of the early visual cortex Laminar profile of visual false percepts. J Neurosci:JN-RM-0998-23. https://doi.org/10.1523/JNEUROSCI.0998-23.2023

Han S, Son JP, Cho H, Park J, Kim S (2019) Gradient-echo and spin-echo blood oxygenation level–dependent functional MRI at ultrahigh fields of 9.4 and 15.2 Tesla. Magn Reson Med 81(2): 1237–1246. https://doi.org/10.1002/mrm.27457

Hassanpour MS, Simmons WK, Feinstein JS, Luo Q, Lapidus RC, Bodurka J, Paulus MP, Khalsa SS (2018) The insular cortex dynamically maps changes in cardiorespiratory interoception. Neuropsychopharmacology 43(2):426–434

Haufler D, Liran O, Buchanan RJ, Pare D (2022) Human anterior insula signals salience and deviations from expectations via bursts of beta oscillations. J Neurophysiol 128(1):160–180. https://doi.org/10.1152/jn.00106.2022

Higley MJ, Picciotto MR (2014) Neuromodulation by acetylcholine: examples from schizophrenia and depression. Curr Opin Neurobiol 29:88–95. https://doi.org/10.1016/j.conb.2014.06.004

Hindy NC, Ng FY, Turk-Browne NB (2016) Linking pattern completion in the hippocampus to predictive coding in visual cortex. Nat Neurosci 19(5):665–667. https://doi.org/10.1038/nn.4284

Hodono S, Rideaux R, van Kerkoerle T, Cloos MA (2023) Initial experiences with Direct Imaging of Neuronal Activity (DIANA) in humans. arXiv:2303.00161:arXiv. http://arxiv.org/abs/2303.00161

Howes OD, Kapur S (2009) The dopamine hypothesis of schizophrenia: version III—the final common pathway. Schizophr Bull 35(3):549–562

Huber L, Goense J, Kennerley AJ, Trampel R, Guidi M, Reimer E, Ivanov D, Neef N, Gauthier CJ, Turner R, Möller HE (2015) Cortical lamina-dependent blood volume changes in human brain at 7 T. NeuroImage 107:23–33. https://doi.org/10.1016/j.neuroimage.2014.11.046

Huber L, Uludağ K, Möller HE (2019) Non-BOLD contrast for laminar fMRI in humans: CBF, CBV, and CMRO2. NeuroImage 197:742–760. https://doi.org/10.1016/j.neuroimage.2017.07.041

Huot P, Johnston TH, Darr T, Hazrati L-N, Visanji NP, Pires D, Brotchie JM, Fox SH (2010) Increased 5-HT $_{2A}$ receptors in the temporal cortex of parkinsonian patients with visual hallucinations: 5-HT $_{2A}$ receptors in visual hallucinations. Mov Disord 25(10):1399–1408. https://doi.org/10.1002/mds.23083

Jeganathan J, Campbell MEJ, Legrand N, Allen M, Breakspear M (2024) Aberrant cardiac interoception in psychosis. Schizophr Bull 51(1):208–216. https://doi.org/10.1093/schbul/sbae078

Kafadar E, Mittal VA, Strauss GP, Chapman HC, Ellman LM, Bansal S, Gold JM, Alderson-Day B, Evans S, Moffatt J, Silverstein SM, Walker EF, Woods SW, Corlett PR, Powers AR (2020) Modeling perception and behavior in individuals at clinical high risk for psychosis: support for the predictive processing framework. Schizophr Res 226:167–175. https://doi.org/10.1016/j.schres.2020.04.017

Kay K, Jamison KW, Vizioli L, Zhang R, Margalit E, Ugurbil K (2019) A critical assessment of data quality and venous effects in sub-millimeter fMRI. NeuroImage 189:847–869. https://doi.org/10.1016/j.neuroimage.2019.02.006

Keller GB, Mrsic-Flogel TD (2018) Predictive processing:a canonical cortical computation. Neuron 100(2):424–435. https://doi.org/10.1016/j.neuron.2018.10.003

Keller GB, Bonhoeffer T, Hübener M (2012) Sensorimotor mismatch signals in primary visual cortex of the behaving mouse. Neuron 74(5):809–815. https://doi.org/10.1016/j.neuron.2012.03.040

Khalsa SS, Craske MG, Li W, Vangala S, Strober M, Feusner JD (2015) Altered interoceptive awareness in anorexia nervosa: effects of meal anticipation, consumption and bodily arousal. Int J Eat Disord 48(7):889–897. https://doi.org/10.1002/eat.22387

Khalsa SS, Adolphs R, Cameron OG, Critchley HD, Davenport PW, Feinstein JS, Feusner JD, Garfinkel SN, Lane RD, Mehling WE, Meuret AE, Nemeroff CB, Oppenheimer S, Petzschner FH, Pollatos O, Rhudy JL, Schramm LP, Simmons WK, Stein MB et al (2018) Interoception and mental health: a roadmap. Biol Psychiatry Cognit Neurosci Neuroimaging 3(6):501–513. https://doi.org/10.1016/j.bpsc.2017.12.004

Kim J, Matney CJ, Blankenship A, Hestrin S, Brown SP (2014) Layer 6 Corticothalamic neurons activate a cortical output layer, layer 5a. J Neurosci 34(29):9656–9664. https://doi.org/10.1523/JNEUROSCI.1325-14.2014

Klein BP, Fracasso A, Van Dijk JA, Paffen CLE, Te Pas SF, Dumoulin SO (2018) Cortical depth dependent population receptive field attraction by spatial attention in human V1. NeuroImage 176:301–312. https://doi.org/10.1016/j.neuroimage.2018.04.055

Kok P, Turk-Browne NB (2018) Associative prediction of visual shape in the hippocampus. J Neurosci 38(31):6888–6899. https://doi.org/10.1523/JNEUROSCI.0163-18.2018

Kok P, Failing MF, de Lange FP (2014) Prior expectations evoke stimulus templates in the primary visual cortex. J Cogn Neurosci 26(7):1546–1554. https://doi.org/10.1162/jocn_a_00562

Kok P, Bains LJ, Van Mourik T, Norris DG, de Lange FP (2016) Selective activation of the deep layers of the human primary visual cortex by top-down feedback. Curr Biol 26(3):371–376

Kok P, Mostert P, De Lange FP (2017) Prior expectations induce prestimulus sensory templates. Proc Natl Acad Sci 114(39):10473–10478

Kometer M, Cahn BR, Andel D, Carter OL, Vollenweider FX (2011) The 5-HT2A/1A agonist psilocybin disrupts modal object completion associated with visual hallucinations. Biol Psychiatry 69(5):399–406. https://doi.org/10.1016/j.biopsych.2010.10.002

Kometer M, Schmidt A, Jancke L, Vollenweider FX (2013) Activation of serotonin 2A receptors underlies the psilocybin-induced effects on oscillations, N170 visual-evoked potentials, and visual hallucinations. J Neurosci 33(25):10544–10551. https://doi.org/10.1523/JNEUROSCI.3007-12.2013

Koopmans PJ, Barth M, Norris DG (2010) Layer-specific BOLD activation in human V1. Hum Brain Mapp 31(9):1297–1304. https://doi.org/10.1002/hbm.20936

Koopmans PJ, Barth M, Orzada S, Norris DG (2011) Multi-echo fMRI of the cortical laminae in humans at 7T. NeuroImage 56(3):1276–1285. https://doi.org/10.1016/j.neuroimage.2011.02.042

Koreki A, Funayama M, Terasawa Y, Onaya M, Mimura M (2021) Aberrant interoceptive accuracy in patients with schizophrenia performing a heartbeat counting task. Schizophrenia Bulletin Open 2(1):sgaa067. https://doi.org/10.1093/schizbullopen/sgaa067

Koster R, Chadwick MJ, Chen Y, Berron D, Banino A, Düzel E, Hassabis D, Kumaran D (2018) Big-loop recurrence within the hippocampal system supports integration of information across episodes. Neuron 99(6):1342–1354.e6. https://doi.org/10.1016/j.neuron.2018.08.009

Larkum M (2013) A cellular mechanism for cortical associations: an organizing principle for the cerebral cortex. Trends Neurosci 36(3):141–151. https://doi.org/10.1016/j.tins.2012.11.006

Lavenex P, Amaral DG (2000) Hippocampal-neocortical interaction: a hierarchy of associativity. Hippocampus 10(4):420–430. https://doi.org/10.1002/1098-1063(2000)10:4<420::AID-HIPO8>3.0.CO;2-5

Lawrence SJD, van Mourik T, Kok P, Koopmans PJ, Norris DG, de Lange FP (2018) Laminar organization of working memory signals in human visual cortex. Curr Biol 28(21):3435–3440.e4. https://doi.org/10.1016/j.cub.2018.08.043

Lawrence SJD, Formisano E, Muckli L, de Lange FP (2019a) Laminar fMRI: applications for cognitive neuroscience. NeuroImage 197:785–791. https://doi.org/10.1016/j.neuroimage.2017.07.004

Lawrence SJ, Norris DG, de Lange FP (2019b) Dissociable laminar profiles of concurrent bottom-up and top-down modulation in the human visual cortex. eLife 8:e44422. https://doi.org/10.7554/eLife.44422

Lee TS, Nguyen M (2001) Dynamics of subjective contour formation in the early visual cortex. Proc Natl Acad Sci 98(4):5

Leinweber M, Ward DR, Sobczak JM, Attinger A, Keller GB (2017) A sensorimotor circuit in mouse cortex for visual flow predictions. Neuron 95(6):1420–1432.e5. https://doi.org/10.1016/j.neuron.2017.08.036

Lennox BR, Palmer-Cooper EC, Pollak T, Hainsworth J, Marks J, Jacobson L, Lang B, Fox H, Ferry B, Scoriels L, Crowley H, Jones PB, Harrison PJ, Vincent A (2017) Prevalence and clinical characteristics of serum neuronal cell surface antibodies in first-episode psychosis: A case-control study. Lancet Psychiatry 4(1):42–48. https://doi.org/10.1016/S2215-0366(16)30375-3

Lewandowski KE, DePaola J, Camsari GB, Cohen BM, Öngür D (2009) Tactile, olfactory, and gustatory hallucinations in psychotic disorders: a descriptive study. Ann Acad Med Singap 38(5):383–387. https://doi.org/10.47102/annals-acadmedsg.V38N5p383

Lim A, Hoek HW, Deen ML, Blom JD, Bruggeman R, Cahn W, De Haan L, Kahn RS, Meijer CJ, Myin-Germeys I, Van Os J, Wiersma D (2016) Prevalence and classification of hallucinations in multiple sensory modalities in schizophrenia spectrum disorders. Schizophr Res 176(2–3): 493–499. https://doi.org/10.1016/j.schres.2016.06.010

Liuzzi L, Gascoyne LE, Tewarie PK, Barratt EL, Boto E, Brookes MJ (2017) Optimising experimental design for MEG resting state functional connectivity measurement. NeuroImage 155: 565–576. https://doi.org/10.1016/j.neuroimage.2016.11.064

Lu H, Golay X, Pekar JJ, Van Zijl PCM (2003) Functional magnetic resonance imaging based on changes in vascular space occupancy. Magn Reson Med 50(2):263–274. https://doi.org/10.1002/mrm.10519

Maass A, Schütze H, Speck O, Yonelinas A, Tempelmann C, Heinze H-J, Berron D, Cardenas-Blanco A, Brodersen KH, Enno Stephan K, Düzel E (2014) Laminar activity in the hippocampus and entorhinal cortex related to novelty and episodic encoding. Nat Commun 5(1):5547. https://doi.org/10.1038/ncomms6547

Major G, Larkum ME, Schiller J (2013) Active properties of neocortical pyramidal neuron dendrites. Annu Rev Neurosci 36(1):1–24. https://doi.org/10.1146/annurev-neuro-062111-150343

Malekian V, Graedel NN, Hickling A, Aghaeifar A, Dymerska B, Corbin N, Josephs O, Maguire EA, Callaghan MF (2023) Mitigating susceptibility-induced distortions in high-resolution 3DEPI fMRI at 7T. NeuroImage 279:120294. https://doi.org/10.1016/j.neuroimage.2023.120294

Markov NT, Vezoli J, Chameau P, Falchier A, Quilodran R, Huissoud C, Lamy C, Misery P, Giroud P, Ullman S, Barone P, Dehay C, Knoblauch K, Kennedy H (2014) Anatomy of hierarchy: feedforward and feedback pathways in macaque visual cortex. J Comp Neurol 522(1):225–259. https://doi.org/10.1002/cne.23458

Marr D (2010) Vision: a computational investigation into the human representation and processing of visual information. MIT Press

McGHIE A, Chapman J (1961) Disorders of attention and perception in early schizophrenia. Br J Med Psychol 34(2):103–116. https://doi.org/10.1111/j.2044-8341.1961.tb00936.x

Meltzer HY, Stahl SM (1976) The dopamine hypothesis of schizophrenia: a review*. Schizophr Bull 2(1):19–76. https://doi.org/10.1093/schbul/2.1.19

Merritt K, Egerton A (2017) 165. Nature of glutamate alterations in schizophrenia: a meta-analysis of proton magnetic resonance spectroscopy studies. Schizophr Bull 43(suppl_1):S84. https://doi.org/10.1093/schbul/sbx021.223

Meyer SS, Bonaiuto J, Lim M, Rossiter H, Waters S, Bradbury D, Bestmann S, Brookes M, Callaghan MF, Weiskopf N, Barnes GR (2017) Flexible head-casts for high spatial precision MEG. J Neurosci Methods 276:38–45. https://doi.org/10.1016/j.jneumeth.2016.11.009

Mocellin R, Walterfang M, Velakoulis D (2006) Neuropsychiatry of complex visual hallucinations. Aust N Z J Psychiatry. https://doi.org/10.1080/j.1440-1614.2006.01878.x

Moerel M, De Martino F, Uğurbil K, Yacoub E, Formisano E (2019) Processing complexity increases in superficial layers of human primary auditory cortex. Sci Rep 9(1):5502. https://doi.org/10.1038/s41598-019-41965-w

Mondino M, Dondé C, Lavallé L, Haesebaert F, Brunelin J (2019) Reality-monitoring deficits and visual hallucinations in schizophrenia. Eur Psychiatry 62:10–14. https://doi.org/10.1016/j.eurpsy.2019.08.010

Moran RJ, Campo P, Symmonds M, Stephan KE, Dolan RJ, Friston KJ (2013) Free energy, precision and learning: the role of cholinergic Neuromodulation. J Neurosci 33(19): 8227–8236. https://doi.org/10.1523/JNEUROSCI.4255-12.2013

Morgan AT, Nothnagel N, Petro LS, Goense J, Muckli L (2020) High-resolution line-scanning reveals distinct visual response properties across human cortical layers [Preprint]. Neuroscience. https://doi.org/10.1101/2020.06.30.179762

Muckli L, De Martino F, Vizioli L, Petro LS, Smith FW, Ugurbil K, Goebel R, Yacoub E (2015) Contextual feedback to superficial layers of V1. Curr Biol 25(20):2690–2695. https://doi.org/10.1016/j.cub.2015.08.057

Muzzu T, Saleem AB (2021) Feature selectivity can explain mismatch signals in mouse visual cortex. Cell Rep 37(1):109772. https://doi.org/10.1016/j.celrep.2021.109772

Nord CL, Lawson RP, Dalgleish T (2021) Disrupted dorsal mid-insula activation during Interoception across psychiatric disorders. Am J Psychiatry 178(8):761–770. https://doi.org/10.1176/appi.ajp.2020.20091340

Norris DG (2012) Spin-echo fMRI: the poor relation? NeuroImage 62(2):1109–1115. https://doi.org/10.1016/j.neuroimage.2012.01.003

O'Brien J, Taylor JP, Ballard C, Barker RA, Bradley C, Burns A, Collerton D, Dave S, Dudley R, Francis P, Gibbons A, Harris K, Lawrence V, Leroi I, McKeith I, Michaelides M, Naik C, O'Callaghan C, Olsen K et al (2020) Visual hallucinations in neurological and ophthalmological disease: pathophysiology and management. J Neurol Neurosurg Psychiatry 91(5):512–519. https://doi.org/10.1136/jnnp-2019-322702

O'Callaghan C, Hall JM, Tomassini A, Muller AJ, Walpola IC, Moustafa AA, Shine JM, Lewis SJG (2017) Visual hallucinations are characterized by impaired sensory evidence accumulation: insights from hierarchical drift diffusion Modeling in Parkinson's disease. Biol Psychiatry Cognit Neurosci Neuroimaging 2(8):680–688. https://doi.org/10.1016/j.bpsc.2017.04.007

Olman CA, Harel N, Feinberg DA, He S, Zhang P, Ugurbil K, Yacoub E (2012) Layer-specific fMRI reflects different neuronal computations at different depths in human V1. PLoS One 7(3): e32536. https://doi.org/10.1371/journal.pone.0032536

Olney JW, Newcomer JW, Farber NB (1999) NMDA receptor hypofunction model of schizophrenia. J Psychiatr Res 33(6):523–533. https://doi.org/10.1016/S0022-3956(99)00029-1

Pagonabarraga J, Martinez-Horta S, Fernández de Bobadilla R, Pérez J, Ribosa-Nogué R, Marín J, Pascual-Sedano B, García C, Gironell A, Kulisevsky J (2016) Minor hallucinations occur in drug-naive Parkinson's disease patients, even from the premotor phase: minor hallucinations in untreated PD patients. Mov Disord 31(1):45–52. https://doi.org/10.1002/mds.26432

Pak A, Ryu E, Li C, Chubykin AA (2020) Top-down feedback controls the cortical representation of illusory contours in mouse primary visual cortex. J Neurosci 40(3):648–660. https://doi.org/10.1523/JNEUROSCI.1998-19.2019

Paulus MP, Stein MB (2010) Interoception in anxiety and depression. Brain Struct Funct 214(5–6): 451–463. https://doi.org/10.1007/s00429-010-0258-9

Pearl D, Yodashkin-Porat D, Katz N, Valevski A, Aizenberg D, Sigler M, Weizman A, Kikinzon L (2009) Differences in audiovisual integration, as measured by McGurk phenomenon, among adult and adolescent patients with schizophrenia and age-matched healthy control groups. Compr Psychiatry 50(2):186–192. https://doi.org/10.1016/j.comppsych.2008.06.004

Persichetti AS, Avery JA, Huber L, Merriam EP, Martin A (2020) Layer-specific contributions to imagined and executed hand movements in human primary motor cortex. Curr Biol 30(9): 1721–1725.e3. https://doi.org/10.1016/j.cub.2020.02.046

Polimeni JR, Fischl B, Greve DN, Wald LL (2010) Laminar analysis of 7T BOLD using an imposed spatial activation pattern in human V1. NeuroImage 52(4):1334–1346. https://doi.org/10.1016/j.neuroimage.2010.05.005

Powers AR, Kelley M, Corlett PR (2016) Hallucinations as top-down effects on perception. Biol Psychiatry Cognit Neurosci Neuroimaging 1(5):393–400. https://doi.org/10.1016/j.bpsc.2016.04.003

Powers AR, Mathys C, Corlett PR (2017) Pavlovian conditioning–induced hallucinations result from overweighting of perceptual priors. Science 357(6351):596–600. https://doi.org/10.1126/science.aan3458

Reichert DP, Seriès P, Storkey AJ (2013) Charles Bonnet syndrome: evidence for a generative model in the cortex? PLoS Comput Biol 9(7):e1003134. https://doi.org/10.1371/journal.pcbi.1003134

Rosier AM, Arckens L, Orban GA, Vandesande F (1993) Laminar distribution of NMDA receptors in cat and monkey visual cortex visualized by [3H]-MK-801 binding. J Comp Neurol 335(3):369–380. https://doi.org/10.1002/cne.903350307

Schapiro AC, Kustner LV, Turk-Browne NB (2012) Shaping of object representations in the human medial temporal lobe based on temporal regularities. Curr Biol 22(17):1622–1627. https://doi.org/10.1016/j.cub.2012.06.056

Schizophrenia Working Group of the Psychiatric Genomics Consortium (2014) Biological insights from 108 schizophrenia-associated genetic loci. Nature 511(7510):421–427. https://doi.org/10.1038/nature13595

Schmack K, Gomez-Carrillo de Castro A, Rothkirch M, Sekutowicz M, Rossler H, Haynes J-D, Heinz A, Petrovic P, Sterzer P (2013) Delusions and the role of beliefs in perceptual inference. J Neurosci 33(34):13701–13712. https://doi.org/10.1523/JNEUROSCI.1778-13.2013

Schmack K, Bosc M, Ott T, Sturgill JF, Kepecs A (2021) Striatal dopamine mediates hallucination-like perception in mice. Science 372(6537):eabf4740. https://doi.org/10.1126/science.abf4740

Self MW, van Kerkoerle T, Goebel R, Roelfsema PR (2019) Benchmarking laminar fMRI: neuronal spiking and synaptic activity during top-down and bottom-up processing in the different layers of cortex. NeuroImage 197:806–817. https://doi.org/10.1016/j.neuroimage.2017.06.045

Serre T, Oliva A, Poggio T (2007) A feedforward architecture accounts for rapid categorization. Proc Natl Acad Sci 104(15):6424–6429. https://doi.org/10.1073/pnas.0700622104

Seth AK, Tsakiris M (2018) Being a beast machine: the somatic basis of selfhood. Trends Cogn Sci 22(11):969–981. https://doi.org/10.1016/j.tics.2018.08.008

Sharoh D, Van Mourik T, Bains LJ, Segaert K, Weber K, Hagoort P, Norris DG (2019) Laminar specific fMRI reveals directed interactions in distributed networks during language processing. Proc Natl Acad Sci 116(42):21185–21190. https://doi.org/10.1073/pnas.1907858116

Shergill SS, Samson G, Bays PM, Frith CD, Wolpert DM (2005) Evidence for sensory prediction deficits in schizophrenia. Am J Psychiatry 162(12):2384–2386. https://doi.org/10.1176/appi.ajp.162.12.2384

Simmons WK, Avery JA, Barcalow JC, Bodurka J, Drevets WC, Bellgowan P (2013) Keeping the body in mind: insula functional organization and functional connectivity integrate interoceptive, exteroceptive, and emotional awareness. Hum Brain Mapp 34(11):2944–2958. https://doi.org/10.1002/hbm.22113

Simons JS, Garrison JR, Johnson MK (2017) Brain mechanisms of reality monitoring. Trends Cogn Sci 21(6):462–473. https://doi.org/10.1016/j.tics.2017.03.012

Spratling MW (2017) A review of predictive coding algorithms. Brain Cogn 112:92–97. https://doi.org/10.1016/j.bandc.2015.11.003

Spruston N (2008) Pyramidal neurons: dendritic structure and synaptic integration. Nat Rev Neurosci 9(3):206–221. https://doi.org/10.1038/nrn2286

Stachenfeld KL, Botvinick MM, Gershman SJ (2017) The hippocampus as a predictive map. Nat Neurosci 20(11):1643–1653. https://doi.org/10.1038/nn.4650

Stephan KE, Baldeweg T, Friston KJ (2006) Synaptic plasticity and Dysconnection in schizophrenia. Biol Psychiatry 59(10):929–939. https://doi.org/10.1016/j.biopsych.2005.10.005

Stephan KE, Petzschner FH, Kasper L, Bayer J, Wellstein KV, Stefanics G, Pruessmann KP, Heinzle J (2019) Laminar fMRI and computational theories of brain function. NeuroImage 197:699–706. https://doi.org/10.1016/j.neuroimage.2017.11.001

Sterzer P, Adams RA, Fletcher P, Frith C, Lawrie SM, Muckli L, Petrovic P, Uhlhaas P, Voss M, Corlett PR (2018) The predictive coding account of psychosis. Biol Psychiatry 84(9):634–643. https://doi.org/10.1016/j.biopsych.2018.05.015

Stuke H, Kress E, Weilnhammer VA, Sterzer P, Schmack K (2021) Overly strong priors for socially meaningful visual signals are linked to psychosis proneness in healthy individuals. Front Psychol 12:583637. https://doi.org/10.3389/fpsyg.2021.583637

Teufel C, Subramaniam N, Dobler V, Perez J, Finnemann J, Mehta PR, Goodyer IM, Fletcher PC (2015) Shift toward prior knowledge confers a perceptual advantage in early psychosis and psychosis-prone healthy individuals. Proc Natl Acad Sci 112(43):13401–13406. https://doi.org/10.1073/pnas.1503916112

Thomas ER, Haarsma J, Nicholson J, Yon D, Kok P, Press C (2024) Predictions and errors are distinctly represented across V1 layers. Curr Biol 34(10):2265–2271.e4. https://doi.org/10.1016/j.cub.2024.04.036

Timmermann C, Spriggs MJ, Kaelen M, Leech R, Nutt DJ, Moran RJ, Carhart-Harris RL, Muthukumaraswamy SD (2018) LSD modulates effective connectivity and neural adaptation mechanisms in an auditory oddball paradigm. Neuropharmacology 142:251–262. https://doi.org/10.1016/j.neuropharm.2017.10.039

Toi PT, Jang HJ, Min K, Kim S-P, Lee S-K, Lee J, Kwag J, Park J-Y (2021) In vivo direct imaging of neuronal activity at high temporo-spatial resolution [Preprint]. Neuroscience. https://doi.org/10.1101/2021.05.21.444581

Torregrossa LJ, Amedy A, Roig J, Prada A, Park S (2022) Interoceptive functioning in schizophrenia and schizotypy. Schizophr Res 239:151–159. https://doi.org/10.1016/j.schres.2021.11.046

Trampel R, Bazin P-L, Pine K, Weiskopf N (2019) In-vivo magnetic resonance imaging (MRI) of laminae in the human cortex. NeuroImage 197:707–715. https://doi.org/10.1016/j.neuroimage.2017.09.037

Troebinger L, López JD, Lutti A, Bestmann S, Barnes G (2014a) Discrimination of cortical laminae using MEG. NeuroImage 102:885–893. https://doi.org/10.1016/j.neuroimage.2014.07.015

Troebinger L, López JD, Lutti A, Bradbury D, Bestmann S, Barnes G (2014b) High precision anatomy for MEG. NeuroImage 86:583–591. https://doi.org/10.1016/j.neuroimage.2013.07.065

Uludağ K, Blinder P (2018) Linking brain vascular physiology to hemodynamic response in ultra-high field MRI. NeuroImage 168:279–295. https://doi.org/10.1016/j.neuroimage.2017.02.063

Van Boxtel JJA, Lu H (2013) A predictive coding perspective on autism spectrum disorders. Front Psychol 4. https://doi.org/10.3389/fpsyg.2013.00019

Van De Cruys S, Evers K, Van Der Hallen R, Van Eylen L, Boets B, de Wit L, Wagemans J (2014) Precise minds in uncertain worlds: predictive coding in autism. Psychol Rev 121(4):649–675. https://doi.org/10.1037/a0037665

van Kerkoerle T, Self MW, Roelfsema PR (2017) Layer-specificity in the effects of attention and working memory on activity in primary visual cortex. Nat Commun 8(1):13804. https://doi.org/10.1038/ncomms13804

van Loon AM, Fahrenfort JJ, van der Velde B, Lirk PB, Vulink NCC, Hollmann MW, Steven Scholte H, Lamme VAF (2016) NMDA receptor antagonist ketamine distorts object recognition by reducing feedback to early visual cortex. Cereb Cortex 26(5):1986–1996. https://doi.org/10.1093/cercor/bhv018

Vicario CM, Nitsche MA, Salehinejad MA, Avanzino L, Martino G (2020) Time processing, interoception, and insula activation: a mini-review on clinical disorders. Front Psychol 11:1893

Wagstyl K, Larocque S, Cucurull G, Lepage C, Cohen JP, Bludau S, Palomero-Gallagher N, Lewis LB, Funck T, Spitzer H, Dickscheid T, Fletcher PC, Romero A, Zilles K, Amunts K, Bengio Y, Evans AC (2020) BigBrain 3D atlas of cortical layers: cortical and laminar thickness gradients diverge in sensory and motor cortices. PLoS Biol 18(4):e3000678. https://doi.org/10.1371/journal.pbio.3000678

Walsh KS, McGovern DP, Clark A, O'Connell RG (2020) Evaluating the neurophysiological evidence for predictive processing as a model of perception. Ann N Y Acad Sci 1464(1): 242–268. https://doi.org/10.1111/nyas.14321

Warburton DM, Wesnes K, Edwards J, Larrad D (1985) Scopolamine and the sensory conditioning of hallucinations. Neuropsychobiology 14(4):198–202

Weber LA, Diaconescu AO, Mathys C, Schmidt A, Kometer M, Vollenweider F, Stephan KE (2020) Ketamine affects prediction errors about statistical regularities: a computational single-trial analysis of the mismatch negativity. J Neurosci 40(29):5658–5668

White TP, Wigton RL, Joyce DW, Bobin T, Ferragamo C, Wasim N, Lisk S, Shergill SS (2014) Eluding the illusion? Schizophrenia, dopamine and the McGurk effect. Front Hum Neurosci 8. https://doi.org/10.3389/fnhum.2014.00565

Whittington JCR, Muller TH, Mark S, Chen G, Barry C, Burgess N, Behrens TEJ (2020) The Tolman-Eichenbaum machine: unifying space and relational memory through generalization in the hippocampal formation. Cell 183(5):1249–1263.e23. https://doi.org/10.1016/j.cell.2020.10.024

Woelk SP, Garfinkel SN (2024) Dissociative symptoms and interoceptive integration. Springer, pp 1–33. https://doi.org/10.1007/7854_2024_480

Yon D, Frith CD (2021) Precision and the Bayesian brain. Curr Biol 31(17):R1026–R1032. https://doi.org/10.1016/j.cub.2021.07.044

Yu Y, Huber L, Yang J, Jangraw DC, Handwerker DA, Molfese PJ, Chen G, Ejima Y, Wu J, Bandettini PA (2019) Layer-specific activation of sensory input and predictive feedback in the human primary somatosensory cortex. Sci Adv 5(5):eaav9053. https://doi.org/10.1126/sciadv.aav9053

Zarkali A, Adams RA, Psarras S, Leyland L-A, Rees G, Weil RS (2019) Increased weighting on prior knowledge in Lewy body-associated visual hallucinations. Brain Commun 1(1):fcz007. https://doi.org/10.1093/braincomms/fcz007

Sensory and Multisensory Processing Changes and Their Contributions to Autism and Schizophrenia

Sarah G. Vassall and Mark T. Wallace

Contents

Abstract Natural environments are typically multisensory, comprising information from multiple sensory modalities. It is in the integration of these incoming sensory signals that we form our perceptual gestalt that allows us to navigate through the world with relative ease. However, differences in multisensory integration (MSI) ability are found in a number of clinical conditions. Throughout this chapter, we

S. G. Vassall
Vanderbilt Brain Institute, Vanderbilt University, Nashville, TN, USA

M. T. Wallace (✉)
Department of Psychology, Vanderbilt University, Nashville, TN, USA

Department of Hearing and Speech, Vanderbilt University Medical Center, Nashville, TN, USA

Vanderbilt Vision Research Center, Nashville, TN, USA

Department of Psychiatry and Behavioral Sciences, Vanderbilt University Medical Center, Nashville, TN, USA

Department of Pharmacology, Vanderbilt University, Nashville, TN, USA

Vanderbilt Kennedy Center, Vanderbilt University Medical Center, Nashville, TN, USA
e-mail: mark.wallace@Vanderbilt.Edu

Curr Topics Behav Neurosci (2025) 74: 73–103
https://doi.org/10.1007/7854_2025_589
Published Online: 11 May 2025

discuss how MSI differences contribute to phenotypic characterization of autism and schizophrenia. Although these clinical populations are often described as opposite each other on a number of spectra, we describe similarities in behavioral performance and neural functions between the two conditions. Understanding the shared features of autism and schizophrenia through the lens of MSI research allows us to better understand the neural and behavioral underpinnings of both disorders. We provide potential avenues for remediation of MSI function in these populations.

Keywords Autism · Multisensory · Schizophrenia · Transdiagnostic

1 Introduction

Natural environments are typically multisensory, comprising information from multiple sensory modalities. Multisensory integration (MSI) – the process by which this information from the different sensory modalities is combined by the brain – not only creates our perceptual gestalt but also results in profound behavioral benefits, including during spatial localization (Fleming et al. 2020; Neil et al. 2006), orientation and navigation through the environment (Nardini and Cowie 2012), and complex social behaviors such as speech comprehension (Sumby and Pollack 1954; Grant and Seitz 2000; van Wassenhove et al. 2005) and social orienting (Bahrick 2010). Furthermore, integration of information across the different sensory modalities can result in distinct perceptual changes, which can be best illustrated through a multitude of illusions (e.g., McGurk illusion (McGurk and MacDonald 1976), stream-bounce effect (Sekuler et al. 1997), and rubber hand illusion (Noel et al. 2020; Crespi and Dinsdale 2019; Cascio et al. 2012)).

Importantly, differences in multisensory integration have been shown in a number of clinical conditions. For the purposes of this review, we will focus specifically on autism and schizophrenia, both of which have accompanying changes in unisensory and multisensory function. Studies of MSI differences in these clinical populations are useful for providing not only pathophysiological insights but also important clues as to the mechanisms that may underlie both normal and disrupted MSI. The advent of the National Institute of Mental Health's Research Domain Criteria (RDoC (National Institute of Mental Health n.d.)) has led to a significant paradigm shift in psychiatric research, with the aim of elucidating the neurophysiological mechanisms underlying phenotypic traits across disorders rather than focusing upon diagnoses defined solely by combinations of symptoms.

Such an approach makes it possible to identify common physiological changes across disorders, assess physiological effects on symptomology, and compare psychiatric disorders to other non-psychiatric disorders and diseases. Here, we aim to discuss some of the shared features of autism and schizophrenia that may have their roots in altered MSI, as well as some of their divergent characteristics. Following this discussion, we will highlight some key findings across both clinical groups that shed light on target brain regions in which altered MSI is seen, and how these regions may impact the clinical phenotype. Finally, we will discuss potential experimental paradigms and training approaches grounded in MSI with an eye toward remediation.

2 Autism and Schizophrenia: Phenotypic Characterization

Autism spectrum disorder (ASD) is a neurodevelopmental disorder that is characterized by social communication impairment (SCI) and patterns of restricted and repetitive behavior (RRBs) (American Psychiatric Association 2013). In contrast, schizophrenia is a psychotic disorder typically diagnosed in adolescence or early adulthood and is characterized primarily by cognitive (e.g., delusions) and sensory (e.g., hallucinations) disturbances (American Psychiatric Association 2013). On the surface it might seem unlikely that these two disorders share common mechanistic features. However, it has been suggested by some that autism and schizophrenia may be characterized by social and cognitive profiles that look clinically distinct but are underpinned by shared mechanisms (Noel et al. 2020). Furthermore, a recent paper sampling 3,837 individuals with autism and 38,370 comparators estimated 32.7 per 1,000 (i.e., 3.27%) of individuals with ASD will have a comorbid schizophrenia diagnosis, with this number being larger in females (5.93%) than in males (2.74%) (Chien et al. 2021). Upon further inspection, the diagnostic criteria for autism and schizophrenia share a surprising degree of overlap (Fig. 1), although the clinical presentation of each differs in several ways. It is through the examination of where clinical features and physiological differences converge – and perhaps more importantly, diverge – that the role of sensory and multisensory function in the emergence of specific symptomology can be unraveled. In the following sections, we focus upon some of the key impacted domains.

2.1 Non-speech Social Communication

Both individuals with ASD and schizophrenia exhibit social and social communication challenges, with SCI being a core component of the autism diagnosis. As with most features of the condition, in autism, SCI presents along a spectrum but is generally characterized by difficulty in social-emotional reciprocity and understanding non-verbal communication (eye contact, body language, etc.). Some individual with ASD also show an absence of or delayed language acquisition. However, many individuals with ASD who are able to speak may experience challenges understanding not only non-verbal communication cues but also prosody and intonation changes in the communication signal that convey emotional content.

In schizophrenia, social functioning is diminished and often presents as social withdrawal, although it has been observed that people with schizophrenia also exhibit difficulty categorizing emotional faces and voices (de Gelder et al. 2005;

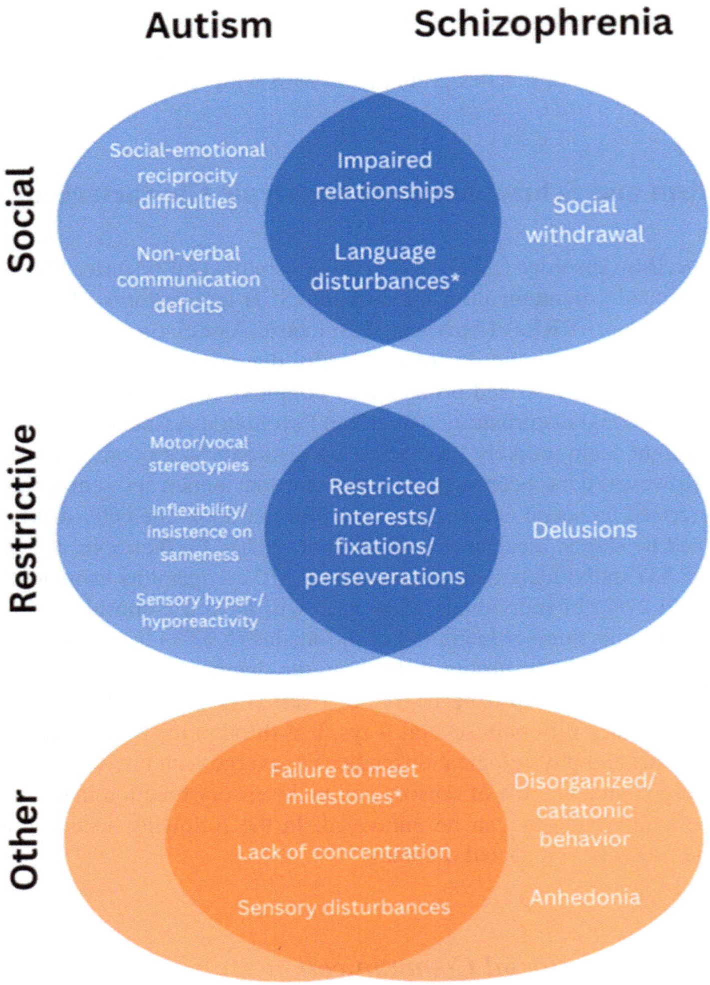

Fig. 1 Schematic of shared phenotypes in autism and schizophrenia. Both groups exhibit social (top), restrictive (middle), and additional features (bottom). While autism and schizophrenia are unique diagnoses, they share a number of behavioral phenotypes shown in the Venn diagrams above. In the *social domain*, both groups exhibit difficulty maintaining social relationships and may show language disturbances. Note, however, that language impairment is not required for either diagnosis and does not affect all individuals in these groups. In the *restrictive* domain, both groups show intense fixations on certain ideas or elements in the environment. Finally, *other shared features* include sensory disturbances, difficulty concentrating, and, in the case of childhood-onset schizophrenia, failure to meet milestones

de Jong et al. 2009; Castagna et al. 2013; Martínez et al. 2019), suggesting that the integration of complex auditory and visual stimuli that make up social interactions may be impaired in this population (much like has been suggested in autism). The following section discusses in detail several of the key non-speech social communication features in autism and schizophrenia.

2.1.1 Changes in Visual Information Integration

When examining audiovisual integration of social stimuli, both people with ASD and schizophrenia utilize different visual features when making judgments about social information. However, while individuals with ASD appear to utilize visual information *less* than that of controls, the opposite appears true in schizophrenia.

For instance, a multitude of studies have revealed that when gazing at faces, people with ASD tend to look significantly less at the eyes (Trepagnier et al. 2002; Spezio et al. 2007; Jones et al. 2008; Jones and Klin 2013; Baron-Cohen 2017; Hotier et al. 2017) and more at peripheral features (Trepagnier et al. 2002) and background (Parish-Morris et al. 2019).

In dynamic social tasks, those with ASD attend less to biological motion (Annaz et al. 2012), and gaze patterns reveal no preference for upright compared to inverted biological stimuli, as is observed in non-individuals with ASD (Klin et al. 2009). Furthermore, individuals with ASD have been shown to be less accurate at recognizing emotions present from visual and audiovisual facial expressions (Charbonneau et al. 2013) and biological motion (Atkinson 2009; Mazzoni et al. 2021). This may be in part related to a reduction in activation of striate and dorsal extrastriate cortical regions in response to visual emotional cues (Martínez et al. 2019).

Individuals with schizophrenia also show changes in gaze patterns toward salient social features; these differences have long been hypothesized to be driven by an ineffective strategy for processing social and emotional information. For instance, in face emotion categorization tasks, people with schizophrenia show increased fixation duration (Shimizu et al. 2000; Zhu et al. 2013; Navalón et al. 2021) but decreased fixation count (Shimizu et al. 2000; Zhu et al. 2013; Loughland et al. 2002; Nikolaides et al. 2016; Wade et al. 2018; Li et al. 2020) and scanpath length (Nikolaides et al. 2016; Li et al. 2020) toward salient facial features (e.g., eyes) regardless of the emotion being displayed. Likely as a result of these atypical gaze patterns, people with schizophrenia are regularly outperformed by control participants in the accuracy of emotion categorization (Kington et al. 2000; Tso et al. 2014; Nikolaides et al. 2016; Hillmann et al. 2018; Wade et al. 2018), with more pronounced deficits being seen when only eye information is available (Kington et al. 2000) and when target faces are embedded in social images (Nikolaides et al. 2016; Li et al. 2020; Green et al. 2008). Thus, like autism, schizophrenia is characterized by distinctive changes in visual gaze patterns; however, in direct contrast to autism, viewing of visual emotional faces is coupled with *increased* activation in primary visual cortex and visual attention areas relative to control groups (Spilka et al. 2019).

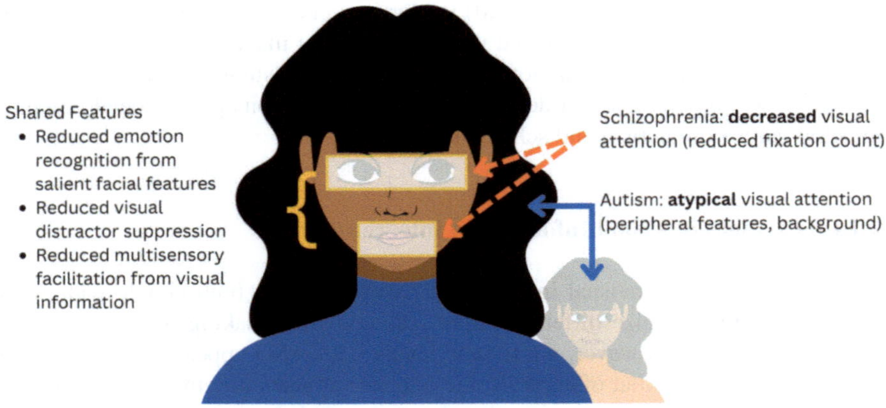

Fig. 2 Summary schematic of similarities and differences between social visual processing in autism and schizophrenia. Both populations exhibit reduced visual distractor suppression and difficulty categorizing emotions from salient facial features (i.e., eyes and mouth). It is also theorized that these changes in visual processing may result in reduced multisensory facilitation from the visual information available, particularly for complex social behaviors like speech comprehension. However, the mechanisms by which these behaviors occur is likely to be somewhat different. Specifically, while autistic individuals exhibit increased visual attention to non-salient features such as peripheral features (e.g., hair) and background (e.g., other people or objects), schizophrenia is characterized by attention, but reduced fixation count, to salient facial features. In both cases, individuals are likely receiving less relevant visual information to inform social behaviors

These visual processing differences observed in both populations are thought to stem from changes in the balance between local and global feature extraction. This is observed even in the case of non-social stimuli. For example, the Enhanced Perceptual Functioning theory (EPF) (Mottron and Burack 2001) of autism posits that individuals with ASD show superior performance in extracting locally oriented features relative to individuals without ASD, but often underperform in tasks that require global scene analysis (Happé and Frith 2006; Pellicano et al. 2005; Manjaly et al. 2007). For instance, studies of global dot motion showed that children with ASD show higher motion detection thresholds compared to non-ASD controls (Pellicano et al. 2005; Milne et al. 2002), suggesting a deficit in global visual perception. Individuals with schizophrenia show an opposite pattern of visual integration (i.e., reduced local feature extraction), as shown in tasks of foreground-background discrimination (Tso et al. 2014).

Despite these differing profiles of social behavior, both autism and schizophrenia result in social deficits, wherein salient social features are not optimally extracted (Fig. 2). Indeed, changes in the balance of local and global processing in these clinical populations may be expected to have two key effects on social function. First, information conveyed primarily by visual features (e.g., emotion) is lost. Second, in naturalistic settings in which visual information is complementary and facilitates information processing (e.g., conversing in a noisy room), we would expect the limited processing of visual stimuli to result in reduced multisensory facilitation.

2.1.2 Relations Between Temporal Integration and Social Function

Differences in temporal integration, particularly for audiovisual stimuli, is also associated with the social communication differences seen in both autism and schizophrenia. The temporal binding window (TBW) – the window of time in which sensory stimuli are likely to be bound into a single perceptual entity – is known to be altered in both populations. Research has largely shown an extended TBW across a variety of audiovisual tasks in both individuals with ASD (Foss-Feig et al. 2010; Stevenson et al. 2014a; Noel et al. 2016, 2017; Zhou et al. 2018; Kawakami et al. 2020) and schizophrenia (Foucher et al. 2007; Stevenson et al. 2017a; Zvyagintsev et al. 2017). Interestingly, while the increased TBW width observed in autism has been shown to relate to the complexity of the stimuli (e.g., temporal integration differences are more pronounced for face-voice stimuli than for simple flashes and beeps (Stevenson et al. 2014b)), people with schizophrenia appear to have a broad deficit in temporal integration across a variety of audiovisual stimuli, including in the sound-induced flash illusion (Haß et al. 2017; de Boer-Schellekens et al. 2014; Vanes et al. 2016), stream-bounce illusion (Zvyagintsev et al. 2017), and asynchronous McGurk illusion (Martin et al. 2013). An enlarged TBW has also been shown to correlate with SCI in both people with ASD (Stevenson et al. 2014a) and schizophrenia (Zhou et al. 2018).

2.1.3 Relations Between Spatial Integration and Social Function

An additional feature of disrupted social communication observed in both groups is the understanding and utilization of peripersonal space (PPS). PPS – the actionable space surrounding the body, which is created through multisensory spatial interactions – can be measured by evaluating performance on psychophysical tasks at different distances from the body. Using a dynamic visuotactile stimulus in which stimuli approached or receded from participants, Noel et al. (2020) evaluated the difference in response times (RTs) to approaching versus receding stimuli; it is known that RTs get faster as stimuli approach the body within the PPS but are relatively stable outside the PPS boundary (Canzoneri et al. 2012; Serino et al. 2015). Furthermore, the researchers defined the PPS flexibility as the slope of this psychometric function, wherein a steeper slope indicates a sharper boundary and vice versa. Noel et al. (2020) found that, relative to the control group, the group with ASD showed a decrease in response time much closer to the body and a steeper psychometric curve, whereas the opposite was true for the group with schizophrenia (Fig. 3a) (Noel et al. 2020). These results suggest that the PPS boundary is smaller and more sharply defined in autism, but more amorphous in schizophrenia (Fig. 3b).

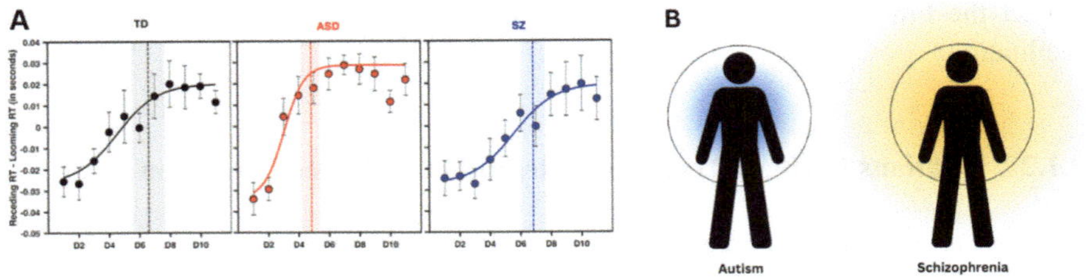

Fig. 3 Peripersonal space (PPS) as measured by visuotactile integration in ASD and schizophrenia samples. (**a**) Data adapted with permission from Noel et al. (2020). Difference in receding vs. approaching visuotactile detection response times (RTs; *y*-axis) among control (black), autistic (blue), and schizophrenic (orange) individuals at discrete distances from the body (*x*-axis) presented in Noel et al. (2020). PPS can be measured as the distance from the body at which RTs begin to decrease; the slope of the curve can be thought of as the flexibility of PPS. Individuals with ASD show a sharper decrease in RT significantly closer to the body relative to controls. In contrast, people with schizophrenia individuals show a shallower change in RT relative to controls. (**b**) Schematic of PPS in each group relative to controls (black circle). People with ASD show a smaller PPS with a sharper boundary, whereas individuals with schizophrenia show a larger PPS with a more diffuse boundary

In addition to PPS size, modulation of PPS based on the social context is an important factor for social function, as it informs how an individual interacts with others in social settings. It has been found that PPS typically shrinks in response to the presence of an unknown person, whereas PPS boundaries expand to incorporate the other person when the interaction is friendly and cooperative (Teneggi et al. 2013). In contrast, invasion of PPS by aversive stimuli and threatening social encounters (e.g., being followed) increases PPS boundaries as well as physiological stress responses (Åhs et al. 2015). Thus, PPS acts as a flexible boundary not only for the comfort of those in one's immediate environment but also for one's own comfort with individuals, given the social context. Interestingly, in direct contrast to controls, it has been found that PPS size does not change in autism regardless of whether the participant is alone or in the presence of other people (Noel et al. 2021). An opposite pattern is observed in schizophrenia: when comparing social and non-social stimuli, individuals with schizophrenia showed a wider PPS boundary in the social condition relative to the control group (Lee et al. 2021). Therefore, it appears that both autistic and individuals with schizophrenia show alterations in their perception of self-other relationships in space, which points to a direct link between MSI and social features of each disorder.

Changes to PPS flexibility have also been linked with distinct clinical features in each group. One well-studied paradigm to evaluate this link is the rubber hand illusion (RHI). The RHI is induced when a prosthetic hand, placed close to a participant's real hand (which is occluded from view), is stroked. As the participant watches the prosthetic hand, they begin to perceive stroking on their own hand by integrating the visual stimulus with their own body. The RHI only occurs within a region of space that could be occupied by the real hand (no more than about 30 cm) (Lloyd 2007), indicating that there must be a plausible relationship between the visual stimulus (i.e., the stroking of the fake hand) and the body. While individuals with ASD do experience the RHI, it has been shown to take longer to occur relative to controls, with the delay being correlated with overall autism severity scores (Cascio et al. 2012). In contrast, individuals with schizophrenia show earlier onset of the RHI (Peled et al. 2000) and a stronger tactile perception (Peled et al. 2000; Thakkar et al. 2011; He et al. 2022) compared to controls. These findings suggest that, relative to control groups, while individuals with ASD are less likely to bind the stimuli into a single percept, individuals with schizophrenia are more biased toward fusion of visuotactile stimuli. Furthermore, individuals with schizophrenia show a reduced sensitivity to temporal asynchrony between the stroking of the fake hand and their own hand (Zopf et al. 2021), indicating a wider window of visuotactile integration in this population. The strength of the RHI has been positively linked to severity of hallucinations (Peled et al. 2000; Thakkar et al. 2011) and disturbances in body perception in schizophrenia (Zopf et al. 2021).

In addition to the RHI, it has been shown more broadly that alterations to the perceived relationship between the self and the environment are correlated with clinical symptomology in schizophrenia. For instance, individuals with schizophrenia have shown an extended perception of their own reaching distance, the degree of which is correlated with the presence and severity of both positive and negative symptoms (Delevoye-Turrell et al. 2011).

We hypothesize that in multisensory social environments, the relative weighting of visual stimuli against other stimuli may be at the heart of these PPS and RHI differences in each group. If, as discussed above, individuals with ASD and schizophrenia show relative decreases and increases in the influence of social visual stimuli, respectively, it is possible that these changes may be a primary factor causing changes to perceived self-other and self-environment relationships. Specifically, failure to appropriately integrate the visual cues with tactile information would lead to an incomplete representation of the salient stimulus in space, thus generating an aberrant representation of the body relative to that stimulus. For instance, if individuals with ASD show reduced visual attention to biological stimuli (Annaz et al. 2012; Klin et al. 2009), perception of the RHI may be delayed due to the relative decrease in salience of one's own hand. Similarly, decreased attention to informative facial features (e.g., eyes (Trepagnier et al. 2002; Spezio et al. 2007; Jones et al. 2008; Jones and Klin 2013; Baron-Cohen 2017; Hotier et al. 2017)) may lead to decreased salience of social information, thus limiting the size and flexibility of PPS to accommodate various social situations. In schizophrenia, we might see the opposite pattern arise, wherein increased influence of social visual information would lead to increased salience of one's own body; in conjunction with enhanced salience of others' social visual cues, this may lead to a poorly delineated self-other divide.

Although each population shows distinct PPS alterations that appear to oppose one another in some ways, we and others argue that the features of autism and schizophrenia present along different functional continua, with certain brain regions acting at the center of specific behaviors. Indeed, there may be similar mechanistic changes in both groups giving rise to the respective PPS profiles in autism and schizophrenia. For example, the intraparietal sulcus (IPS) is a region of the posterior parietal cortex that facilitates visuomotor transformations of the arms, hands, and eyes, among other processes. IPS is heavily involved in the social aspects of PPS, such as differential activation to social and nonsocial stimuli (Holt et al. 2014; Vieira et al. 2017, 2020). Decreased IPS in autism (Massaccesi et al. 2021; Webster et al. 2020) and increased IPS activity in schizophrenia (Holt et al. 2015) in response to social images may indicate a functional difference that contributes to atypical extraction and synthesis of social information, rather than the processing of incoming streams of sensory information per se. Additionally, as discussed in Sect. 2.1.1, both populations exhibit decreased gaze toward social stimuli, such as faces. Atypical IPS function may also be implicated in this cross-diagnostic behavior.

2.2 *Speech and Language*

2.2.1 Characterizing Expressive and Receptive Language Differences

Speech and language changes are often present in both autism and schizophrenia, although not required for either diagnosis. Differences in expressive language are fairly well characterized in both groups. While a number of individuals with ASD never acquire fluent speech, research has found that many autistic individuals who do acquire speech exhibit differences in prosody (Peppé et al. 2007; Fusaroli et al. 2017) and syntax (Peppé et al. 2007; Brien et al. 2021) (also see (Arunachalam and Luyster 2016) for a comprehensive review), and do not adapt to others' speech rate as would be expected in a typical social situation (Wynn et al. 2018).

While the same known motor challenges are not necessarily observed in schizophrenia, individuals with schizophrenia experiencing their first onset of psychosis exhibit less complex syntax (Thomas et al. 1996), more syntactic errors (Thomas et al. 1996; Perlini et al. 2012) during free speech, and reduced verbal fluency as measured by clinical fluency evaluations (Lundin et al. 2022). Additionally, reduced verbal fluency is associated with more global deficits in social function (Lundin et al. 2022). Furthermore, these language changes observed in schizophrenia are likely to develop long before onset of positive and negative symptoms; adolescents at high risk of psychosis show atypical prosody, the degree of which is correlated with the severity of negative psychotic symptoms that appear later in first-onset psychosis (Agurto et al. 2020).

Receptive language is a somewhat more difficult function to disentangle from other clinical features. For instance, while receptive language scores may be reduced due to true differences in comprehension, deficits may also wrongly be concluded due to a host of other factors such as reduced attention, cooperation, or task engagement. In autism, characterizing differences in receptive language is made more difficult by the extreme phenotypic heterogeneity observed in this population. While one study showed that decreased receptive language in autism is correlated with increased expression of RRBs (Ray-Subramanian and Ellis Weismer 2012), another study showed that severity of SCI, but not RRBs, negatively correlated with receptive language ability (Kim et al. 2020). Yet others still have shown that individuals with ASD with and without language impairments do not show differences in overall scores of autism severity (Loucas et al. 2008), nor is receptive language trajectory different between children with and without ASD when accounting for baseline ability (Brignell et al. 2019). Furthermore, in many individuals, nonverbal IQ and language outcomes – both expressive and receptive – are unrelated (Slušná et al. 2021). However, despite the heterogeneity, receptive language differences have been linked to auditory processing differences, differences in visual attention, and changes in audiovisual integration (see below). For example, poorer receptive language outcomes have been linked with impaired prosodic affective discrimination (Peppé et al. 2007) and attention to speaker (McDaniel et al. 2018), respectively.

Receptive language differences are even less well understood in schizophrenia, although we are aware of one somewhat recent study that has characterized the relationship between receptive language and schizophrenia features. When accounting for verbal knowledge, individuals with schizophrenia have been shown to perform comparably to healthy controls at discriminating real and pseudo-words, as well as identifying synonyms, although individuals with schizophrenia trend toward selecting homophone distractors rather than true synonyms more frequently than controls (Tan et al. 2016). However, when asked to judge whether syntactically different sentences (i.e., different grammatical arrangement of words) were the same in meaning, individuals with schizophrenia make significantly more errors than healthy controls (Tan et al. 2016), suggesting that syntactic comprehension is impaired in this population. Furthermore, syntactic errors are significantly positively correlated with disorganized expressive speech features, such as tangentiality and derailment (Tan et al. 2016). Importantly, this study focused solely on language and did not include any metrics of sensory or other social communication function, so it is difficult to draw conclusions about the nature of receptive language deficits in this population. Nonetheless, it is likely that aberrant sensory processing discussed above – particularly differences in utilization of visual social information – contributes to changes in the comprehension of complex, nuanced linguistic features.

Examining the language deficits shared by these two populations, we suggest that changes to the function of multisensory regions that support language function (e.g., posterior temporal cortex) may underlie prosodic, syntactic, and fluency deficits. Importantly, we must point out that it is likely to be these multipurpose multisensory areas, rather than expressive language-specific brain regions (e.g., inferior frontal cortex), that give rise to the shared language phenotypes. Disruptions to speech- and language-specific regions lead to various aphasias. While some autistic individuals do not develop speech, lack of speech and atypical speech patterns do not meet clinical criteria for this condition. Rather, it is more likely that the cognitive representations of audiovisual features of speech and language, and not the motor output itself, are disrupted. We discuss the role of posterior temporal cortex more thoroughly in Sect. 3.1. Additionally, more general multisensory areas – such as angular gyrus in inferior parietal lobe – should also be considered when evaluating potential pathophysiology in autism and schizophrenia, given the breadth of multisensory disruptions, particularly in the audiovisual domain, discussed here and throughout the chapter.

2.2.2 Characterizing Audiovisual Emotion Recognition Differences

As with non-speech stimuli, research shows divergent patterns of difficulties extracting meaningful social information from speech in these clinical groups. For instance, when asked to judge audiovisual emotional expressions, some research has found that individuals with ASD are able to reach comparable performance to controls only when the signal-to-noise ratio is increased (Charbonneau et al. 2013) or when the presentation rate is slowed (Tardif et al. 2007), suggesting reduced multisensory facilitation of social emotional information. In contrast, studies of emotion categorization in schizophrenia have revealed that while non-emotional prosody and congruent audiovisual emotional prosody discrimination are unimpaired, individuals with schizophrenia show difficulty with this task in conflicting audiovisual conditions. When asked to discriminate voice emotion, individuals with schizophrenia show *increased* influence of the emotional content of the face (de Gelder et al. 2005; de Jong et al. 2009; Castagna et al. 2013), an effect that is correlated with executive function, attention, and presence of negative symptoms (e.g., catatonia) (Castagna et al. 2013). Similarly, when asked to categorize emotional faces, schizophrenic show relatively *decreased* influence of conflicting emotional voices relative to controls (de Gelder et al. 2005). Combined, this evidence suggests that, in direct opposition to the pattern seen in individuals with ASD, individuals with schizophrenia may overestimate the importance of visual social information in audiovisual contexts.

Importantly, to date, it is unclear whether autistic and schizophrenic individuals exhibit differences in audiovisual emotional recognition resulting from pathophysiology of areas that support social function, low-level sensory processing differences leading to deficits in more complex areas, or the presentation of other features which inhibit efficient social processing. In reality, it is likely a combination of all three, and we can posit shared mechanisms which might hinder audiovisual emotion recognition in both groups. We have already delineated some potential brain regions that are implicated in various aspects of social processing in these populations – posterior temporal cortex, intraparietal sulcus, inferior parietal lobe – and it has been established, specifically in autism, that audiovisual processing deficits increase as stimulus complexity increases (Stevenson et al. 2014b, 2017b, 2018).

But could other features of each condition contribute to atypical emotion recognition, or indeed any of the speech and non-speech deficits described so far? For instance, we have previously discussed the relationship between eye gaze and social function in both groups. There are also motivational changes that are associated with both autism and schizophrenia. For example, alterations in oxytocin levels – a hormone associated with social bonding, prosocial behavior, and social adaptability – have been linked to features of both autism and schizophrenia. Specifically, decreased plasma oxytocin levels are correlated with increased RRBs in autism and increased severity of negative symptoms in schizophrenia (see Florea et al. (2022) for a review), both of which can indirectly impair social function. Additionally, because of oxytocin's role in the development of adaptive social behaviors, atypical levels in childhood could lead to cascading, amplified deficits in adulthood.

2.2.3 Relation Between MSI and Language Abilities

Given that, for individuals with normal vision and hearing, speech is typically an audiovisual experience, integration of auditory and visual cues is critical for understanding and articulating speech; gleaning information about both the temporal relationship between acoustic and visual speech cues and the spatial relationship between a target speaker and their speech output is enhanced in multisensory contexts. However, changes to audiovisual integration need not be restricted to speech stimuli to interfere with language outcomes. Importantly, there is significant evidence that individuals with ASD exhibit differences in MSI, with complex multisensory stimuli (e.g., speech) yielding more pronounced group-level differences than simple stimuli (Stevenson et al. 2014b, 2017b, 2018). Although this particular framework is not necessarily adopted in schizophrenia research, the principle should hold true that low-level integration differences (e.g., differences in perception of the flash-beep illusion) have cascading and amplified effects on integration of more complex stimuli (e.g., speech cues).

That said, there is significant evidence that difficulties integrating auditory and visual social information – as well as interoceptive and tactile information – can have serious implications for real-world scenarios. For instance, speech-in-noise comprehension benefits profoundly from visual cues (Sumby and Pollack 1954; Grant and Seitz 2000; van Wassenhove et al. 2005), and this ability may be compromised in clinical populations who exhibit suboptimal weighting of one target modality (i.e., audition or vision). Indeed, both individuals with autism (Borgolte et al. 2021; Foxe et al. 2015; Russo et al. 2009; Schwartz et al. 2020) and schizophrenia (Senkowski and Moran 2022; Ross et al. 2007) show speech-in-noise comprehension deficits, with reduced multisensory facilitation relative to control groups. We speculate that if individuals with ASD exhibit a reduced influence of social visual cues (e.g., visual articulatory movements) (Fig. 4a), the target speech perception will be compromised. This may occur due to not only reduced facilitation of phoneme discrimination by visual speech but also increased influence of distracting auditory stimuli that might fall into the external gaze path. We expect the opposite pattern of changes in schizophrenia (Fig. 4b), with increased social visual weight to cause relatively reduced use of the target visual speech cues, coupled with increased influence of distracting visual stimuli in the environment. Nonetheless, both these patterns of deficits are likely to result in challenges in noisy social environments.

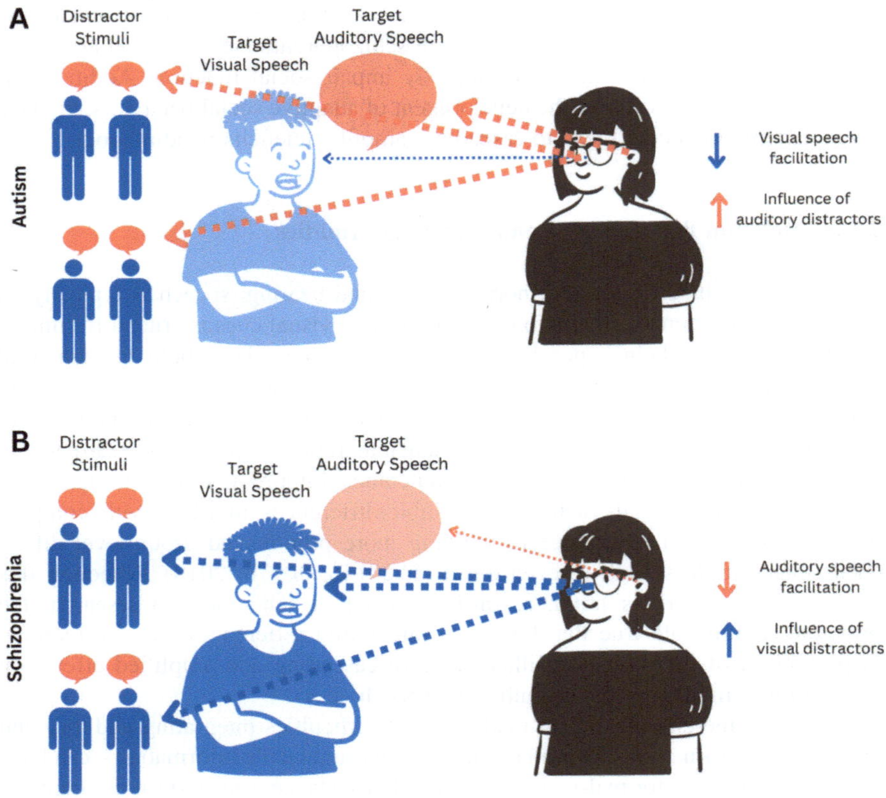

Fig. 4 Schematic of social audiovisual integration differences according to the atypical weighting hypothesis. (**a**) In autism, a reduction in the weight of social visual stimuli would present as decreased facilitation by target visual speech in noisy conditions, with increased influence of the distractor auditory speech. (**b**) In schizophrenia, an increased weight of social visual stimuli would present as decreased facilitation by target auditory speech, but increased influence of visual distractors

2.3 *Restricted Interests, Fixations, and Perseverations*

As demonstrated in Fig. 1, both individuals with ASD and schizophrenia show restrictive behavioral patterns, although presentation is quite different. Individuals with ASD' RRBs are often driven by specific hyper- or hyposensitivities to various sensory stimuli and may present as sensory-seeking behaviors (i.e., "stimming," through vocal or motor stereotypies), restricted interests and fixations, and insistence on sameness (American Psychiatric Association 2013). In contrast, the restrictive behaviors in schizophrenia are more abstract, presenting in the form of delusions – false beliefs that are retained despite evidence to the contrary – as well as restricted affect and catatonia (American Psychiatric Association 2013). Research to date is scarce concerning the relationship between MSI and RRBs in autism and schizophrenia. However, there is some evidence from other clinical groups that changes to MSI may underlie certain cognitive patterns that give rise to restrictive patterns of thinking in both groups.

For example, as described above, RHI perception has been shown to be stronger in schizophrenia, with a wider tolerance for asynchronous visual and tactile stimuli. Similarly, individuals with Parkinson's disease (PD), which features affective changes and impaired initiation of movement (catatonia) akin to the negative symptoms of schizophrenia, show stronger RHI perception compared to control groups, with a wider tolerance for asynchronous visual and tactile stimuli (Waldmann et al. 2020; Ding et al. 2017). These remarkably similar findings coupled with shared clinical features among schizophrenia and PD indicate that restricted motor behaviors and emotional affect may share common mechanisms dependent on intact visuotactile integration ability. Visuotactile integration differences may also be related to RRBs observed in autism. For instance, individuals with Tourette's syndrome – characterized by uncontrollable vocal and/or motor tics (American Psychiatric Association 2013) that may look similar to vocal and motor stereotypies observed in autism – show a relationship between RHI susceptibility and premonitory sensations that precede tics (Rae et al. 2018). Additionally, attention deficit/hyperactivity disorder (ADHD), which is commonly comorbid with autism, is characterized by reduced susceptibility to the RHI (Boehme et al. 2020). Combined, this evidence suggests that changes to multisensory integration which gives rise to PPS may be related to the interoceptive sensations that elicit atypical vocal and motor behaviors, as well as impaired regulation or inhibition of behavioral responses to these sensations.

Because there is quite limited evidence concerning the relationship between MSI and RRBs in these disorders, we believe it is important to note the role of sensory responsivity more broadly, not restricted to multisensory interactions, in the emergence of restricted patterns of behavior and thinking. Specifically, sensory hypersensitivity appears to link strongly with the RRBs observed in autism. Not only do most individuals with ASD exhibit sensory hypersensitivity, but also do many individuals with shared RRB-like clinical features, including OCD (Isaacs et al. 2022), tic disorders (Isaacs et al. 2020, 2022), and ADHD (Lucker et al. 1996). Furthermore, individuals with tic disorders show increased sensory hyperresponsivity, the degree of which is correlated with severity of OCD-like behaviors and thinking patterns, as well as ADHD-like symptoms and severity of premonitory urges (Isaacs et al. 2020). Across all these disorders, sensory hypersensitivity and hyper-responsivity may contribute to impulsive or compulsive behaviors aimed at alleviating discomfort associated with an overstimulating sensory environment.

3 Brain Regions of Interest for Shared Multisensory Phenotypes Across Schizophrenia and Autism

The purpose of this review is, in the spirit of RDoC, to identify common pathophysiology that may underlie MSI differences across clinical populations. Thus far, we have discussed behavioral evidence that might point to common or disparate changes in MSI function that give rise to the phenotypes specific to each disorder. In addition to these behavioral and perceptual differences, both autism and schizophrenia are known to be associated with a variety of changes to brain structures and connectivity that are relevant to sensory and multisensory processing. We believe that understanding how these critical multisensory hubs may be altered in autism and schizophrenia will offer further understanding of how clinical phenotypes arise.

3.1 Posterior Superior Temporal Sulcus (pSTS)

One cortical region that has been implicated in both autism and schizophrenia is posterior superior temporal sulcus (pSTS). Not only is pSTS a site of convergence for auditory and visual pathways, but it is also particularly active during the processing of social information (e.g., facial features (Hotier et al. 2017; Isik et al. 2017)) and language (e.g., syntax (Friederici et al. 2009)). Furthermore, stronger within-network functional connectivity (FC) of pSTS – and resting-state FC between pSTS and occipital cortex – is associated with better face emotion recognition in healthy adults (Wang et al. 2016). Thus, research examining pSTS may offer insight into not only the broader audiovisual integration differences observed in the autism and schizophrenia populations but also into the social-specific changes in behavior and cognition.

We can look to performance on various speech and language tasks to better illustrate potential changes in pSTS activity. For instance, in healthy adults, it has been shown that FC between pSTS and auditory and visual cortices changes such that pSTS is more strongly connected to the sensory cortex for which there is the most reliable speech information (Nath and Beauchamp 2011). In other words, pSTS activity and connectivity are altered to accommodate the noisiness of the environment, and has also been shown to affect the perception of fused syllables in modified McGurk tasks (Nath and Beauchamp 2011). As described above, it is well documented that autistic and individuals with schizophrenia both exhibit difficulties comprehending speech and noise and show altered rates of McGurk perception relative to control groups.

We can also examine metrics of excitation and inhibition to explore the role of pSTS in social communication outcomes and MSI. It has been observed that the concentration of the inhibitory neurotransmitter γ-aminobutyric acid (GABA) in STS significantly correlates with gamma-band oscillation power (Balz et al. 2016), suggesting that the balance of excitation and inhibition (E/I) is critical for regulating cortical oscillations. Importantly, gamma-band oscillations are involved in synchronization within and across cortical regions and serve to facilitate feature binding – one important aspect of MSI (Stone et al. 2014). An altered E/I balance has been proposed in numerous models of autism (Foss-Feig et al. 2017; Seymour et al. 2019), and there is evidence to suggest that auditory- and visual-evoked gamma activity is reduced throughout the cortex in autism (Wilson et al. 2007; Port et al. 2016; Safar et al. 2021). Thus, increased E/I balance in pSTS may be a target for further investigation in the affective and linguistic challenges experienced by individuals with ASD.

Like autism, features of schizophrenia may also be associated with atypical pSTS activity. While increased within-network and decreased between-network FC have been observed in pSTS in autism (see Wallace and Stevenson (2014) for a comprehensive review), individuals with schizophrenia exhibit hyperactivity of pSTS while viewing neutral faces but hypoactivity when categorizing emotional faces relative to healthy controls (Mier et al. 2017). Additionally, like for autism, research has demonstrated increased baseline and stimulus-evoked gamma-band power in schizophrenia throughout the cortex (see (Foss-Feig et al. 2017) for a review), supporting the hypothesis of E/I imbalance in the schizophrenic population as well.

While the focus of this section has been pSTS, it should also be noted that aberrations throughout the posterior temporal cortex have been noted in both populations as well. Specifically, when considering changes to language comprehension and production, posterior superior temporal gyrus (pSTG) – the canonical Wernicke's area – should be considered an additional research target. Not only does pSTG support phonological and semantic information processing (Chang et al. 2015), but also speech perception in noise (Ozker et al. 2017). Furthermore, left lateralization of the core language areas in temporal cortex – including pSTG – is a known structural correlate of language processing; lack of left lateralization has been linked to various deficits in language comprehension and production (de Guibert et al. 2011). A subset of both autistic (Baum Miller and Wallace 2019) and schizophrenic (Ocklenburg et al. 2013) individuals show reduced or non-existent lateralization of these canonical language regions, which may contribute to the deficits in speech-in-noise comprehension as well as changes to prosody and semantic and syntactic errors observed in free speech in these populations.

3.2 Posterior Parietal Cortex (PPC)

Posterior parietal cortex (PPC) is another brain region known for its multisensory properties. In this section, we will discuss the two primary PPC regions known to be related to clinical phenotypes in both autism and schizophrenia: the first is inferior parietal lobule (IPL), which contains angular gyrus (AG) and supramarginal gyrus (SMG), and the second is intraparietal sulcus (IPS) which contains lateral, ventral, medial, and anterior intraparietal regions (LIP, VIP, MIP, and AIP, respectively).

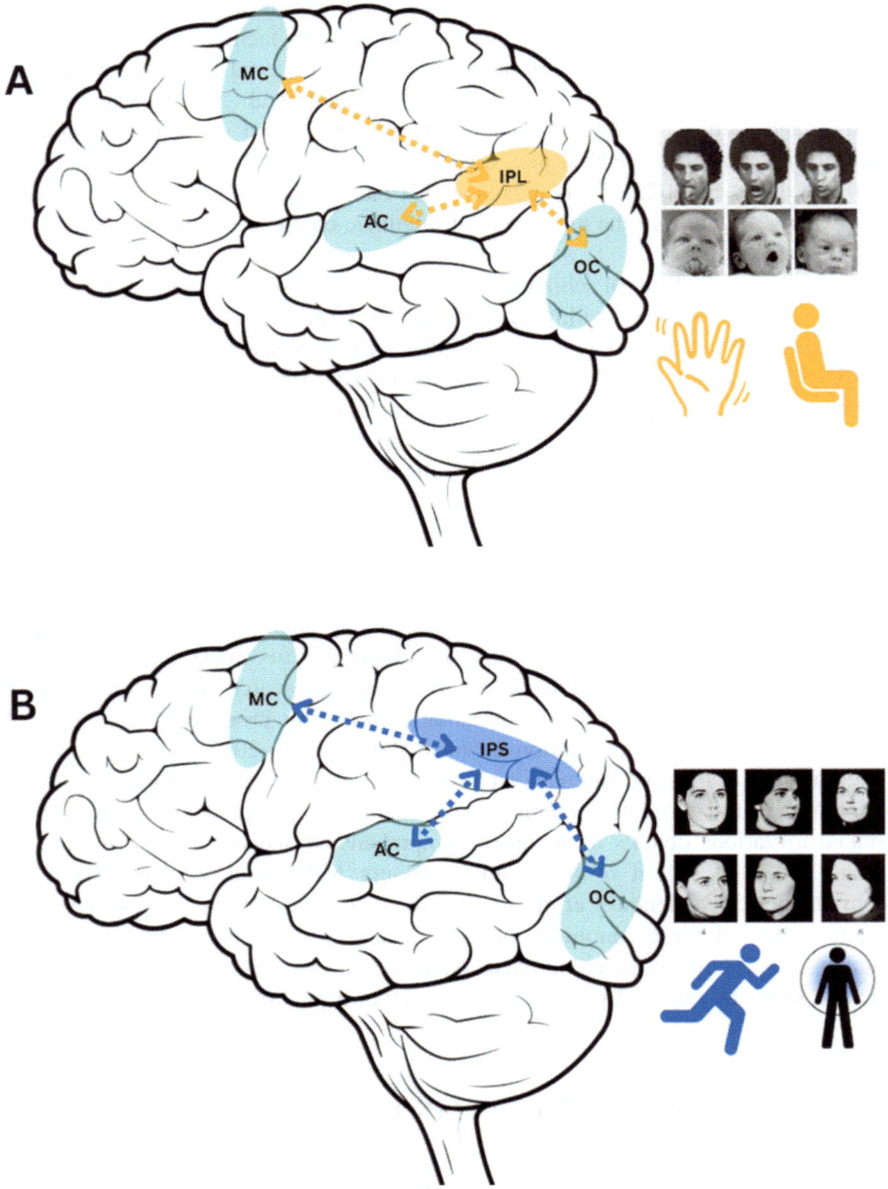

Fig. 5 Schematic detailing the role of inferior parietal lobule (IPL) and intraparietal sulcus (IPS) in the social and motor features of autism and schizophrenia. Functional connectivity (FC) with motor cortex (MC), auditory cortex (AC), and occipital cortex (OC), as well as activity of IPL and IPS independently, are shown to be altered in both clinical groups. (**a**) Structures of AG and SMG receive their primary input from AC and OC and are known to be associated with both verbal and non-verbal social communication as well as observation and imitation of others. The famous Meltzoff experiment, in which infants have been shown to mimic adults' facial expressions soon after birth, is one of the best-known examples of innate facial imitation skills very often affected in both populations. FC between IPL and other areas (e.g., MC) are involved in the execution and

While AG and SMG are more involved in audiovisual integration, the areas in IPS are largely involved in various aspects of visuotactile and visuomotor integration.

3.2.1 Inferior Parietal Lobule

AG and SMG (Fig. 5a) are multisensory regions that are highly important for language comprehension, non-verbal communication, and social interaction more broadly. In autism, research has shown that decreased gray matter volume in both AG and SMG (Duan et al. 2022; Wallace et al. 2010) is correlated with poorer non-verbal communication abilities (Duan et al. 2022). Additionally, research points to atypical functional connectivity (FC) between IPL and other brain regions that support language function. For instance, in individuals with speech praxis (i.e., motor execution impairments), the severities of the disorder and of SCI are negatively correlated with cross-hemispheric IPL connectivity to language centers and pre-motor areas (Wymbs et al. 2021).

In addition to speech and language abilities, IPL appears to be critically important for more nuanced aspects of social interaction. In a series of dyadic play tasks, autistic children show increased IPL activity relative to non-autistic controls during conditions in which they are asked to lead or compete with their peer, but decreased IPL activity during cooperative play (Su et al. 2022). This may suggest a role of IPL in distinguishing self- and other-initiated actions.

Although IPL is largely known for its role in language and social communication, it is also linked to motor features of autism. For instance, an analysis of the Autism Brain Imaging Data Exchange (ABIDE) II dataset found that while most RRBs decreased with age, there remained a significant negative relationship between AG-superior occipital FC and RRB severity (Zhang et al. 2022). This suggests that IPL is a key player in motor urges and initiation in autism. However, the role of IPL may extend to mentalizing of motor behaviors as well. For instance, during hand gesture imitation, individuals with ASD have been shown to perform comparably to non-autistic controls, but accompanied by significantly reduced AG activity (Wadsworth et al. 2017); AG activation during motor imitation is also negatively correlated with overall endorsement of autism features (Wadsworth et al. 2017). Furthermore, when asked to either observe or judge the intent of (i.e., mentalize) others' actions, while non-autistic controls show an increase in fronto-parietal FC, individuals with ASD show no change in FC (Cole et al. 2019); this indiscriminate pattern of FC is also linked with overall autism severity (Cole et al. 2019). Taken together, these findings suggest that the role of IPL in visuomotor integration may be

Fig. 5 (continued) initiation of gross and fine motor behaviors. (**b**) Like IPL, IPS receives a significant amount of input from AC and OC but is more involved in complex social judgments of unfamiliar faces (e.g., approachability; see faces from the Benton Facial Recognition Test) as well as attention to biological motion and social features. Additionally, IPS is heavily involved in visuomotor integration, playing a key role in the formation of PPS

compromised in autism as task demands increase. Specifically, while simple visuomotor imitation may be intact in autism, the attribution of complex social meaning – largely derived from auditory, visual, and audiovisual information (speech, facial expressions, tone of voice, etc.) – may be impaired.

In schizophrenia, structural and functional differences in IPL are debated but have been implicated in studies of sensory integration, concept of the self and body, and executive functioning (see Torrey (2007) for review). While there is somewhat less research describing IPL differences in schizophrenia relative to autism, the existing literature supports a relationship between IPL function, social function, and motor behaviors, in line with what is observed in autism. Generally speaking, the research points to reduced AG volume (Nierenberg et al. 2005), left IPL thickness and gyrification (Danivas et al. 2009; Venkatasubramanian et al. 2011; Palaniyappan and Liddle 2012; Otte et al. 2022), and resting-state IPL activation (Soni et al. 2018). Additionally, like autism, IPL activity is reduced during observation of others' actions relative to controls, a feature that is correlated with overall clinical severity and social impairments in this population (Kato et al. 2011; Thakkar et al. 2014). Taken together, this research suggests that IPL may be functioning at a somewhat reduced capacity in schizophrenia.

This is perhaps best illustrated by studies of FC between IPL and other cortical areas. For instance, reduced feedforward connectivity between IPL and dorsolateral prefrontal cortex is correlated with increased endorsement of delusions (Greenman et al. 2020), whereas reduced connectivity between SMG and post-central gyrus is associated with increased severity of negative symptoms (e.g., catatonia) (Ji et al. 2019). Thus, IPL may be a critical relay for higher-order cognition and thought patterns as well as lower-order functions such as motor initiation in schizophrenia. Furthermore, dysregulation of within-network IPL connectivity is observed in schizophrenia during the presentation of audiovisual emotion stimuli (Müller et al. 2013). This may indicate that there is a reduction in the specificity of information conveyed by IPL, perhaps as a result of non-specific or reduced population encoding.

3.2.2 Intraparietal Sulcus

IPS (Fig. 5b) is an additional posterior parietal area heavily implicated in MSI, particularly visuomotor and visuotactile integration important for the representation of PPS. The two major parameters governing PPS are its permeability – the space at which entry by another is perceived as intrusive – and its flexibility – the ability to adapt PPS based on the social context. In addition to ventral premotor cortex, activity in dorsal IPS is not only increased in response to social stimuli intruding on PPS relative to nonsocial stimuli (Holt et al. 2014; Vieira et al. 2017, 2020), but is also correlated with broad metrics of self-reported social function, such as time spent with others (Holt et al. 2014).

In conjunction with self-reported feelings of increased discomfort, individuals with ASD exhibit decreased bilateral IPS activity when watching videos of approaching individuals (Massaccesi et al. 2021). This suggests that decreased IPS activity may be linked with a narrower window of permeability in this population. Decreased left IPS activation in autism has also been observed when making social judgments (e.g., approachable), but not physical judgments (e.g., gender) about unfamiliar faces, as well as when viewing social biological motion (Webster et al. 2020). Importantly, viewing non-social motion (e.g., random dot kinematogram) is *not* accompanied by atypical IPS activity relative to controls (Robertson et al. 2014). Combined, this evidence suggests that altered PPS in autism is associated with not only atypical visuomotor and visuotactile integration (see Fig. 3) in IPS but also with differences in the synthesis of social information in this cortical area. Indeed, social information may be represented in IPS with less specificity in autism. An fMRI study investigating cortical responses to faces and houses found reduced decoding accuracy in IPS in the autistic group relative to non-autistic controls (Nunes et al. 2021).

Similarly, IPS has been implicated in social functioning in schizophrenia, but in an opposite pattern to that seen in autism. Approaching faces elicit increased dorsal IPS activity in individuals with schizophrenia relative to healthy controls, the degree of which is correlated with extended PPS boundaries and severity of negative symptoms (e.g., social withdrawal) (Holt et al. 2015). Importantly, lower-order face-responsive regions (e.g., fusiform face area) show normal activation in schizophrenia (Holt et al. 2015), suggesting that, like autism, the synthesis of higher-order social information at the level of IPS may be impaired rather than the processing of the stimulus characteristics themselves. Additionally, in individuals with schizophrenia relative to healthy controls, lateral intraparietal area LIP – a region of IPS largely dedicated to oculomotor transformations – shows a reduction in velocity-dependent modulation of activity in response to tracked moving objects (Nagel et al. 2012). Like autism, this latter finding may suggest a relatively indiscriminate activation of IPS where object motion-related modulation would be expected. However, changes to motion-evoked IPS activation may also be a result of reduced feedforward connectivity to this region, as has been observed between the motion-sensitive extrastriate area V5 and various regions of IPS (Lencer et al. 2011).

Although it remains unclear the degree to which IPS function itself contributes to features of schizophrenia, it is apparent that this cortical area warrants further study. Furthermore, PPC more broadly should be investigated, not only for its role in PPS and social function in both autism and schizophrenia, but also for its role in more explicit social communication skills, such as syntax, praxis, and social and non-social action judgments.

3.3 Temporo-Parietal Junction (TPJ)

To this point, we have discussed findings of changes in the structure and function of temporal and parietal cortical areas in autism and schizophrenia. We will briefly discuss a cortical intersection between these areas, temporo-parietal junction (TPJ). TPJ is a hub that links IPL, pSTS/G, and lateral occipital cortex. Activity of both ventral posterior (vpTPJ) and anterior (vaTJP) TPJ has been shown to act as moderators of autistic traits and proneness to psychosis in healthy populations (Abu-Akel et al. 2017). Specifically, decreased baseline vpTPJ activity correlated with increased autistic traits, whereas increased vpTPJ activity correlated with increased psychosis proneness; the reverse pattern of activity was seen in vaTPJ (Abu-Akel et al. 2017). In autistic groups, reduced TPJ activity and within-network FC have been associated with poorer judgments of intentionality (Kana et al. 2014); similarly, reduced temporo-parietal FC has been linked to reduced humor comprehension and increased positive and disorganization symptoms in schizophrenia (Jáni et al. 2021). As discussed in prior sections, this evidence points to disruptions in common cortical regions and neural pathways, which may result in similar or opposing feature presentation in autism and schizophrenia.

4 In Pursuit of Novel Therapeutic Interventions

This review has characterized some of the shared behavioral phenotypes and cortical structures implicated in both autism and schizophrenia, with the goal of identifying common mechanisms that may give rise to different patterns of clinical features. The purpose of using this RDoC-inspired model is to better understand the physiological roots of a given phenotype *across* clinical populations rather than within the context of a *single* diagnostic group. With this in mind, we will discuss the potential for remediation of MSI impairments using non-invasive and non-pharmacological behavioral approaches. As perception and cognition are scaffolded on a (multi)-sensory foundation, the hope is that improvements in MSI will cascade into improvements in these higher-order domains. Additionally, by characterizing individuals' clinical profiles and performance on MSI tasks across diagnoses – and in conjunction with non-clinical samples – we are better able to elucidate the links between MSI disruption and the presentation of certain cross-diagnostic symptomology. As noted above, both autism and schizophrenia share clinical features not only with one another but also with other well-studied disorders such as OCD, PD, and ADHD. By identifying not only the behaviors and features common across these disorders but also patterns of similarities in MSI function, we may be able to better understand how MSI function underlies maladaptive behavior and which physiological mechanisms may be beneficial to target.

Behavioral training is likely to be an important adjuvant to any novel therapeutic intervention in either of these populations; not only may immediate perceptual experience be shifted using such paradigms, but training could also result in significant cross-modal plasticity. Furthermore, there is already evidence that perceptual shifts are accomplished faster and more effectively using multisensory paradigms compared to those that employ only a single modality (Seitz et al. 2006; Kim et al. 2008; Shams and Seitz 2008; Opoku-Baah et al. 2020). This latter finding underscores not only the behavioral benefits gleaned from MSI but also the potential therapeutic use for a variety of perceptual deficits.

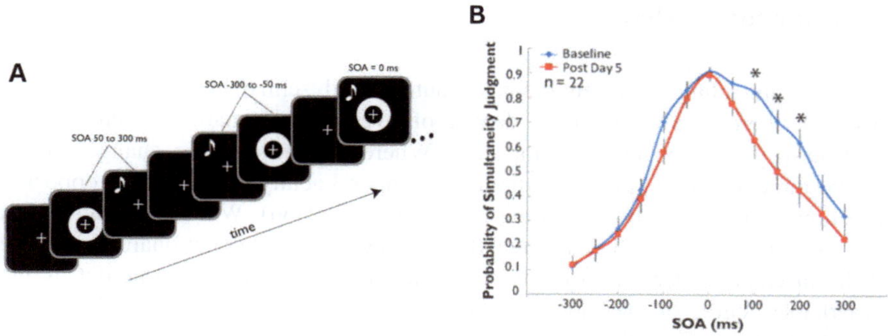

Fig. 6 Audiovisual temporal binding window is sensitive to perceptual training and feedback. (Data adapted with permission from Powers et al. 2009). (**a**) When asked to determine whether simple flashes and beeps are presented simultaneously, participants are then provided with feedback concerning the accuracy of their response. A TBW is estimated from their simultaneity judgments. (**b**) After repeated feedback, participants show a narrowing of their TBW, particularly in visual-leading (SOA > 0) asynchronies

For example, it has been long understood that in healthy samples, audiovisual TBW is plastic and can be trained with perceptual feedback; indeed, in 2-alternative forced choice simultaneity judgment tasks (Fig. 6a), providing individuals with feedback over a repeated course of training narrows their TBW (Fig. 6b), such that they become more accurate in their simultaneity judgment, particularly for visual-leading stimuli (Powers et al. 2009; Stevenson et al. 2013). Importantly, these studies have also shown that the effects last at least a week after the training, suggesting that this may be an effective way of producing long-term changes in multisensory precision in clinical groups. Indeed, preliminary evidence has shown that in children with autism who are verbally fluent and require little support to function, audiovisual training can in some cases improve temporal acuity temporarily (Feldman et al. 2020) (although see Feldman et al. (2022)). Although these results are to date controversial, there is promise that at least some individuals with ASD may benefit from specific multisensory perceptual training. While we are unaware of any similar findings in schizophrenia, multisensory training in emotion recognition has demonstrated temporary improvement in this group in one study (Bechi et al. 2018). Thus, we believe that both individuals with autism and with schizophrenia are likely to benefit from sensory training as a part of a comprehensive intervention plan and that improved (multi)sensory acuity is likely to have positive downstream effects on complex sensory-dependent functions (e.g., speech-in-noise comprehension).

5 Concluding Remarks

Although the clinical presentations of autism and schizophrenia are strikingly different, there is a surprisingly degree of overlap in domains of sensory and multisensory function that are impacted. Whereas some of these changes are in similar directions (e.g., poorer audiovisual temporal acuity), many are in opposing directions (e.g., susceptibility to the rubber hand illusion). When thinking from a dimensional perspective (i.e., RDoC), these changes suggest commonalities in some of the mechanistic targets in these conditions, a suggestion also borne out by the remarkable degree of overlap in affected brain regions and circuits. Collectively, these findings reinforce the basic premise that sensory and multisensory circuits and their functional correlates scaffold the development and maintenance of higher-order cognitive representations. In the context of autism and schizophrenia, the most impacted of these representations are in the arenas and social function and language and communication. Much more work is needed in examining these links, but we believe that there is promise in leveraging the information learned about (multi)-sensory functional differences in these conditions to develop non-pharmacological and non-invasive treatment options.

References

Abu-Akel AM, Apperly IA, Wood SJ, Hansen PC (2017) Autism and psychosis expressions diametrically modulate the right temporoparietal junction. Soc Neurosci 12:506–518

Agurto C et al (2020) Analyzing acoustic and prosodic fluctuations in free speech to predict psychosis onset in high-risk youths. In: 2020 42nd Annual International Conference of the IEEE Engineering in Medicine & Biology Society (EMBC), pp 5575–5579. https://doi.org/10.1109/EMBC44109.2020.9176841

Åhs F, Dunsmoor JE, Zielinski D, LaBar KS (2015) Spatial proximity amplifies valence in emotional memory and defensive approach-avoidance. Neuropsychologia 70:476–485

American Psychiatric Association (2013) Diagnostic and statistical manual of mental disorders. 5th ed.

Annaz D, Campbell R, Coleman M, Milne E, Swettenham J (2012) Young children with autism spectrum disorder do not preferentially attend to biological motion. J Autism Dev Disord 42:401–408

Arunachalam S, Luyster RJ (2016) The integrity of lexical acquisition mechanisms in autism spectrum disorders: a research review. Autism Res 9:810–828

Atkinson AP (2009) Impaired recognition of emotions from body movements is associated with elevated motion coherence thresholds in autism spectrum disorders. Neuropsychologia 47:3023–3029

Bahrick LE (2010) Intermodal perception and selective attention to intersensory redundancy: implications for typical social development and autism. In: Blackwell handbook of infant development, vol 1, pp 120–165

Balz J et al (2016) GABA concentration in superior temporal sulcus predicts gamma power and perception in the sound-induced flash illusion. NeuroImage 125:724–730

Baron-Cohen S (2017) The eyes as window to the mind. Am J Psychiatry 174:1–2

Baum Miller SH, Wallace MT (2019) Multisensory processing differences in individuals with autism spectrum disorder. In: Multisensory processes: the auditory perspective, vol 68. ASA Press, pp 243–272
Bechi M et al (2018) Visual and audio emotion processing training for outpatients with schizophrenia: an integrated multisensory approach. Neuropsychol Rehabil 28:1131–1144
Boehme R, Frost Karlsson M, Heilig M, Olausson H, Capusan AJ (2020) Sharpened self-other distinction in attention deficit hyperactivity disorder. NeuroImage Clin 27:102317
Borgolte A et al (2021) Enhanced attentional processing during speech perception in adult high-functioning autism spectrum disorder: an ERP-study. Neuropsychologia 161:108022
Brien A, Hutchins TL, Westby C (2021) Autobiographical memory in autism spectrum disorder, attention-deficit/hyperactivity disorder, hearing loss, and childhood trauma: implications for social communication intervention. Lang Speech Hear Serv Sch 52:239–259
Brignell A, May T, Morgan AT, Williams K (2019) Predictors and growth in receptive vocabulary from 4 to 8 years in children with and without autism spectrum disorder: a population-based study. Autism 23:1322–1334
Canzoneri E, Magosso E, Serino A (2012) Dynamic sounds capture the boundaries of peripersonal space representation in humans. PLoS One 7:e44306
Cascio CJ, Foss-Feig JH, Burnette CP, Heacock JL, Cosby AA (2012) The rubber hand illusion in children with autism spectrum disorders: delayed influence of combined tactile and visual input on proprioception. Autism 16:406–419
Castagna F et al (2013) Prosody recognition and audiovisual emotion matching in schizophrenia: the contribution of cognition and psychopathology. Psychiatry Res 205:192–198
Chang EF, Raygor KP, Berger MS (2015) Contemporary model of language organization: an overview for neurosurgeons. J Neurosurg 122:250
Charbonneau G et al (2013) Multilevel alterations in the processing of audio–visual emotion expressions in autism spectrum disorders. Neuropsychologia 51:1002–1010
Chien Y-L, Wu C-S, Tsai H-J (2021) The comorbidity of schizophrenia spectrum and mood disorders in autism spectrum disorder. Autism Res 14:571–581
Cole EJ, Barraclough NE, Andrews TJ (2019) Reduced connectivity between mentalizing and mirror systems in autism spectrum condition. Neuropsychologia 122:88–97
Crespi B, Dinsdale N (2019) Autism and psychosis as diametrical disorders of embodiment. Evol Med Public Health 2019:121–138
Danivas V et al (2009) Inferior parietal lobule volume and schneiderian first-rank symptoms in antipsychotic-naïve schizophrenia: a 3-Tesla MRI Study. Indian J Psychol Med 31:82–87
de Boer-Schellekens L, Stekelenburg JJ, Maes JP, Van Gool AR, Vroomen J (2014) Sound improves diminished visual temporal sensitivity in schizophrenia. Acta Psychol 147:136–142
de Gelder B et al (2005) Multisensory integration of emotional faces and voices in schizophrenics. Schizophr Res 72:195–203
de Guibert C et al (2011) Abnormal functional lateralization and activity of language brain areas in typical specific language impairment (developmental dysphasia). Brain 134:3044–3058
de Jong JJ, Hodiamont PPG, Van den Stock J, de Gelder B (2009) Audiovisual emotion recognition in schizophrenia: reduced integration of facial and vocal affect. Schizophr Res 107:286–293
Delevoye-Turrell Y, Vienne C, Coello Y (2011) Space boundaries in schizophrenia: voluntary action for improved judgments of social distances. Soc Psychol 42:193
Ding C et al (2017) Parkinson's disease alters multisensory perception: insights from the Rubber Hand Illusion. Neuropsychologia 97:38–45
Duan Y et al (2022) Identifying and predicting autism spectrum disorder based on multi-site structural MRI with machine learning. Front Hum Neurosci 15:765517
Feldman JI et al (2020) Plasticity of temporal binding in children with autism spectrum disorder: a single case experimental design perceptual training study. Res Autism Spectr Disord 74:101555
Feldman JI et al (2022) A randomized controlled trial for audiovisual multisensory perception in autistic youth. J Autism Dev Disord 53:4318. https://doi.org/10.1007/s10803-022-05709-6

Fleming JT, Noyce AL, Shinn-Cunningham BG (2020) Audio-visual spatial alignment improves integration in the presence of a competing audio-visual stimulus. Neuropsychologia 146:107530
Florea T et al (2022) Oxytocin: narrative expert review of current perspectives on the relationship with other neurotransmitters and the impact on the main psychiatric disorders. Medicina 58:923
Foss-Feig J et al (2010) An extended multisensory temporal binding window in autism spectrum disorders. Exp Brain Res 203:381
Foss-Feig JH et al (2017) Searching for cross-diagnostic convergence: neural mechanisms governing excitation and inhibition balance in schizophrenia and autism spectrum disorders. Biol Psychiatry 81:848–861
Foucher JR, Lacambre M, Pham B-T, Giersch A, Elliott MA (2007) Low time resolution in schizophrenia: lengthened windows of simultaneity for visual, auditory and bimodal stimuli. Schizophr Res 97:118–127
Foxe JJ et al (2015) Severe multisensory speech integration deficits in high-functioning school-aged children with Autism Spectrum Disorder (ASD) and their resolution during early adolescence. Cereb Cortex 25:298–312
Friederici AD, Makuuchi M, Bahlmann J (2009) The role of the posterior superior temporal cortex in sentence comprehension. Neuroreport 20:563
Fusaroli R, Lambrechts A, Bang D, Bowler DM, Gaigg SB (2017) Is voice a marker for Autism spectrum disorder? A systematic review and meta-analysis. Autism Res 10:384–407
Grant KW, Seitz P-F (2000) The use of visible speech cues for improving auditory detection of spoken sentences. J Acoust Soc Am 108:1197–1208
Green M, Waldron JH, Simpson I, Coltheart M (2008) Visual processing of social context during mental state perception in schizophrenia. J Psychiatry Neurosci 33:34
Greenman DLB et al (2020) Parietal-prefrontal feedforward connectivity in association with schizophrenia genetic risk and delusions. Am J Psychiatry 177:1151–1158
Happé F, Frith U (2006) The weak coherence account: detail-focused cognitive style in autism spectrum disorders. J Autism Dev Disord 36:5–25
Haß K et al (2017) Enlarged temporal integration window in schizophrenia indicated by the double-flash illusion. Cogn Neuropsychiatry 22:145–158
He J et al (2022) Deficits in sense of body ownership, sensory processing, and temporal perception in schizophrenia patients with/without auditory verbal hallucinations. Front Neurosci 16: 831714
Hillmann TE, Riehle M, Lincoln TM (2018) Reduced scanning of salient facial features mediates the association between paranoia and emotion recognition. Psychiatry Res 269:430–436
Holt DJ et al (2014) Neural correlates of personal space intrusion. J Neurosci 34:4123
Holt DJ et al (2015) Abnormalities in personal space and parietal–frontal function in schizophrenia. NeuroImage Clin 9:233–243
Hotier S et al (2017) Social cognition in autism is associated with the neurodevelopment of the posterior superior temporal sulcus. Acta Psychiatr Scand 136:517–525
Isaacs D et al (2020) Sensory hypersensitivity severity and association with obsessive-compulsive symptoms in adults with tic disorder. Neuropsychiatr Dis Treat 16:2591
Isaacs D et al (2022) Cross-disorder comparison of sensory over-responsivity in chronic tic disorders and obsessive-compulsive disorder. Compr Psychiatry 113:152291
Isik L, Koldewyn K, Beeler D, Kanwisher N (2017) Perceiving social interactions in the posterior superior temporal sulcus. Proc Natl Acad Sci USA 114:E9145
Jáni M et al (2021) The neural substrates of diminished humor comprehension in schizophrenia and its relationship with psychopathology. Acta Neurobiol Exp 81:181–190
Ji E et al (2019) Increased and decreased superficial white matter structural connectivity in schizophrenia and bipolar disorder. Schizophr Bull 45:1367–1378
Jones W, Klin A (2013) Attention to eyes is present but in decline in 2–6-month-old infants later diagnosed with autism. Nature 504:427–431

Jones W, Carr K, Klin A (2008) Absence of preferential looking to the eyes of approaching adults predicts level of social disability in 2-year-old toddlers with autism spectrum disorder. Arch Gen Psychiatry 65:946–954

Kana RK, Libero LE, Hu CP, Deshpande HD, Colburn JS (2014) Functional brain networks and white matter underlying theory-of-mind in autism. Soc Cogn Affect Neurosci 9:98–105

Kato Y et al (2011) Magnetoencephalography study of right parietal lobe dysfunction of the evoked mirror neuron system in antipsychotic-free schizophrenia. PLoS One 6:e28087

Kawakami S et al (2020) Atypical multisensory integration and the temporal binding window in autism spectrum disorder. J Autism Dev Disord 50:3944–3956

Kim RS, Seitz AR, Shams L (2008) Benefits of stimulus congruency for multisensory facilitation of visual learning. PLoS One 3:e1532

Kim H, Ahn J, Lee H, Ha S, Cheon K-A (2020) Differences in language ability and emotional-behavioral problems according to symptom severity in children with autism spectrum disorder. Yonsei Med J 61:880–890

Kington JM, Jones LA, Watt AA, Hopkin EJ, Williams J (2000) Impaired eye expression recognition in schizophrenia. J Psychiatr Res 34:341–347

Klin A, Lin DJ, Gorrindo P, Ramsay G, Jones W (2009) Two-year-olds with autism orient to non-social contingencies rather than biological motion. Nature 459:257–261

Lee H-S et al (2021) Altered peripersonal space and the bodily self in schizophrenia: a virtual reality study. Schizophr Bull 47:927–937

Lencer R et al (2011) Altered transfer of visual motion information to parietal association cortex in untreated first-episode psychosis: implications for pursuit eye tracking. Psychiatry Res Neuroimaging 194:30–38

Li X-B et al (2020) The attenuated visual scanpaths of patients with schizophrenia whilst recognizing emotional facial expressions are worsened in natural social scenes. Schizophr Res 220: 155–163

Lloyd DM (2007) Spatial limits on referred touch to an alien limb may reflect boundaries of visuo-tactile peripersonal space surrounding the hand. Brain Cogn 64:104–109

Loucas T et al (2008) Autistic symptomatology and language ability in autism spectrum disorder and specific language impairment. J Child Psychol Psychiatry 49:1184–1192

Loughland CM, Williams LM, Gordon E (2002) Visual scanpaths to positive and negative facial emotions in an outpatient schizophrenia sample. Schizophr Res 55:159–170

Lucker JR, Geffner D, Koch W (1996) Perception of loudness in children with ADD and without ADD. Child Psychiatry Hum Dev 26:181–190

Lundin NB, Jones MN, Myers EJ, Breier A, Minor KS (2022) Semantic and phonetic similarity of verbal fluency responses in early-stage psychosis. Psychiatry Res 309:114404

Manjaly ZM et al (2007) Neurophysiological correlates of relatively enhanced local visual search in autistic adolescents. NeuroImage 35:283–291

Martin B, Giersch A, Huron C, van Wassenhove V (2013) Temporal event structure and timing in schizophrenia: preserved binding in a longer "now". Neuropsychologia 51:358–371

Martínez A et al (2019) Differential patterns of visual sensory alteration underlying face emotion recognition impairment and motion perception deficits in schizophrenia and autism spectrum disorder. Biol Psychiatry 86:557–567

Massaccesi C et al (2021) Neural correlates of interpersonal space permeability and flexibility in autism spectrum disorder. Cereb Cortex 31:2968–2979

Mazzoni N, Ricciardelli P, Actis-Grosso R, Venuti P (2021) Difficulties in recognising dynamic but not static emotional body movements in autism spectrum disorder. J Autism Dev Disord 52: 1092. https://doi.org/10.1007/s10803-021-05015-7

McDaniel J, Yoder P, Woynaroski T, Watson L (2018) Predicting receptive–expressive vocabulary discrepancies in preschool children with autism spectrum disorder. J Speech Lang Hear Res 61: 1426–1439

McGurk H, MacDonald J (1976) Hearing lips and seeing voices. Nature 264:746–748

Mier D et al (2017) Aberrant activity and connectivity of the posterior superior temporal sulcus during social cognition in schizophrenia. Eur Arch Psychiatry Clin Neurosci 267:597–610

Milne E et al (2002) High motion coherence thresholds in children with autism. J Child Psychol Psychiatry 43:255–263

Mottron L, Burack JA (2001) Enhanced perceptual functioning in the development of autism. In J. A. Burack, T. Charman, N. Yirmiya, & P. R. Zelazo (Eds.), The development of autism: Perspectives from theory and research (pp. 131–148). Lawrence Erlbaum Associates Publishers.

Müller V, Cieslik E, Laird A, Fox P, Eickhoff S (2013) Dysregulated left inferior parietal activity in schizophrenia and depression: functional connectivity and characterization. Front Hum Neurosci 7:268

Nagel M, Sprenger A, Steinlechner S, Binkofski F, Lencer R (2012) Altered velocity processing in schizophrenia during pursuit eye tracking. PLoS One 7:e38494

Nardini M, Cowie D (2012) The development of multisensory balance, locomotion, orientation, and navigation. In: Multisensory development. Oxford University Press, pp 137–158

Nath AR, Beauchamp MS (2011) Dynamic changes in superior temporal sulcus connectivity during perception of noisy audiovisual speech. J Neurosci 31:1704

National Institute of Mental Health. About RDoC. https://www.nimh.nih.gov/research/research-funded-bynimh/rdoc/about-rdoc.shtml

Navalón P et al (2021) Attentional processing biases to threat in schizophrenia: evidence from a free-viewing task with emotional scenes. J Psychiatr Res 144:80–86

Neil PA, Chee-Ruiter C, Scheier C, Lewkowicz DJ, Shimojo S (2006) Development of multisensory spatial integration and perception in humans. Dev Sci 9:454–464

Nierenberg J et al (2005) Reduced left angular gyrus volume in first-episode schizophrenia. Am J Psychiatry 162:1539–1541

Nikolaides A et al (2016) Restricted attention to social cues in schizophrenia patients. Eur Arch Psychiatry Clin Neurosci 266:649–661

Noel J-P, Łukowska M, Wallace M, Serino A (2016) Multisensory simultaneity judgment and proximity to the body. J Vis 16:21–21

Noel J-P, De Niear MA, Stevenson R, Alais D, Wallace MT (2017) Atypical rapid audio-visual temporal recalibration in autism spectrum disorders. Autism Res 10:121–129

Noel J-P et al (2020) Visual-tactile spatial multisensory interaction in adults with autism and schizophrenia. Front Psych 11:1099

Noel J-P et al (2021) Inflexible updating of the self-other divide during a social context in autism: psychophysical, electrophysiological, and neural network modeling evidence. Biol Psychiatry Cogn Neurosci Neuroimaging 7:756. https://doi.org/10.1016/j.bpsc.2021.03.013

Nunes AS et al (2021) Classification of evoked responses to inverted faces reveals both spatial and temporal cortical response abnormalities in Autism spectrum disorder. NeuroImage Clin 29: 102501

Ocklenburg S, Westerhausen R, Hirnstein M, Hugdahl K (2013) Auditory hallucinations and reduced language lateralization in schizophrenia: a meta-analysis of dichotic listening studies. J Int Neuropsychol Soc 19:410–418

Opoku-Baah C, Hou B, Wallace M (2020) A multisensory perspective on low vision rehabilitation: audition facilitates visual learning of contrast detection. Invest Ophthalmol Vis Sci 61:4269–4269

Otte M-L et al (2022) Cortical morphology and illness insight in patients with schizophrenia. Eur Arch Psychiatry Clin Neurosci 272:985–995

Ozker M, Schepers IM, Magnotti JF, Yoshor D, Beauchamp MS (2017) A double dissociation between anterior and posterior superior temporal gyrus for processing audiovisual speech demonstrated by electrocorticography. J Cogn Neurosci 29:1044–1060

Palaniyappan L, Liddle PF (2012) Dissociable morphometric differences of the inferior parietal lobule in schizophrenia. Eur Arch Psychiatry Clin Neurosci 262:579–587

Parish-Morris J et al (2019) Adaptation to different communicative contexts: an eye tracking study of autistic adults. J Neurodev Disord 11:5

Peled A, Ritsner M, Hirschmann S, Geva AB, Modai I (2000) Touch feel illusion in schizophrenic patients. Biol Psychiatry 48:1105–1108
Pellicano E, Gibson L, Maybery M, Durkin K, Badcock DR (2005) Abnormal global processing along the dorsal visual pathway in autism: a possible mechanism for weak visuospatial coherence? Neuropsychologia 43:1044–1053
Peppé S, McCann J, Gibbon F, O'Hare A, Rutherford M (2007) Receptive and expressive prosodic ability in children with high-functioning autism. J Speech Lang Hear Res 50:1015–1028
Perlini C et al (2012) Linguistic production and syntactic comprehension in schizophrenia and bipolar disorder. Acta Psychiatr Scand 126:363–376
Port RG et al (2016) Maturation of auditory neural processes in autism spectrum disorder – A longitudinal MEG study. NeuroImage Clin 11:566–577
Powers AR, Hillock AR, Wallace MT (2009) Perceptual training narrows the temporal window of multisensory binding. J Neurosci 29:12265
Rae CL, Larsson DEO, Eccles JA, Ward J, Critchley HD (2018) Subjective embodiment during the rubber hand illusion predicts severity of premonitory sensations and tics in Tourette Syndrome. Conscious Cogn 65:368–377
Ray-Subramanian CE, Ellis Weismer S (2012) Receptive and expressive language as predictors of restricted and repetitive behaviors in young children with autism spectrum disorders. J Autism Dev Disord 42:2113–2120
Robertson CE et al (2014) Global motion perception deficits in autism are reflected as early as primary visual cortex. Brain 137:2588–2599
Ross LA et al (2007) Impaired multisensory processing in schizophrenia: deficits in the visual enhancement of speech comprehension under noisy environmental conditions. Schizophr Res 97:173–183
Russo N, Zecker S, Trommer B, Chen J, Kraus N (2009) Effects of background noise on cortical encoding of speech in autism spectrum disorders. J Autism Dev Disord 39:1185–1196
Safar K, Vandewouw MM, Taylor MJ (2021) Atypical development of emotional face processing networks in autism spectrum disorder from childhood through to adulthood. Dev Cogn Neurosci 51:101003
Schwartz S, Wang L, Shinn-Cunningham BG, Tager-Flusberg H (2020) Neural evidence for speech processing deficits during a cocktail party scenario in minimally and low verbal adolescents and young adults with autism. Autism Res 13:1828–1842
Seitz AR, Kim R, Shams L (2006) Sound facilitates visual learning. Curr Biol 16:1422–1427
Sekuler R, Sekuler AB, Lau R (1997) Sound alters visual motion perception. Nature 385:308–308
Senkowski D, Moran JK (2022) Early evoked brain activity underlies auditory and audiovisual speech recognition deficits in schizophrenia. NeuroImage Clin 33:102909
Serino A et al (2015) Body part-centered and full body-centered peripersonal space representations. Sci Rep 5:18603
Seymour RA, Rippon G, Gooding-Williams G, Schoffelen JM, Kessler K (2019) Dysregulated oscillatory connectivity in the visual system in autism spectrum disorder. Brain 142:3294–3305
Shams L, Seitz AR (2008) Benefits of multisensory learning. Trends Cogn Sci 12:411–417
Shimizu T et al (2000) Comparison of eye-movement patterns in schizophrenic and normal adults during examination of facial affect displays. Percept Mot Skills 91:1045–1056
Slušná D, Rodríguez A, Salvadó B, Vicente A, Hinzen W (2021) Relations between language, non-verbal cognition, and conceptualization in non- or minimally verbal individuals with ASD across the lifespan. Autism Dev Lang Impair 6:23969415211053264
Soni S, Muthukrishnan SP, Sood M, Kaur S, Sharma R (2018) Hyperactivation of left inferior parietal lobule and left temporal gyri shortens resting EEG microstate in schizophrenia. Schizophr Res 201:204–207
Spezio ML, Adolphs R, Hurley RSE, Piven J (2007) Analysis of face gaze in autism using "Bubbles". Neuropsychologia 45:144–151

Spilka MJ, Pittman DJ, Bray SL, Goghari VM (2019) Manipulating visual scanpaths during facial emotion perception modulates functional brain activation in schizophrenia patients and controls. J Abnorm Psychol 128:855–866
Stevenson RA, Wilson MM, Powers AR, Wallace MT (2013) The effects of visual training on multisensory temporal processing. Exp Brain Res 225:479–489
Stevenson RA et al (2014a) Multisensory temporal integration in autism spectrum disorders. J Neurosci 34:691
Stevenson RA et al (2014b) Brief report: arrested development of audiovisual speech perception in autism spectrum disorders. J Autism Dev Disord 44:1470–1477
Stevenson RA et al (2017a) The associations between multisensory temporal processing and symptoms of schizophrenia. Schizophr Res 179:97–103
Stevenson RA et al (2017b) Multisensory speech perception in autism spectrum disorder: from phoneme to whole-word perception. Autism Res 10:1280–1290
Stevenson RA et al (2018) The cascading influence of multisensory processing on speech perception in autism. Autism 22:609–624
Stone DB, Coffman BA, Bustillo JR, Aine CJ, Stephen JM (2014) Multisensory stimuli elicit altered oscillatory brain responses at gamma frequencies in patients with schizophrenia. Front Hum Neurosci 8:788
Su W-C, Culotta M, Tsuzuki D, Bhat A (2022) Cortical activation during cooperative joint actions and competition in children with and without an autism spectrum condition (ASC): an fNIRS study. Sci Rep 12:5177
Sumby WH, Pollack I (1954) Visual contribution to speech intelligibility in noise. J Acoust Soc Am 26:212–215
Tan EJ, Yelland GW, Rossell SL (2016) Characterising receptive language processing in schizophrenia using word and sentence tasks. Cogn Neuropsychiatry 21:14–31
Tardif C, Lainé F, Rodriguez M, Gepner B (2007) Slowing down presentation of facial movements and vocal sounds enhances facial expression recognition and induces facial–vocal imitation in children with autism. J Autism Dev Disord 37:1469–1484
Teneggi C, Canzoneri E, di Pellegrino G, Serino A (2013) Social modulation of peripersonal space boundaries. Curr Biol 23:406–411
Thakkar KN, Nichols HS, McIntosh LG, Park S (2011) Disturbances in body ownership in schizophrenia: evidence from the rubber hand illusion and case study of a spontaneous out-of-body experience. PLoS One 6:e27089
Thakkar KN, Peterman JS, Park S (2014) Altered brain activation during action imitation and observation in schizophrenia: a translational approach to investigating social dysfunction in schizophrenia. Am J Psychiatry 171:539–548
Thomas P et al (1996) Speech and language in first onset psychosis differences between people with schizophrenia, mania, and controls. Br J Psychiatry 168:337–343
Torrey EF (2007) Schizophrenia and the inferior parietal lobule. Schizophr Res 97:215–225
Trepagnier C, Sebrechts MM, Peterson R (2002) Atypical face gaze in autism. Cyberpsychol Behav 5:213–217
Tso IF, Carp J, Taylor SF, Deldin PJ (2014) Role of visual integration in gaze perception and emotional intelligence in schizophrenia. Schizophr Bull 40:617–625
van Wassenhove V, Grant KW, Poeppel D (2005) Visual speech speeds up the neural processing of auditory speech. Proc Natl Acad Sci USA 102:1181
Vanes LD et al (2016) Reduced susceptibility to the sound-induced flash fusion illusion in schizophrenia. Psychiatry Res 245:58–65
Venkatasubramanian G, Jayakumar PN, Keshavan MS, Gangadhar BN (2011) Schneiderian first rank symptoms and inferior parietal lobule cortical thickness in antipsychotic-naïve schizophrenia. Prog Neuro-Psychopharmacol Biol Psychiatry 35:40–46
Vieira JB, Tavares TP, Marsh AA, Mitchell DGV (2017) Emotion and personal space: neural correlates of approach-avoidance tendencies to different facial expressions as a function of coldhearted psychopathic traits. Hum Brain Mapp 38:1492–1506

Vieira JB, Pierzchajlo SR, Mitchell DGV (2020) Neural correlates of social and non-social personal space intrusions: role of defensive and peripersonal space systems in interpersonal distance regulation. Soc Neurosci 15:36–51

Wade J et al (2018) Extraction of emotional information via visual scanning patterns: a feasibility study of participants with schizophrenia and neurotypical individuals. ACM Trans Access Comput 11:1

Wadsworth HM et al (2017) The Action Imitation network and motor imitation in children and adolescents with autism. Neuroscience 343:147–156

Waldmann A, Volkmann J, Zeller D (2020) Parkinson's disease may reduce sensitivity to visual-tactile asynchrony irrespective of dopaminergic treatment: evidence from the rubber hand illusion. Parkinsonism Relat Disord 78:100–104

Wallace MT, Stevenson RA (2014) The construct of the multisensory temporal binding window and its dysregulation in developmental disabilities. Neuropsychologia 64:105–123

Wallace GL, Dankner N, Kenworthy L, Giedd JN, Martin A (2010) Age-related temporal and parietal cortical thinning in autism spectrum disorders. Brain 133:3745–3754

Wang X, Song Y, Zhen Z, Liu J (2016) Functional integration of the posterior superior temporal sulcus correlates with facial expression recognition. Hum Brain Mapp 37:1930–1940

Webster PJ et al (2020) Processing of real-world, dynamic natural stimuli in autism is linked to corticobasal function. Autism Res 13:539–549

Wilson TW, Rojas DC, Reite ML, Teale PD, Rogers SJ (2007) Children and adolescents with autism exhibit reduced MEG steady-state gamma responses. Biol Psychiatry 62:192–197

Wymbs NF, Nebel MB, Ewen JB, Mostofsky SH (2021) Altered inferior parietal functional connectivity is correlated with praxis and social skill performance in children with autism spectrum disorder. Cereb Cortex 31:2639–2652

Wynn CJ, Borrie SA, Sellers TP (2018) Speech rate entrainment in children and adults with and without autism spectrum disorder. Am J Speech Lang Pathol 27:965–974

Zhang A et al (2022) Connectivity-based brain network supports restricted and repetitive behaviors in autism spectrum disorder across development. Front Psych 13:874090

Zhou H et al (2018) Multisensory temporal binding window in autism spectrum disorders and schizophrenia spectrum disorders: a systematic review and meta-analysis. Neurosci Biobehav Rev 86:66–76

Zhu XL et al (2013) Visual scanning of emotional faces in schizophrenia. Neurosci Lett 552:46–51

Zopf R, Boulton K, Langdon R, Rich AN (2021) Perception of visual-tactile asynchrony, bodily perceptual aberrations, and bodily illusions in schizophrenia. Schizophr Res 228:534–540

Zvyagintsev M, Parisi C, Mathiak K (2017) Temporal processing deficit leads to impaired multisensory binding in schizophrenia. Cogn Neuropsychiatry 22:361–372

Visual Perceptual Processing Abnormalities in Body Dysmorphic Disorder

Joel P. Diaz-Fong and Jamie D. Feusner

Contents

J. P. Diaz-Fong
Institute of Medical Science, Temerty Faculty of Medicine, University of Toronto, Toronto, ON, Canada

Centre for Addiction and Mental Health, Toronto, ON, Canada

J. D. Feusner (✉)
Institute of Medical Science, Temerty Faculty of Medicine, University of Toronto, Toronto, ON, Canada

Centre for Addiction and Mental Health, Toronto, ON, Canada

Department of Psychiatry, Temerty Faculty of Medicine, University of Toronto, Toronto, ON, Canada

Department of Women's and Children's Health, Karolinska Institutet, Stockholm, Sweden
e-mail: jamie.feusner@utoronto.ca

Curr Topics Behav Neurosci (2025) 74: 105–127
https://doi.org/10.1007/7854_2024_472
Published Online: 02 May 2024

Abstract Phenomenological observations of individuals with body dysmorphic disorder (BDD), coupled with evidence from neuropsychological, psychophysical, and neuroimaging studies, support a model of aberrant visual perception characterized by deficient global/holistic, enhanced detail/local processing, and selective visual-attentional biases. These features may contribute to the core symptomatology of distorted perception of their appearance, in addition to misinterpretation of others' facial expressions and poor insight regarding their misperceived appearance defects. Insights from visual processing studies can contribute to the development of novel interventions, such as perceptual retraining and non-invasive neuromodulation. However, much remains to be understood about visual perception in BDD. Future research should leverage brain imaging modalities with high temporal resolutions and employ study designs that induce conflicts in multisensory integration, thereby advancing our mechanistic understanding of distorted visual perception observed in BDD.

Keywords Attention · Body dysmorphic disorder · Insight · Visual perception

1 Introduction

Body dysmorphic disorder (BDD) is a prevalent and debilitating yet understudied psychiatric disorder. Individuals with BDD experience obsessive thoughts regarding the perceived defects of their appearance that are minor or not noticeable to others, causing considerable distress and impairment of their daily functioning (American Psychiatric Association 2013, p. 991). Although any appearance feature can be the focus of these misperceptions and preoccupations, the most common features involve the face and head areas, such as concerns about skin, hair, and nose (Phillips 2005, p. 412). Some examples include misperceptions that one has acne scarring on their face, that one's hair is thinning, or that one's eyes are asymmetric, accompanied by beliefs that these are prominent, noticeable by others, and make the person look ugly or unattractive. Preoccupations related to the body are also common, including concerns about body weight and body shape (Phillips and Diaz 1997). Appearance distortions are often delusional in nature – they are fixed, false beliefs (American Psychiatric Association 2013) – such that some individuals with BDD are fully convinced that their appearance is deformed or abnormal and appears unattractive (Phillips et al. 2014). BDD has a high prevalence in the general population of 1.7–2.9% (Buhlmann et al. 2010; Koran et al. 2008; Rief et al. 2006; Schieber et al. 2015) (5–7.5 million in the USA alone) and in 13% of psychiatric inpatients (Mufaddel et al. 2013). Poor BDD-related insight could also contribute to seeking cosmetic procedures rather than mental health treatment. A recent meta-analysis estimates a prevalence of BDD of 19.2% in patients presenting for cosmetic surgery (Salari et al. 2022) and commonly undergo rhinoplasty, lipectomy, liposuction, and breast surgery (Lee et al. 2023). In addition, the lifetime rate of suicide attempts in

BDD is remarkably high: approximately 25% (Phillips and Menard 2006). Despite this, it is often misdiagnosed, or the diagnosis is missed, and it remains vastly understudied. The field still knows relatively little about the neurobiology of BDD.

The etiology of BDD has not been fully elucidated, although it is likely multifactorial; developmental, social, and neurobiological factors presumably interact to shape its characteristic features (Feusner et al. 2010d). Of the neurobiological determinants, evidence exists for genetic factors (including family heritability (Bienvenu et al. 2000; Phillips et al. 1998), twin (Monzani et al. 2012), and a candidate gene study (Phillips et al. 2015)), dysfunction of frontostriatal systems (Atmaca et al. 2010; Buchanan et al. 2014; Deckersbach et al. 2000; Feusner et al. 2010c; Kerwin et al. 2014; Rauch et al. 2003), and widespread aberrant structural white matter connectivity (Buchanan et al. 2013) and whole-brain network organization (Arienzo et al. 2013).

Numerous neuropsychological and psychophysical studies have revealed abnormalities in visual processing and visuospatial organization among individuals with BDD. These abnormalities manifest as deficiencies in global perception and feature integration, accompanied by a selective attentional bias to details (Deckersbach et al. 2000; Feusner et al. 2010b; Jefferies et al. 2012; Mundy and Sadusky 2014; Stangier et al. 2008), highlighting a potentially crucial clinical phenotype in BDD: perceptual distortion of appearance. This supports a role in the pathophysiology for aberrant neural systems underpinning disturbed visual information processing (Li et al. 2013).

Disturbances in neural systems may also exist for other perceptual modalities related to self-body experiences, such as somatoperception (egocentric mapping of information from the body surface, exteroceptive perception of external objects contacting the body, and interoception about the nature and state of the body (Longo et al. 2010)), proprioception, and equilibrioception, as well as their multisensory integration (Kaplan et al. 2013). Yet, as the functioning of these other sensory modalities has received little investigation to this point in BDD, the focus of this chapter is on visual perception and how disturbances may produce the experiences of distorted appearance.

This chapter highlights brain and behavioral evidence of disturbances of visual processing in BDD, and how these contribute to clinical phenomenology. In addition, it puts forth a working model of how consciously-experienced distortions of appearance in those with BDD may form as a result of disturbances in lower-level sensory processing combined with attentional biases and influence from emotional arousal. Finally, the chapter will discuss the translation of existing research findings to novel treatments and – as the field is in relatively early stages of understanding these complex experiences in BDD and other disorders involving body image disturbance – potential future directions for new lines of research.

2 Visual Perceptual Abnormalities

2.1 Visuospatial Deficits

Some of the earliest investigations of visuospatial function in BDD utilized tasks to probe global-local processing. Two studies examined this using the Rey-Osterrieth Complex Figure Test; when participants were asked to recall a portion of the test, those with BDD, to a greater extent than healthy controls, recalled specific details of the figure rather than the overall organizational features (Deckersbach et al. 2000; Greenberg et al. 2018) suggesting deficits in organizational strategies for visuospatial processing. One study, however, found no differences between groups on this test (Hanes 1998). In the embedded figures task (Oltman et al. 1971), which requires participants to identify a target shape embedded within one of three complex shapes, individuals with BDD responded slower and less accurately than controls (Kerwin et al. 2014).

2.2 Reduced Global Processing of Visual Signals

To investigate potential functional neural underpinnings of distorted perceptual experiences for appearance in BDD, and following up on earlier studies of potential visuospatial deficits in BDD, several studies using functional magnetic resonance imaging (fMRI) probed responses to viewing faces, bodies, and non-appearance-related objects. Specifically, these task-based fMRI studies used spatial frequency-filtered own face (Feusner et al. 2010c), other-face (Feusner et al. 2007; Li et al. 2013), and house (Feusner et al. 2011; Li et al. 2013) stimuli to selectively convey configural and holistic information. This task manipulation uncovered abnormalities in both primary and secondary visual processing systems. In a similar study using body stimuli, abnormal hypoactivity in dorsal visual stream (DVS) regions was found in those with BDD, including the lateral occipital cortex and superior parietal lobule (Moody et al. 2021). Further, activity in these regions was associated with subjective appearance ratings of these bodies in terms of attractiveness, aversiveness, and how over- and underweight they perceived them to be. A joint electroencephalogram (EEG) and functional magnetic resonance imaging (fMRI) analysis, allowing for estimating spatial activation patterns tied to specific time-locked event-related potential components, found hypoactivity in the DVS in BDD for viewing low-detail faces and houses associated with the N170 component (170 ms after stimulus presentation). There was also enhanced ventral visual stream (VVS) activity for high-detail images; further, the magnitude of this activity was correlated with how unattractive they perceived faces to be (Li et al. 2013).

These findings point to disturbances in the perception of visual stimuli characterized by the under-utilization of brain systems responsible for processing low levels of detail (global processing). This suggests that individuals with BDD have difficulties in perceiving and integrating the overall features of visual information. This phenotype may explain the peculiar, increased attention given to minuscule defects and the inability to contextualize these perceptions within the larger context of the face or body (Li et al. 2013). In other words, individuals with BDD may struggle to recognize the relative insignificance of these perceived defects. In sum, these functional neuroimaging studies in BDD demonstrate reduced DVS and (although less consistently across studies and stimuli) enhanced VVS activity, which are linked to the subjective perception of appearance.

2.3 *Over-Attention to Visual Details*

A useful psychophysical probe of global/local visual processing is a test of a phenomenon referred to as the face inversion effect (FIE) (Valentine 1988). The FIE, observed in most individuals, consists of lower recognition accuracy and slower performance when viewing upside-down faces compared to upright faces on the face inversion task (Fig. 1). The FIE is one of the most robust behavioral effects in the field of face perception. When faces are inverted, it disrupts the holistic processing of facial information. Thus, our brains may have a specific template for processing upright faces but not inverted faces (Freire et al. 2000), given that faces appear more frequently in nature in this orientation. Inversion necessitates feature-based or detailed strategies, typically resulting in delayed response latencies and/or decreased recognition accuracy. Reduced FIE has been demonstrated in three studies in BDD (Feusner et al. 2010b; Jefferies et al. 2012; Toh et al. 2017b), two studies in high body image disturbance individuals (Beilharz et al. 2016; Mundy and Sadusky 2014)

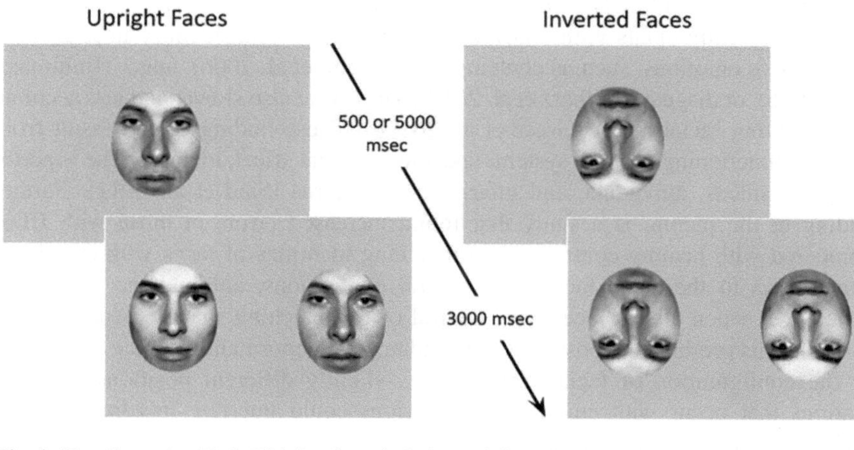

Fig. 1 Face Inversion Task. This is a forced-choice task featuring upright and inverted faces. In this version (Feusner et al. 2010b), stimuli were presented for short (500 ms) or long (5,000 ms) durations

(the majority who likely exceeded clinical cutoffs for BDD), and a study in high body image disturbance individuals, although using body stimuli (Dhir et al. 2018). One study did not find evidence for abnormal FIE in BDD (Monzani et al. 2013). Reduced FIE in BDD may be attributed to similar visual processing strategies for upright and inverted faces due to greater reliance on feature-based/detailed strategies regardless of orientation. In one study, reduced FIE was found in individuals with BDD compared with healthy controls on response times during long duration, but not short duration, face viewing (Feusner et al. 2010b). The reduced FIE was accounted for by faster response times for inverted faces in the BDD group compared with controls, suggesting an abnormal propensity for detailed processing that may have conferred an advantage. However, there was no significant difference between BDD and controls for short-duration (500 ms) stimuli. This differential response based on stimuli duration was reproduced using body stimuli in individuals with high body image disturbance (Dhir et al. 2018).

These findings seem to indicate that enhanced local visual processing may not be due to early perceptual abnormalities but instead driven by a selective attentional bias to local details, which requires a longer time to unfold. This heightened selective attention to local visual details combined with aberrant global visual processing may, in turn, lead to detecting minor imperfections in mirror reflections or photos of their own faces that are not noticeable to others (Buhlmann et al. 2014; Yaryura-Tobias et al. 2002).

2.4 *Emotional Face Misinterpretation*

In addition to global and local visual processing deficits, several studies have shown that individuals with BDD may have deficiencies in facial affect recognition. Specifically, individuals with BDD were more likely to misidentify neutral faces for negative emotions, such as contempt (Buhlmann et al. 2006), anger (Buhlmann et al. 2006), or disgust (Beilharz et al. 2017). They were also slower and less accurate at identifying facial affect (Grace et al. 2019b). These findings could result from aberrant functioning of brain systems specific for facial affect, including the superior temporal sulcus, amygdala, and inferior frontal gyrus (Said et al. 2011). Further adding to the picture is a study that found increased errors in those with BDD compared with healthy controls when matching identities of faces with emotional expressions to the same faces with neutral expressions, although no significant differences when all the faces were neutral or for matching shapes (Feusner et al. 2010a). This presents the possibility of fundamental abnormalities in the processing of the configuration of facial features; i.e., slightly different positions of facial features that occur with emotional expressions could interfere inordinately with their ability to recognize faces. Alternatively, the emotional content conveyed by the faces could have interfered with performance, although the effect did not differ by positive or negative facial expressions.

Furthermore, several studies have provided evidence to support the notion that individuals with BDD have a general bias toward self-referential or emotionally-provoking content. For instance, these individuals have been found to display delayed responses in a Stroop task when salient words are presented as distractors (Buhlmann et al. 2002a; Toh et al. 2017a; Borgers et al. 2022). They have also been found to have a stronger tendency to perceive ambiguous body-related scenarios and ambiguous social situations as threatening compared to healthy controls (Buhlmann et al. 2002b). These observations, along with neuroimaging findings (Bohon et al. 2012; Grace et al. 2019a), suggest that the emotional brain systems in individuals with BDD may exert a stronger influence on the visual systems compared to healthy controls. This could result in a tendency to engage in enhanced detail processing of small imperfections on their own face or body and to demonstrate biased attention toward relevant features in the faces or bodies of others (Grocholewski et al. 2012). In sum, the research to date suggests the possibility that aberrant facial emotional recognition in BDD could be a function of an interaction of impaired facial affect processing and imbalances in configural vs. detail processing, which could worsen under conditions of high emotional arousal.

3 Relationship to Other Clinical Phenomenology

3.1 Level of Insight

Research to date in BDD, as described, suggests the presence of abnormal holistic and configural visual processing. The brains of those with BDD may experience a disrupted ability to generate global visual "templates" to allow them to effectively integrate and contextualize details, including minor flaws or imperfections. These difficulties could contribute to distortions in their subjective perception of their appearance. Given the general tendency of humans to accept what they see without questioning it, individuals with BDD frequently assume that not only what they subjectively perceive when they look in the mirror or in photographs is an "objective reality" of their appearance, but that others perceive them the same way that they do. These misperceptions may contribute to limited insight into their condition, reinforced by repetitive thinking (obsessions) and behaviors (compulsions).

Clinically, the concept of insight in BDD refers to the degree of conviction in one's disorder-relevant belief (Phillips et al. 2014). At the extreme end of poor insight is delusionality. About 30–60% of those with BDD are delusional (Eisen et al. 2004; Mancuso et al. 2010; Phillips et al. 1994, 2012). Whether there is a similar underlying pathophysiology in BDD as in other disorders involving delusions such as schizophrenia is not completely clear at this point, although a single study comparing them on tasks measuring perceptual organization found abnormalities in schizophrenia that were not present in those with BDD (Silverstein et al. 2015). Further, comorbidity rates between BDD and schizophrenia have not been directly examined. Those with BDD with poorer insight are less likely to seek mental health treatment (Phillips et al. 2005) – as they believe their problems, even if they include psychological distress, stem from their reaction to an underlying severe physical appearance defect. This represents a considerable barrier to initiating treatment and, often, to adherence to treatment. Further, it underscores the importance of understanding the neural underpinnings of poor insight in BDD and ultimately developing and applying treatments to improve it.

In addition to this theoretical model of how the neuroimaging, psychophysical, and neuropsychological findings in BDD might produce distorted subjective perception that in turn could lead to poor insight, there are several findings from neuroimaging and psychophysical studies in BDD that demonstrate more direct associations between visuospatial task performance, brain structure, and brain function with poor insight. In a psychophysical study of global/local processing using the embedded figure test, those with BDD with slower performance had poorer insight (Kerwin et al. 2014). Using the Navon task (Navon 1997) involving recognizing larger letters made up of smaller letters to study global-local processing and set-shifting, those with BDD with poorer insight were less accurate when having to shift from identifying the global stimulus to identifying the local stimulus (Kerwin et al. 2014).

A study of white matter tracts using diffusion tensor imaging (DTI) found that individuals with BDD with worse insight had lower fractional anisotropy (FA) and linear anisotropy and higher mean diffusivity (MD) in the inferior longitudinal fasciculus (ILF), which connects the anterior temporal cortex (including the amygdala and hippocampus) to the occipital lobe (Feusner et al. 2013). The ILF is involved in visually guided behaviors as well as in facial recognition, visual emotional recognition, reading, and memory. The findings in BDD suggest the possibility of associations between worse insight/delusionality and reduced fiber organization in pathways that integrate emotional signals with visual perception. Further, when individuals with BDD assess their appearance, disruptions in feedforward (bottom-up observations from the external environment) and/or feedback (top-down contextualization based on prior beliefs or expectations) transfer of information within the ILF (Schmahmann and Pandya 2006; Shinoura et al. 2007; Ross 2008) could lead to difficulties in accurately updating visual memories. Consequently, they may develop persistent, distorted mental images of appearance flaws. For example, prior experiences such as exposure to extreme lighting conditions or past blemishes like acne that have since resolved could have produced distorted visual "templates," (akin to influential priors in Bayesian terms (Ghaderinezhad and Ley 2019; Limanowski and Friston 2020)) the visual memories of which were reinforced by high emotional states. Subsequently, when regarding one's appearance, these templates may pre-color subsequent viewing such that the old memories dominate over new information. An imbalance in visual tolerance (invariance, or stable representation over different viewing conditions) over visual discrimination (explicitness, or discerning new information) (Schurgin 2018) could lead to the persistence of a limited number of percepts of "magnified" defects whenever viewing oneself.

Supporting this model of disrupted feed-forward of information comes from an EEG study examining responses to others' faces and houses in individuals with BDD (Li et al. 2013). The amplitudes of the N170 signals were smaller in those with BDD than in healthy controls when viewing both faces and houses, and those with lower amplitudes for normal and low spatial frequency images had worse insight. This suggests an abnormal encoding of faces (or perhaps stimuli in general), particularly the holistic and configural elements conveyed by low spatial frequency information, which could contribute to perceptual distortions and subsequent poor insight in BDD. Reduced fiber organization in white matter tracts conveying feed-forward information to update visual memories would be most consistent with this model, although it remains unclear if these disruptions are direction-specific since bidirectional signals are conveyed in this tract.

In sum, abnormal visual processing in BDD could lead affected individuals to experience the belief that their subjective point of view is an objective reality, resulting in limited insight. Poor insight has been demonstrated to be associated with slower performance in global/local processing tasks as well as electrophysiological signatures of abnormal encoding of holistic and configural elements. In addition, poor insight is associated with disruptions in white matter tracts integrating emotional signals and visual perception and involved in visual memory. This could lead to difficulties in those with BDD in updating visual memories resulting in persistent distorted visual images of flaws, and subsequently increasing the fixity of beliefs that their appearance is defective, which could hinder mental health-treatment-seeking.

3.2 Social and Affective Misinterpretation

Misinterpretation of the facial emotional expression of others in BDD can contribute to the misperceptions of one's appearance. As this occurs prominently in self-referential scenarios (Buhlmann et al. 2006), those with BDD often misinterpret that others are taking notice of what they themselves misperceive to be defects and flaws, and reacting emotionally to them – for example, with disgust or contempt. This reinforces the belief that one's appearance is not only defective but repulsive, reducing their ability to consciously override the misperceptions. Clinically, this also occurs with misinterpretations of others' actions, such as averting gaze, walking away, etc. Thus, misinterpretation of others' facial emotional expressions and social behaviors prevent the accumulation of corrective social information that others, in fact, are not perceiving or taking special notice of their appearance defects. This could lead not only to the fixity of their beliefs and poor insight but also to social avoidance in order to avoid the distress they experience in these situations.

4 Theoretical Model

4.1 A General Model of Visual Processing

After light is converted into neural signals by the retina, visual information travels through the optic tract to reach the striate cortex (V1), where it is thought to be processed by two principal neuronal pathways: the dorsal and ventral visual streams (Goodale and Milner 1992). The first extends dorsally into the parietal cortex and rapidly processes coarse, low spatial frequency (LSF) information. This type of information represents the overall "whole" of an object, as well as its motion and spatial location. It includes V1, V2, V5, and MT and primarily receives inputs from the magnocellular cell layers (Laycock et al. 2007). Because of its rapid action, the dorsal stream is believed to play a crucial role in engaging instinctive defensive behaviors by enhancing the detection of stimuli, even when they are presented in unattended locations or fall below the threshold for conscious recognition. The other stream lies ventrally and involves the occipital fusiform gyrus and temporal cortex – V1, V2, V4, and inferior temporal cortex – and receives inputs from both magnocellular and parvocellular cell layers. It represents a more recent evolutionary development in the visual system and is responsible for slower, piecemeal processing of an object's high-fidelity attributes (featural or "local" processing of form and color). Configural processing involves sensitivity to the relative positions of features (first-order relations), holistic integration of features into a gestalt, and awareness of relations between the features (Maurer et al. 2002). In configural processing, the parts of an object are not individually represented but instead recognized as "templates" (Tanaka and Farah 1991). Featural processing is the modular analysis of features that can be measured independently from each other, are local in their spatial extent, and are characterized by discontinuities (Bartlett et al. 2003). This is also referred to as local part-based or fragmented-based processing (Schwaninger et al. 2002). Information from the dorsal and ventral pathways converge in inferior temporal regions such that the more rapidly-processed global and configural information provides a template for contextualization of the more slowly-processed details arriving from the ventral stream (Tanaka and Farah 1991; Takahashi et al. 2013; Ungerleider and Haxby 1994). Of note, the fusiform face area, involved in the recognition of own and others' faces (Kircher et al. 2001), is in the ventral visual pathway (Kanwisher et al. 1997) in inferior temporal regions that additionally receive dorsal pathway input (Kravitz et al. 2013). The integration of the dorsal and ventral pathways is crucial for a coherent visual representation. When these inputs fail to harmonize, such as when one pathway dominates or the system cannot adapt to changing visual demands, perceptual capacity is compromised.

4.2 A Model of Abnormal Visual Processing

Human evolution has preserved intricate neural functions that facilitate adaptation and survival in complex and potentially hazardous environments. This is evident in the assortment of visually guided mechanisms that aid the detection and appraisal of threats, such as the subcortical visual pathway mediated by the superior colliculus (McFadyen et al. 2020). Additionally, there are extensive connections between the visual cortex and other brain regions involved in modulating attention, negative emotions, motivation, and motor responses. These connections reflect the evolutionary link between the visual cortex and circuits that regulate various physiological, cognitive, and affective states, both conscious and unconscious (Mobbs et al. 2015). Thus, there are multiple brain pathways in which aberrancies could impact perceptual accuracy and overall perceptual experience (Feusner et al. 2017).

Observations of aberrant visual object recognition have been characterized not only in BDD, as described, but in obsessive-compulsive disorder, social anxiety, and schizophrenia (Laycock et al. 2007) which point to impaired early visual processing in the dorsal stream, leading not only to distorted perceptions and difficulties in categorizing images but also biases in social perception. Knowledge of these functionally and anatomically distinct – although interconnected – systems, in addition to the existing brain and behavioural research findings related to visual and visuospatial processing in BDD, contribute to our model of distorted perception in BDD (Fig. 2)

In this model, "bottom-up" deficiencies in global/configural perception originating from visual cortical abnormalities in low spatial frequency processing contribute to an imbalance in global/local processing. These interact with "top-down"

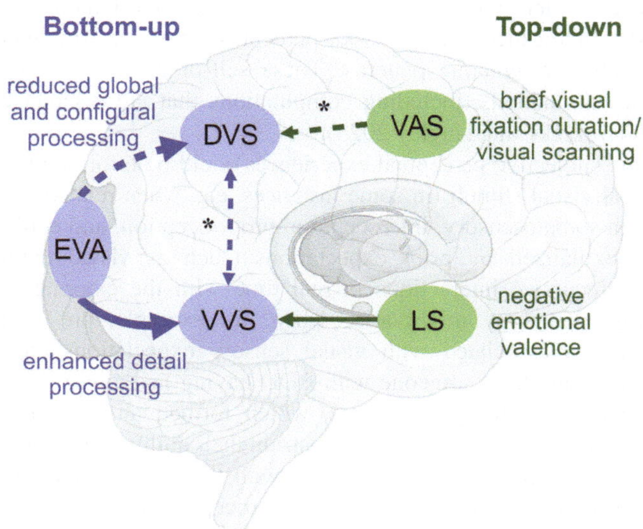

Fig. 2 Top-down and bottom-up (highly simplified) model of abnormal visual processing in body dysmorphic disorder. EVA = early visual areas; DVS = dorsal visual stream; VVS = ventral visual stream; LS = limbic system, e.g. amygdala, hippocampus; VAS = visual-attention system, e.g. frontal eye fields; Purple = bottom-up; Green = top-down; Solid arrows = aberrantly strong influence; Dashed arrows = aberrantly weak influence; * = hypothesized activations with no prior studies in BDD

contributions from enhanced attention to visual flaws and may interact as well with aberrant frontostriatal systems (Feusner et al. 2010c), particularly related to set-shifting (Kerwin et al. 2014) and cognitive inflexibility (Deckersbach et al. 2000; Dunai et al. 2010; Hanes 1998) resulting in distorted perception. Enhanced attention to visual flaws could contribute to increased scanning of appearance features, the latter of which results in shorter visual fixations (Toh et al. 2017c). Enhanced detail processing may thus have contributions from aberrant "top-down" attentional allocation and eye gaze behaviors, which, interacting with "bottom-up" deficiencies in global/configural processing, contribute to a subjective perceptual experience in which details are not adequately integrated and contextualized into a whole. The result is an overall heightened detection of "flaws" and an inability to determine that most "flaws" are minuscule relative to the whole.

This is an oversimplified model – one that is put forth to integrate extant neurobiological and behavioral evidence, providing a starting point to refine and build upon as research in visual processing in BDD progresses. Relatedly, the terms "top-down" and "bottom-up" are usually applied within an "outside-in" conceptualization of how living organisms function: i.e., an external stimulus is inputted through a "bottom-up" (typically, primary sensory) system, causing a "top-down" (usually from prefrontal and/or limbic systems) reaction to that extrinsic stimulus. Yet, a competing framework is that the brain does not merely "process" signals externally from the environment. Rather, there is abundant evidence that the brain is a self-organized system with pre-existing connectivity and dynamics (reviewed in Buzsáki 2019). In this theory, a conscious percept is largely determined by ongoing, pre-existing intrinsic dynamic patterns. These self-organized systems constantly create internal simulations, including computations that are disengaged from the environment (Buzsáki and Tingley 2023).

In this way, subjective perceptual experiences could be informed by a combination of external visual stimuli (in some instances, e.g. when related to one's body, integrated with somatosensory, interoception, proprioception, and equilibrioception signals), contextualized by prior expectations (such as visual memories), and updated from recursive interactions in real time with the external environment. Hypothetically, pathology might arise when, for example, pre-formed internal intrinsic dynamics associated with visual sensory expectations are excessively rigid. Take the example of someone with BDD having had a past episode of acne, seen under extreme lighting conditions, which formed an emotionally charged memory of it. This could subsequently hamstring their ability to update their percept with information based on new interactions with the environment. An analogous "context rigidity" has been described for interoceptive pathology (Paulus et al. 2019). In Bayesian terminology, within this framework prior expectations or beliefs are judged to be more precise than incoming sensory information (likelihood), the resulting percept (posterior belief) is biased toward these priors, and there is a failure to update existing expectations or beliefs. As the field advances to design experimental investigations of those with BDD that are more naturalistic and incorporate free interactions with observed stimuli, we will begin to build more comprehensive and accurate models to understand visual perception and how it is altered under pathological conditions.

Box 1 Are Visual Perceptual Disturbances in BDD Limited to Their Own Appearance?
In addition to evidence for aberrant visual system activation patterns for own appearance, there is evidence for disturbances in response to others' faces and bodies and for non-appearance-related stimuli, such as houses (Feusner et al. 2007, 2010c, 2011; Li et al. 2013; Moody et al. 2021). Those with BDD also show evidence of impaired visual organizational strategies for complex figures (Deckersbach et al. 2000; Greenberg et al. 2018). Yet, the appearance of own faces and bodies is more highly salient, as they represent (in an over-valued way) threats to one's sense of self-worth and that others might judge them negatively and potentially reject or ostracize them. This is likely why distortions of their own appearance are at the forefront of their preoccupations. The salience of viewing self-images also generates emotional arousal and may result in different attentional allocation patterns and resulting eye gaze behaviors, which may further contribute to perceptual distortions.

5 New Perspectives

5.1 *Translating Research into Practice*

Perceptual Retraining Following from phenomenological observations of abnormal perception in BDD, perceptual retraining techniques have been used in some CBT-based treatment modules (Wilhelm et al. 2014). A pilot study that combined CBT and computerized cognitive remediation was also proposed as a visual training program for BDD (Beilharz et al. 2018). The methods employed have not been empirically tested in isolation from the other aspects of CBT treatment. Moreover, these approaches involve several concomitant perceptual, behavioral, and emotional interventions, so the contribution from each independently is unclear.

For the treatment of anxiety disorders, attention bias modification strategies have demonstrated efficacy (Amir et al. 2009; Eldar et al. 2012). In healthy individuals, perceptual techniques such as enhancing peripheral vision using neurofeedback (Nan et al. 2013) have been tested. In individuals with high body image concerns, a behavioral study demonstrated that using a Navon (1997) visual stimulus (specifically, a large letter made of smaller letters, presented in a way to promote a global bias) can induce global processing and temporarily reverse visual processing biases (Beilharz et al. 2016). In a mechanistic study testing the effects of a visual-attention modulation paradigm on functional neural connectivity and eye gaze patterns, BDD individuals with face concerns were instructed to hold their gaze constant on the center of an image of their face (Wong et al. 2022a, b). This attention modulation resulted in greater connectivity from occipital to parietal regions in the dorsal visual stream during subsequent compared to prior unconstrained natural viewing of their faces. These findings suggest that the attentional modulation effect of reducing visual scanning while viewing one's face results in persistent enhancement effects on DVS connectivity during subsequent unconstrained viewing. While not seen immediately in this study, it is possible that the same intervention could have later effects in the VVS after longer durations, or multiple repetitions, of attention modulation, secondary to reductions in visual scanning and scrutiny of details.

A separate perceptually-based strategy of modifying neural connectivity in visual systems using brief presentation of own faces was also tested in those with BDD (Wong et al. 2022a, b). While this was associated with weaker VVS connectivity in healthy controls, as hypothesized, this effect was not apparent in those with BDD. There was, however, a nonsignificant trend of an increasing ratio of DVS to VVS connectivity (which represents the balance between global and local processing) with shorter viewing durations in those with BDD.

These findings have translational implications for perceptual retraining treatment design for BDD. Yet, in order to develop and plan novel intervention strategies to remediate perceptual abnormalities in BDD, critically important questions remain unanswered.

Neuromodulation Non-invasive brain stimulation may also have the potential to alter distorted self-perception and could be used to augment perceptual retraining. In a sham-controlled pilot study, intermittent theta burst stimulation (iTBS), a type of repetitive transcranial magnetic stimulation (rTMS) was used to modulate dorsal visual system utilization in BDD participants with face concerns to enhance global/holistic processing while viewing images of their own face (Wong et al. 2021). In the active group ($n = 7$), iTBS enhanced dynamic connectivity in dorsal visual stream and was accompanied by significant self-reported improvements in body image. Although the sample size was very small, enhanced connectivity within DVS and reduced VVS connectivity provide an early signal of the potential for stimulation to change the balance between global vs. local visual processing in BDD. Moreover, several transcranial electrical stimulation (tES) methods have demonstrated the ability to modulate face and object perception by applying electrical stimulation over the lateral occipital cortex (Barbieri et al. 2016; Estudillo et al. 2023; Gonzalez-Perez et al. 2019). However, further investigation is warranted to explore how tES could effectively contribute to the remediation of perceptual distortions. These stimulation methods are either yet to be applied to those with BDD (tES) or need to be tested in larger studies (rTMS) to better understand their mechanisms and if they result in causal effects on perception and clinical symptoms.

5.2 Future Areas of Research

Modalities with High Temporal Resolution In order to capture early visual perceptual and attentional abnormalities that may not be detected by fMRI, tools with higher temporal resolutions such as EEG/MEG and functional near-infrared spectroscopy (fNIRS) could be utilized. While these tools have lower spatial resolution (limited to the cortical surface within a few centimeters), they are particularly suitable for studying the regions of interest associated with appearance disturbance, namely the dorsal and ventral visual streams, which are predominantly located on the cortical surface. By employing EEG/MEG and fNIRS, it is possible to investigate the respective contributions of early perceptual processes and potential biased selective attention to enhanced local part-based processing in BDD. These tools provide valuable insights into the temporal dynamics of neural activity and offer a complementary approach to fMRI, enabling a more comprehensive understanding of the underlying mechanisms involved in distorted body image perception.

In addition to EEG/MEG and fNIRS, eye tracking is another valuable tool that can aid in the investigation of cognitive strategies engaged in distorted visual perception. Moreover, eye tracking could be used to augment perceptual and attentional training paradigms. Further, the high temporal resolution and dense data sampling afforded by eye tracking data lends itself to individual-level computational modeling of behavior. This is likely a necessary strategy for the development of effective treatment approaches since these may need to be individually tailored due to different subgroups of BDD individuals who search and hone in on appearance flaws vs. those who avoid viewing these areas (Toh et al. 2017c). From eye tracking data one could develop computational models for prediction of future gaze patterns for the purpose of developing eye-tracking-based biofeedback (e.g. see https://arxiv.org/abs/1702.00714). As gaze behavior in general typically evolves across the time span of visual exploration (as has been observed in those with BDD (Wong et al. 2022a, b)), prediction of future gaze patterns should be modeled with time-dependent estimation of weights (Coutrot and Guyader 2017; Stadler et al. 2021). These could be used in machine learning or Bayesian analyses (with, for example, emotional arousal), to predict if upcoming saccade amplitude patterns (or, for example, fixation duration for face concerns) are likely to produce perceptual distortions at which point the patient could receive feedback to alter them.

Other Sensory Modalities and Multisensory Integration An underexplored avenue in BDD research involves investigating perceptual abnormalities beyond vision, extending to other sensory modalities. Initial evidence suggests abnormalities in interoception related to trait BDD symptoms (Kunstman et al. 2016) and in those with muscle dysmorphia (Grunewald et al. 2023a, b). (See Jenkinson and Rossell (2023), for discussion about a framework for future interoception research in BDD.)

Indeed, future studies should further explore multisensory integration in BDD, employing experimental designs to probe cross-modal conflicts. For instance, a study utilizing the rubber hand illusion, an experimental manipulation that induces conflict between visual, tactile, and proprioceptive information, found that individuals with BDD demonstrated atypical performance in proprioceptive drift, perceptual drift in the perceived location of the hand, indicating abnormalities in multisensory integration (Kaplan et al. 2014). Tasks that probe cross-modal conflicts have found interesting results in eating disorders (Brizzi et al. 2023) and could be a

Table 1 Future areas of research for exploring mechanisms of visual perceptual processing abnormalities in BDD[a]

Future areas of visual perception research in BDD	Example experimental designs or modalities
Contributions of early perceptual processes, biases in selective attention, emotional arousal, and their interactions	• Brain imaging studies with high temporal resolution (EEG/MEG, fNIRS) • Eye tracking to assess attention allocation in experiments utilizing appearance-related and non-appearance-related stimuli • Single and paired pulse TMS for spatially focal and temporally precise manipulation of brain activity to test top-down and bottom-up models, and causal effects
Computational modeling of abnormal visual perception	• Individual-level computational modeling of eye gaze behaviors to predict abnormal allocation of attention • Tasks that probe perceptual decision making, decision making under uncertainty, and interpretation bias (faces or ambiguous scenarios)
Multisensory integration	• Cross-modal conflicts such as the rubber hand illusion (visual-tactile-proprioceptive), aperture task (visual-proprioceptive-vestibular) (Warren and Whang 1987), and size-weight illusion (visual-haptic proprioceptive) (Plaisier and Smeets 2015)

[a] This is a non-exhaustive list

useful tool to inform the relative contributions of other sensory modalities to the visual perceptual abnormalities found in BDD (Table 1).

6 Conclusions

Body dysmorphic disorder (BDD) is a prevalent and debilitating psychiatric disorder characterized by a preoccupation with perceived defects in one's appearance. Despite its significant impact on individuals' lives, BDD remains understudied and often misdiagnosed. The etiology of BDD is likely multifactorial, involving developmental, social, and neurobiological factors.

Neurobiological research has provided insights into the neural correlates of BDD. Evidence of disturbances in visual processing and visuospatial organization has been demonstrated in several neuropsychological, psychophysical, and neuroimaging studies. Reduced global processing, over-attention to local details, and deficiencies in facial affect recognition have been found in individuals with BDD. This may lead to visual perceptual distortions, potentially influenced by emotional arousal and over-attention to details.

These abnormalities in visual perception contribute to the clinical phenomenology of BDD. They can lead to distorted subjective perceptions of one's appearance, as individuals with BDD are unable to integrate and contextualize details effectively. This limited insight into their condition, paired with rigid internal intrinsic dynamics and reinforced by obsessive thinking and compulsive behaviors, further complicates the illness course including seeking mental health treatment. Understanding the neural underpinnings of poor insight in BDD is crucial for the development of effective treatments.

The findings from neuroimaging, psychophysical, and neuropsychological studies can inform the development of novel interventions. However, further research is needed to deepen our understanding of the complex experiences in BDD and other disorders involving body image disturbance such as eating disorders.

Overall, the existing evidence highlights the importance of studying visual processing, as well as the neurobiological basis of other BDD phenotypes, to advance our knowledge, improve diagnosis and treatment, and pave the way for future research in this field. By unraveling the underlying neural mechanisms, we can develop more targeted and effective interventions to help individuals with BDD achieve better outcomes and enhance their quality of life.

Acknowledgments

This work was supported by the National Institute of Mental Health (NIMH) grant numbers: R01MH105662 (Feusner) and R21MH128815 (Feusner).

Conflict of Interest The authors have nothing to disclose.

References

American Psychiatric Association (2013) Diagnostic and statistical manual of mental disorders, 5th edn. https://doi.org/10.1176/appi.books.9780890425596

Amir N, Beard C, Taylor CT, Klumpp H, Elias J, Burns M, Chen X (2009) Attention training in individuals with generalized social phobia: a randomized controlled trial. J Consult Clin Psychol 77(5):961–973. https://doi.org/10.1037/a0016685

Arienzo D, Leow A, Brown JA, Zhan L, Gadelkarim J, Hovav S, Feusner JD (2013) Abnormal brain network organization in body dysmorphic disorder. Neuropsychopharmacology 38(6): 1130–1139. https://doi.org/10.1038/npp.2013.18

Atmaca M, Bingol I, Aydin A, Yildirim H, Okur I, Yildirim MA, Mermi O, Gurok MG (2010) Brain morphology of patients with body dysmorphic disorder. J Affect Disord 123(1–3): 258–263. https://doi.org/10.1016/j.jad.2009.08.012

Barbieri M, Negrini M, Nitsche MA, Rivolta D (2016) Anodal-tDCS over the human right occipital cortex enhances the perception and memory of both faces and objects. Neuropsychologia 81: 238–244. https://doi.org/10.1016/j.neuropsychologia.2015.12.030

Bartlett JC, Searcy JH, Abdi H (2003) What are the routes to face recognition? Perception of faces, objects, and scenes: analytic and holistic processes, p 21

Beilharz FL, Atkins KJ, Duncum AJ, Mundy ME (2016) Altering visual perception abnormalities: a marker for body image concern. PLoS One 11(3):e0151933. https://doi.org/10.1371/journal.pone.0151933

Beilharz F, Castle DJ, Grace S, Rossell SL (2017) A systematic review of visual processing and associated treatments in body dysmorphic disorder. Acta Psychiatr Scand 136(1):16–36. https://doi.org/10.1111/acps.12705

Beilharz F, Castle DJ, Phillipou A, Rossell SL (2018) Visual training program for body dysmorphic disorder: protocol for a novel intervention pilot and feasibility trial. Pilot Feasibility Stud 4:189. https://doi.org/10.1186/s40814-018-0384-3

Bienvenu OJ, Samuels JF, Riddle MA, Hoehn-Saric R, Liang KY, Cullen BA, Grados MA, Nestadt G (2000) The relationship of obsessive-compulsive disorder to possible spectrum disorders: results from a family study. Biol Psychiatry 48(4):287–293. https://doi.org/10.1016/s0006-3223(00)00831-3

Bohon C, Hembacher E, Moller H, Moody TD, Feusner JD (2012) Nonlinear relationships between anxiety and visual processing of own and others' faces in body dysmorphic disorder. Psychiatry Res 204(2–3):132–139. https://doi.org/10.1016/j.pscychresns.2012.09.003

Borgers T, Kürten M, Kappelhoff A, Enneking V, Möllmann A, Schulte J, Klug M, Leehr EJ, Dohm K, Grotegerd D, Krause P, Zwiky E, Dannlowski U, Buhlmann U, Redlich R (2022) Brain functional correlates of emotional face processing in body dysmorphic disorder. J Psychiatr Res 147:103–110. https://doi.org/10.1016/j.jpsychires.2022.01.007

Brizzi G, Sansoni M, Di Lernia D, Frisone F, Tuena C, Riva G (2023) The multisensory mind: a systematic review of multisensory integration processing in Anorexia and Bulimia Nervosa. J Eat Disord 11(1):204. https://doi.org/10.1186/s40337-023-00930-9

Buchanan BG, Rossell SL, Maller JJ, Toh WL, Brennan S, Castle DJ (2013) Brain connectivity in body dysmorphic disorder compared with controls: a diffusion tensor imaging study. Psychol Med 43(12):2513–2521. https://doi.org/10.1017/S0033291713000421

Buchanan B, Rossell S, Maller JJ, Toh WL, Brennan S, Castle D (2014) Regional brain volumes in body dysmorphic disorder compared to controls. Aust N Z J Psychiatry 48(7):654–662. https://doi.org/10.1177/0004867413520253

Buhlmann U, McNally RJ, Wilhelm S, Florin I (2002a) Selective processing of emotional information in body dysmorphic disorder. J Anxiety Disord 16(3):289–298. https://doi.org/10.1016/s0887-6185(02)00100-7

Buhlmann U, Wilhelm S, McNally RJ, Tuschen-Caffier B, Baer L, Jenike MA (2002b) Interpretive biases for ambiguous information in body dysmorphic disorder. CNS Spectr 7(6):435–443. https://doi.org/10.1017/s1092852900017946

Buhlmann U, Etcoff NL, Wilhelm S (2006) Emotion recognition bias for contempt and anger in body dysmorphic disorder. J Psychiatr Res 40(2):105–111. https://doi.org/10.1016/j.jpsychires.2005.03.006

Buhlmann U, Glaesmer H, Mewes R, Fama JM, Wilhelm S, Brähler E, Rief W (2010) Updates on the prevalence of body dysmorphic disorder: a population-based survey. Psychiatry Res 178(1):171–175. https://doi.org/10.1016/j.psychres.2009.05.002

Buhlmann U, Rupf L, Gleiss MJ, Zschenderlein K, Kathmann N (2014) Seeing "changes" that aren't there: facial and object discrimination in body dysmorphic disorder. Compr Psychiatry 55(3):468–474. https://doi.org/10.1016/j.comppsych.2013.09.011

Buzsáki G (2019) The brain from inside out. Oxford University Press

Buzsáki G, Tingley D (2023) Cognition from the body-brain partnership: exaptation of memory. Annu Rev Neurosci 46:191–210. https://doi.org/10.1146/annurev-neuro-101222-110632

Coutrot A, Guyader N (2017) Learning a time-dependent master saliency map from eye-tracking data in videos. http://arxiv.org/abs/1702.00714

Deckersbach T, Savage CR, Phillips KA, Wilhelm S, Buhlmann U, Rauch SL, Baer L, Jenike MA (2000) Characteristics of memory dysfunction in body dysmorphic disorder. J Int Neuropsychol Soc 6(6):673–681. https://doi.org/10.1017/s1355617700666055

Dhir S, Ryan HS, McKay EL, Mundy ME (2018) Parameters of visual processing abnormalities in adults with body image concerns. PLoS One 13(11):e0207585. https://doi.org/10.1371/journal.pone.0207585

Dunai J, Labuschagne I, Castle DJ, Kyrios M, Rossell SL (2010) Executive function in body dysmorphic disorder. Psychol Med 40(9):1541–1548. https://doi.org/10.1017/S003329170999198X

Eisen JL, Phillips KA, Coles ME, Rasmussen SA (2004) Insight in obsessive compulsive disorder and body dysmorphic disorder. Compr Psychiatry 45(1):10–15. https://doi.org/10.1016/j.comppsych.2003.09.010

Eldar S, Apter A, Lotan D, Edgar KP, Naim R, Fox NA, Pine DS, Bar-Haim Y (2012) Attention bias modification treatment for pediatric anxiety disorders: a randomized controlled trial. Am J Psychiatry 169(2):213–220. https://doi.org/10.1176/appi.ajp.2011.11060886

Estudillo AJ, Lee YJ, Álvarez-Montesinos JA, García-Orza J (2023) High-frequency transcranial random noise stimulation enhances unfamiliar face matching of high resolution and pixelated faces. Brain Cogn 165:105937. https://doi.org/10.1016/j.bandc.2022.105937

Feusner JD, Townsend J, Bystritsky A, Bookheimer S (2007) Visual information processing of faces in body dysmorphic disorder. Arch Gen Psychiatry 64(12):1417–1425. https://doi.org/10.1001/archpsyc.64.12.1417

Feusner JD, Bystritsky A, Hellemann G, Bookheimer S (2010a) Impaired identity recognition of faces with emotional expressions in body dysmorphic disorder. Psychiatry Res 179(3):318–323. https://doi.org/10.1016/j.psychres.2009.01.016

Feusner JD, Moller H, Altstein L, Sugar C, Bookheimer S, Yoon J, Hembacher E (2010b) Inverted face processing in body dysmorphic disorder. J Psychiatr Res 44(15):1088–1094. https://doi.org/10.1016/j.jpsychires.2010.03.015

Feusner JD, Moody T, Hembacher E, Townsend J, McKinley M, Moller H, Bookheimer S (2010c) Abnormalities of visual processing and frontostriatal systems in body dysmorphic disorder. Arch Gen Psychiatry 67(2):197–205. https://doi.org/10.1001/archgenpsychiatry.2009.190

Feusner JD, Neziroglu F, Wilhelm S, Mancusi L, Bohon C (2010d) What causes BDD: research findings and a proposed model. Psychiatr Ann 40(7):349–355. https://doi.org/10.3928/00485713-20100701-08

Feusner JD, Hembacher E, Moller H, Moody TD (2011) Abnormalities of object visual processing in body dysmorphic disorder. Psychol Med 41(11):2385–2397. https://doi.org/10.1017/S0033291711000572

Feusner JD, Arienzo D, Li W, Zhan L, Gadelkarim J, Thompson PM, Leow AD (2013) White matter microstructure in body dysmorphic disorder and its clinical correlates. Psychiatry Res 211(2):132–140. https://doi.org/10.1016/j.pscychresns.2012.11.001

Feusner J, Deshpande R, Strober M (2017) A translational neuroscience approach to body image disturbance and its remediation in anorexia nervosa. Int J Eat Disord 50(9):1014–1017. https://doi.org/10.1002/eat.22742

Freire A, Lee K, Symons LA (2000) The face-inversion effect as a deficit in the encoding of configural information: direct evidence. Perception 29(2):159–170. https://doi.org/10.1068/p3012

Ghaderinezhad F, Ley C (2019) On the impact of the choice of the prior in Bayesian statistics. In: Bayesian inference on complicated data. IntechOpen, p 22

Gonzalez-Perez M, Wakui E, Thoma V, Nitsche MA, Rivolta D (2019) Transcranial alternating current stimulation (tACS) at 40 Hz enhances face and object perception. Neuropsychologia 135:107237. https://doi.org/10.1016/j.neuropsychologia.2019.107237

Goodale MA, Milner AD (1992) Separate visual pathways for perception and action. Trends Neurosci 15(1):20–25. https://doi.org/10.1016/0166-2236(92)90344-8

Grace SA, Labuschagne I, Castle DJ, Rossell SL (2019a) Intranasal oxytocin alters amygdala-temporal resting-state functional connectivity in body dysmorphic disorder: a double-blind placebo-controlled randomized trial. Psychoneuroendocrinology 107:179–186. https://doi.org/10.1016/j.psyneuen.2019.05.022

Grace SA, Toh WL, Buchanan B, Castle DJ, Rossell SL (2019b) Impaired recognition of negative facial emotions in body dysmorphic disorder. J Int Neuropsychol Soc 25(8):884–889. https://doi.org/10.1017/S1355617719000419

Greenberg JL, Weingarden H, Reuman L, Abrams D, Mothi SS, Wilhelm S (2018) Set shifting and visuospatial organization deficits in body dysmorphic disorder. Psychiatry Res 260:182–186. https://doi.org/10.1016/j.psychres.2017.11.062

Grocholewski A, Kliem S, Heinrichs N (2012) Selective attention to imagined facial ugliness is specific to body dysmorphic disorder. Body Image 9(2):261–269. https://doi.org/10.1016/j.bodyim.2012.01.002

Grunewald W, Fogelberg S, Ferguson W, Hines S, Fortenberry B, Smith AR (2023a) Longitudinal relationships between specific domains of interoception and muscle dysmorphia symptoms. Eat Behav 48:101686. https://doi.org/10.1016/j.eatbeh.2022.101686

Grunewald W, Gagliano E, Brown TA, Smith AR (2023b) Conformity to masculine norms, interoceptive dysfunction, and changes in muscle dysmorphia symptoms. Body Image 45: 343–354. https://doi.org/10.1016/j.bodyim.2023.03.015

Hanes KR (1998) Neuropsychological performance in body dysmorphic disorder. J Int Neuropsychol Soc 4(2):167–171. https://doi.org/10.1017/s1355617798001672

Jefferies K, Laws KR, Fineberg NA (2012) Superior face recognition in body dysmorphic disorder. J Obsessive Compuls Relat Disord 1(3):175–179. https://doi.org/10.1016/j.jocrd.2012.03.002

Jenkinson PM, Rossell SL (2023) Disturbed interoception in body dysmorphic disorder: a framework for future research. Aust N Z J Psychiatry:48674231215030. https://doi.org/10.1177/00048674231215030

Kanwisher N, McDermott J, Chun MM (1997) The fusiform face area: a module in human extrastriate cortex specialized for face perception. J Neurosci 17(11):4302–4311. https://doi.org/10.1523/JNEUROSCI.17-11-04302.1997

Kaplan RA, Rossell SL, Enticott PG, Castle DJ (2013) Own-body perception in body dysmorphic disorder. Cogn Neuropsychiatry 18(6):594–614. https://doi.org/10.1080/13546805.2012.758878

Kaplan RA, Enticott PG, Hohwy J, Castle DJ, Rossell SL (2014) Is body dysmorphic disorder associated with abnormal bodily self-awareness? A study using the rubber hand illusion. PLoS One 9(6):e99981. https://doi.org/10.1371/journal.pone.0099981

Kerwin L, Hovav S, Hellemann G, Feusner JD (2014) Impairment in local and global processing and set-shifting in body dysmorphic disorder. J Psychiatr Res 57:41–50. https://doi.org/10.1016/j.jpsychires.2014.06.003

Kircher TT, Senior C, Phillips ML, Rabe-Hesketh S, Benson PJ, Bullmore ET, Brammer M, Simmons A, Bartels M, David AS (2001) Recognizing one's own face. Cognition 78(1):B1–B15. https://doi.org/10.1016/s0010-0277(00)00104-9

Koran LM, Abujaoude E, Large MD, Serpe RT (2008) The prevalence of body dysmorphic disorder in the United States adult population. CNS Spectr 13(4):316–322. https://doi.org/10.1017/s1092852900016436

Kravitz DJ, Saleem KS, Baker CI, Ungerleider LG, Mishkin M (2013) The ventral visual pathway: an expanded neural framework for the processing of object quality. Trends Cogn Sci 17(1): 26–49. https://doi.org/10.1016/j.tics.2012.10.011

Kunstman JW, Clerkin EM, Palmer K, Peters MT, Dodd DR, Smith AR (2016) The power within: the experimental manipulation of power interacts with trait BDD symptoms to predict interoceptive accuracy. J Behav Ther Exp Psychiatry 50:178–186. https://doi.org/10.1016/j.jbtep.2015.08.003

Laycock R, Crewther SG, Crewther DP (2007) A role for the 'magnocellular advantage' in visual impairments in neurodevelopmental and psychiatric disorders. Neurosci Biobehav Rev 31(3): 363–376. https://doi.org/10.1016/j.neubiorev.2006.10.003

Lee AD, Hale EW, Mundra L, Le E, Kaoutzanis C, Mathes DW (2023) The heart of it all: body dysmorphic disorder in cosmetic surgery. J Plast Reconstr Aesthet Surg 87:442–448. https://doi.org/10.1016/j.bjps.2023.10.068

Li W, Arienzo D, Feusner JD (2013) Body Dysmorphic Disorder: Neurobiological Features and an Updated Model. Z. Klin. Psychol. Psychother. 42(3):184–191. https://doi.org/10.1026/1616-3443/a000213

Limanowski J, Friston K (2020) Active inference under visuo-proprioceptive conflict: simulation and empirical results. Sci Rep 10(1):4010. https://doi.org/10.1038/s41598-020-61097-w

Longo MR, Azañón E, Haggard P (2010) More than skin deep: body representation beyond primary somatosensory cortex. Neuropsychologia 48(3):655–668. https://doi.org/10.1016/j.neuropsychologia.2009.08.022

Mancuso SG, Knoesen NP, Castle DJ (2010) Delusional versus nondelusional body dysmorphic disorder. Compr Psychiatry 51(2):177–182. https://doi.org/10.1016/j.comppsych.2009.05.001

Maurer D, Grand RL, Mondloch CJ (2002) The many faces of configural processing. Trends Cogn Sci 6(6):255–260. https://doi.org/10.1016/s1364-6613(02)01903-4

McFadyen J, Dolan RJ, Garrido MI (2020) The influence of subcortical shortcuts on disordered sensory and cognitive processing. Nat. Rev. Neurosci. 21(5):264–276. https://doi.org/10.1038/s41583-020-0287-1

Mobbs D, Hagan CC, Dalgleish T, Silston B, Prévost C (2015) The ecology of human fear: survival optimization and the nervous system. Front Neurosci 9:55. https://doi.org/10.3389/fnins.2015.00055

Monzani B, Rijsdijk F, Anson M, Iervolino AC, Cherkas L, Spector T, Mataix-Cols D (2012) A twin study of body dysmorphic concerns. Psychol Med 42(9):1949–1955. https://doi.org/10.1017/S0033291711002741

Monzani B, Krebs G, Anson M, Veale D, Mataix-Cols D (2013) Holistic versus detailed visual processing in body dysmorphic disorder: testing the inversion, composite and global precedence effects. Psychiatry Res 210(3):994–999. https://doi.org/10.1016/j.psychres.2013.08.009

Moody TD, Morfini F, Cheng G, Sheen CL, Kerr WT, Strober M, Feusner JD (2021) Brain activation and connectivity in anorexia nervosa and body dysmorphic disorder when viewing bodies: relationships to clinical symptoms and perception of appearance. Brain Imaging Behav 15(3):1235–1252. https://doi.org/10.1007/s11682-020-00323-5

Mufaddel A, Osman OT, Almugaddam F, Jafferany M (2013) A review of body dysmorphic disorder and its presentation in different clinical settings. Prim Care Companion CNS Disord 15(4):PCC.12r01464. https://doi.org/10.4088/PCC.12r01464

Mundy EM, Sadusky A (2014) Abnormalities in visual processing amongst students with body image concerns. Adv Cogn Psychol 10(2):39–48. https://doi.org/10.5709/acp-0155-4

Nan W, Wan F, Lou CI, Vai MI, Rosa A (2013) Peripheral visual performance enhancement by neurofeedback training. Appl Psychophysiol Biofeedback 38(4):285–291. https://doi.org/10.1007/s10484-013-9233-6

Navon D (1997) Forest before trees: the precedence of global features in visual perception. Cogn Psychol 9:353–383. https://doi.org/10.1016/0010-0285(77)90012-3

Oltman PK, Raskin E, Witkin HA (1971) Group embedded figures test. Consulting Psychologists Press

Paulus MP, Feinstein JS, Khalsa SS (2019) An active inference approach to interoceptive psychopathology. Annu Rev Clin Psychol 15:97–122. https://doi.org/10.1146/annurev-clinpsy-050718-095617

Phillips KA (2005) The broken mirror: understanding and treating body dysmorphic disorder. Oxford University Press

Phillips KA, Diaz SF (1997) Gender differences in body dysmorphic disorder. J Nerv Ment Dis 185(9):570–577. https://doi.org/10.1097/00005053-199709000-00006

Phillips KA, Menard W (2006) Suicidality in body dysmorphic disorder: a prospective study. Am J Psychiatry 163(7):1280–1282. https://doi.org/10.1176/ajp.2006.163.7.1280

Phillips KA, McElroy SL, Keck PE Jr, Hudson JI, Pope HG Jr (1994) A comparison of delusional and nondelusional body dysmorphic disorder in 100 cases. Psychopharmacol Bull 30(2):179–186

Phillips KA, Gunderson CG, Mallya G, McElroy SL, Carter W (1998) A comparison study of body dysmorphic disorder and obsessive-compulsive disorder. J Clin Psychiatry 59(11):568–575. https://doi.org/10.4088/jcp.v59n1102

Phillips KA, Menard W, Fay C, Weisberg R (2005) Demographic characteristics, phenomenology, comorbidity, and family history in 200 individuals with body dysmorphic disorder. Psychosomatics 46(4):317–325. https://doi.org/10.1176/appi.psy.46.4.317

Phillips KA, Pinto A, Hart AS, Coles ME, Eisen JL, Menard W, Rasmussen SA (2012) A comparison of insight in body dysmorphic disorder and obsessive-compulsive disorder. J Psychiatr Res 46(10):1293–1299. https://doi.org/10.1016/j.jpsychires.2012.05.016

Phillips KA, Hart AS, Simpson HB, Stein DJ (2014) Delusional versus nondelusional body dysmorphic disorder: recommendations for DSM-5. CNS Spectr 19(1):10–20. https://doi.org/10.1017/S1092852913000266

Phillips KA, Zai G, King NA, Menard W, Kennedy JL, Richter MA (2015) A preliminary candidate gene study in body dysmorphic disorder. J Obs Compuls Relat Disord 6:72–76. https://doi.org/10.1016/j.jocrd.2015.06.004

Plaisier MA, Smeets JB (2015) Object size can influence perceived weight independent of visual estimates of the volume of material. Sci Rep 5:17719. https://doi.org/10.1038/srep17719

Rauch SL, Phillips KA, Segal E, Makris N, Shin LM, Whalen PJ, Jenike MA, Caviness VS Jr, Kennedy DN (2003) A preliminary morphometric magnetic resonance imaging study of regional brain volumes in body dysmorphic disorder. Psychiatry Res 122(1):13–19. https://doi.org/10.1016/s0925-4927(02)00117-8

Rief W, Buhlmann U, Wilhelm S, Borkenhagen A, Brähler E (2006) The prevalence of body dysmorphic disorder: a population-based survey. Psychol Med 36(6):877–885. https://doi.org/10.1017/S0033291706007264

Ross ED (2008) Sensory-specific amnesia and hypoemotionality in humans and monkeys: gateway for developing a hodology of memory. Cortex 44(8):1010–1022. https://doi.org/10.1016/j.cortex.2008.02.002

Said CP, Haxby JV, Todorov A (2011) Brain systems for assessing the affective value of faces. Philos Trans R Soc Lond B Biol Sci 366(1571):1660–1670. https://doi.org/10.1098/rstb.2010.0351

Salari N, Kazeminia M, Heydari M, Darvishi N, Ghasemi H, Shohaimi S, Mohammadi M (2022) Body dysmorphic disorder in individuals requesting cosmetic surgery: a systematic review and meta-analysis. J Plast Reconstr Aesthet Surg 75(7):2325–2336. https://doi.org/10.1016/j.bjps.2022.04.098

Schieber K, Kollei I, de Zwaan M, Martin A (2015) Classification of body dysmorphic disorder – what is the advantage of the new DSM-5 criteria? J Psychosom Res 78(3):223–227. https://doi.org/10.1016/j.jpsychores.2015.01.002

Schmahmann J, Pandya D (2006) Fiber pathways of the brain. Oxford University Press

Schurgin MW (2018) Visual memory, the long and the short of it: a review of visual working memory and long-term memory. Atten Percept Psychophys 80(5):1035–1056. https://doi.org/10.3758/s13414-018-1522-y

Schwaninger A, Lobmaier JS, Collishaw SM (2002) Role of featural and configural information in familiar and unfamiliar face recognition. In: International workshop on biologically motivated computer vision. Springer, Berlin, pp 643–650. https://doi.org/10.1007/3-540-36181-2_64

Shinoura N, Suzuki Y, Tsukada M, Katsuki S, Yamada R, Tabei Y, Saito K, Yagi K (2007) Impairment of inferior longitudinal fasciculus plays a role in visual memory disturbance. Neurocase 13(2):127–130. https://doi.org/10.1080/13554790701399254

Silverstein SM, Elliott CM, Feusner JD, Keane BP, Mikkilineni D, Hansen N, Hartmann A, Wilhelm S (2015) Comparison of visual perceptual organization in schizophrenia and body dysmorphic disorder. Psychiatry Res 229(1–2):426–433. https://doi.org/10.1016/j.psychres.2015.05.107

Stadler M, Doebler P, Mertins B, Delucchi Danhier R (2021) Statistical modeling of dynamic eye-tracking experiments: relative importance of visual stimulus elements for gaze behavior in the multi-group case. Behav Res Methods 53(6):2650–2667. https://doi.org/10.3758/s13428-021-01576-8

Stangier U, Adam-Schwebe S, Müller T, Wolter M (2008) Discrimination of facial appearance stimuli in body dysmorphic disorder. J Abnorm Psychol 117(2):435–443. https://doi.org/10.1037/0021-843X.117.2.435

Takahashi E, Ohki K, Kim DS (2013) Dissociation and convergence of the dorsal and ventral visual working memory streams in the human prefrontal cortex. Neuroimage 65:488–498. https://doi.org/10.1016/j.neuroimage.2012.10.002

Tanaka JW, Farah MJ (1991) Second-order relational properties and the inversion effect: testing a theory of face perception. Percept Psychophys 50(4):367–372

Toh WL, Castle DJ, Rossell SL (2017a) Attentional biases in body dysmorphic disorder (BDD): eye-tracking using the emotional Stroop task. Compr Psychiatry 74:151–161. https://doi.org/10.1016/j.comppsych.2017.01.014

Toh WL, Castle DJ, Rossell SL (2017b) Face and object perception in body dysmorphic disorder versus obsessive-compulsive disorder: the Mooney faces task. J Int Neuropsychol Soc 23(6): 471–480. https://doi.org/10.1017/S1355617717000327

Toh WL, Castle DJ, Rossell SL (2017c) How individuals with body dysmorphic disorder (BDD) process their own face: a quantitative and qualitative investigation based on an eye-tracking paradigm. Cogn Neuropsychiatry 22(3):213–232. https://doi.org/10.1080/13546805.2017.1300090

Ungerleider LG, Haxby JV (1994) 'What' and 'where' in the human brain. Curr Opin Neurobiol 4(2):157–165. https://doi.org/10.1016/0959-4388(94)90066-3

Valentine T (1988) Upside-down faces: a review of the effect of inversion upon face recognition. Br J Psychol 79(4):471–491. https://doi.org/10.1111/j.2044-8295.1988.tb02747.x

Warren WH Jr, Whang S (1987) Visual guidance of walking through apertures: body-scaled information for affordances. J Exp Psychol Hum Percept Perform 13(3):371–383. https://doi.org/10.1037//0096-1523.13.3.371

Wilhelm S, Phillips KA, Didie E, Buhlmann U, Greenberg JL, Fama JM, Keshaviah A, Steketee G (2014) Modular cognitive-behavioral therapy for body dysmorphic disorder: a randomized controlled trial. Behav Ther 45(3):314–327. https://doi.org/10.1016/j.beth.2013.12.007

Wong WW, Rangaprakash D, Larson MS, Diaz-Fong JP, Tadayonnejad R, Leuchter AF, Feusner JD (2021) Can excitatory neuromodulation change distorted perception of one's appearance? Brain Stimul 14(5):1197–1200. https://doi.org/10.1016/j.brs.2021.07.010

Wong WW, Rangaprakash D, Diaz-Fong JP, Rotstein NM, Hellemann GS, Feusner JD (2022a) Neural and behavioral effects of modification of visual attention in body dysmorphic disorder. Transl Psychiatry 12(1):325. https://doi.org/10.1038/s41398-022-02099-2

Wong WW, Rangaprakash D, Moody TD, Feusner JD (2022b) Dynamic effective connectivity patterns during rapid face stimuli presentation in body dysmorphic disorder. Front Neurosci 16: 890424. https://doi.org/10.3389/fnins.2022.890424

Yaryura-Tobias JA, Neziroglu F, Chang R, Lee S, Pinto A, Donohue L (2002) Computerized perceptual analysis of patients with body dysmorphic disorder: a pilot study. CNS Spectr 7(6): 444–446. https://doi.org/10.1017/s1092852900017958

Computational Approaches for Uncovering Interoceptive Mechanisms in Psychiatric Disorders and Their Biological Basis

Marishka Mehta, Martin P. Paulus, and Ryan Smith

Contents

Abstract Interoception, the process of detecting, perceiving, and interpreting signals from within the body, is essential for physiological regulation and adaptive behavior. A growing body of research underscores important potential links between interoceptive dysfunction and psychiatric disorders. Parallel advancements in the field of computational psychiatry have led to the development of biologically plausible models of information processing in the brain. This review surveys the current state of traditional and computational research approaches to study interoceptive processes in psychiatry. We also provide a foundational description of predominant computational approaches and theoretical models of interoception. Finally, we discuss the potential molecular foundations of interoceptive computation and consider future directions for incorporating computational models to enhance clinical insights and inform personalized treatments. We conclude that combining interoception and computational modeling approaches holds considerable promise

M. Mehta and R. Smith (✉)
Laureate Institute for Brain Research, Tulsa, OK, USA

University of Tulsa, Tulsa, OK, USA
e-mail: RSmith@laureateinstitute.org

M. P. Paulus
Laureate Institute for Brain Research, Tulsa, OK, USA

Curr Topics Behav Neurosci (2025) 74: 129–156
https://doi.org/10.1007/7854_2024_572
Published Online: 26 February 2025

in moving the field forward, both in addressing unresolved mechanistic questions and identifying novel potential therapeutic targets.

Keywords Computational models · Interoception · Molecular mechanisms · Psychiatric disorder

1 Introduction

Adaptive behavior requires the continuous detection and regulation of a wide array of physiological processes within the body. The term interoception refers to the process of detecting, perceiving, and interpreting these bodily signals and maintaining an updated map of the internal state of the body at conscious and unconscious levels (Khalsa et al. 2018). Interoceptive information, such as that regarding the state of the cardiovascular, respiratory, and gastrointestinal systems, is crucial for guiding visceral regulation and behavior, and it can also influence higher-level cognitive functions involving emotions (Craig 2002; Critchley and Garfinkel 2017; Greenwood and Garfinkel 2024), decision-making (Dunn et al. 2010a), and judgment (Garfinkel et al. 2014; Zhao et al. 2022), among others. Interoception therefore has many routes by which it can potentially influence both physical and mental health.

Interoception and *allostasis*, the body's adaptive mechanism to maintain physiological stability through change (McEwen 1998), share a complex, reciprocal relationship. Interoception provides crucial signals about the body's physiological status, enabling the brain to monitor, predict, and respond to bodily needs, thereby forming the basis for allostasis. These internal signals, or interoceptive cues, help guide allostatic control by informing the timing, intensity, and type of physiological responses necessary for maintaining physiological variables within adaptive ranges. This corresponds to the related concept of *allostatic load*, which refers to the demand placed on allostatic systems to manage current and expected future challenges and its cumulative toll on the body (Guidi et al. 2021). When adjusting bodily states to meet these challenges, allostatic processes can in turn influence interoception by modulating the perceived salience of interoceptive signals based on an individual's physiological and psychological context (Stephan et al. 2016). Variables that are important for sensing allostatic load include blood pressure, oxygen, glucose, hydration, and inflammatory cytokine levels, among many others – which can be regulated either directly via autonomic/endocrine control or indirectly through action (e.g., getting a drink in anticipation of future dehydration). Unlike simpler forms of homeostatic regulation, which respond to *unexpected* deviations from adaptive physiological states in a *reactive* manner, allostasis can dynamically adjust physiological states *in advance* of environmental challenges when anticipated. Crucially, evaluating the real-time success of these regulatory processes requires detection of how body states have changed (Petzschner et al. 2021; Smith et al. 2017). Noisy, inflexible, or otherwise disrupted interoceptive signals may therefore hinder effective regulation and lead to heightened allostatic load. As one might expect, chronically elevated allostatic load has also been associated with onset and risk for several physical and mental health problems (Beckie 2012).

Motivated by the potential links between interoception, allostasis, and mental health, a recent line of research has drawn on tools from computational neuroscience and psychiatry (Huys et al. 2016; Smith et al. 2021a) to build and test mathematical models of underlying information processing and control mechanisms. As described below, this line of work holds promise as a means of uncovering individual differences in computational mechanisms and how they may inform clinical theory and treatment. In particular, cognitive computational models afford the unique ability to extract latent measures of specific patterns in perceptual and behavioral dynamics, which are often missed when relying solely on descriptive measures using summary statistics (Wilson and Collins 2019). Furthermore, these latent measures have clear mechanistic interpretations at the information processing level, and identifiable neural underpinnings, which could potentially facilitate the identification of novel cognitive and biological treatment targets. Computational approaches therefore hold unique promise in identifying interpretable mechanisms underlying interoception and interoceptive dysfunction that could be of considerable clinical relevance. Aside from treatment development, it is also hoped that mechanism identification with this approach could help identify subpopulations with similar disorder phenotypes but distinct underlying mechanisms, which might further entail differential clinical trajectories or treatment responses. Thus, when used to better characterize interoceptive processes implicated in psychiatric disorders, computational approaches are considered promising approaches in addressing mechanistic heterogeneity, nosology, and treatment considerations.

In what follows, we first review existing interoception research with traditional (non-computational) paradigms. We then consider computational approaches and how they have recently been applied to interoception research. We also briefly highlight the potential molecular basis of interoceptive computation. Finally, we discuss future research directions that will be important for testing computational models of interoception and their clinical relevance.

2 Interoception Studies

To date, the majority of interoception research paradigms in humans have focused on heartbeat perception. For example, the heartbeat counting task (HCT; Dale and Anderson 1978) asks individuals to report the number of heartbeats they feel in a given period of time and compares this subjective count to the true number of heartbeats in that time period. A related task, the heartbeat discrimination task (HDT; Whitehead et al. 1977), asks participants to judge whether their heartbeats are in synch or out of synch with a sequence of auditory tones. Heartbeat tapping tasks (HTTs; (McFarland 1975)) instead ask individuals to press a button each time they feel a heartbeat. Each of these tasks has recognized strengths and limitations (Desmedt et al. 2018; Hickman et al. 2020).

Previous work using these and other tasks has found evidence for altered cardiac interoception in depression and anxiety disorders (Eggart et al. 2019; Paulus and Stein 2010; Terhaar et al. 2012), eating disorders (Khalsa et al. 2015), and substance use disorders (Stewart et al. 2020). For example, one study showed an inverse relationship between depression severity and interoceptive accuracy on the HCT, as well as a positive association between accuracy and trait anxiety (Pollatos et al. 2009). Positive associations with anxiety levels have also been confirmed in a couple of other studies using similar methods (e.g., see Dunn et al. 2010b; Pollatos et al. 2007). Recent interoception-focused interventions have also shown promise in improving anxiety and eating disorders symptoms (e.g., see Choquette et al. 2023; Flux et al. 2022; Quadt et al. 2021; Sugawara et al. 2020), suggesting an important connection to psychopathology. However, sample sizes and effect sizes have varied between studies, offering variable levels of support (for a more complete, up-to-date review, see Ritz et al. 2024). A recent review of work testing cardiac biofeedback approaches also suggested this may improve interoception in some cases, but with limited support at present for associated improvements in psychiatric symptoms (Wareing et al. 2024). Along these lines, it is important to highlight that the literature linking interoception to psychopathology is mixed overall. For example, other studies have failed to find relationships between anxiety and interoception (Adams et al. 2022). While some work has also suggested that anorexia nervosa patients show lower interoceptive accuracy (Pollatos et al. 2008), this failed to replicate using the HDT in another eating disorders sample (Eshkevari et al. 2014). A more recent meta-analysis also found that changes in interoception after targeted interventions varied across disorders (Heim et al. 2023), with inconclusive support for links to symptom improvement.

Despite their widespread use, such tasks have also recently come under considerable scrutiny. Counting accuracy, for example, can be driven by expected heart rate as opposed to accurate moment-to-moment detection (Corneille et al. 2020; Desmedt et al. 2018). One study using the HCT also reported only moderate stability in scores with repeated testing (Khalsa et al. 2009b; Murphy et al. 2019). Further, scores on the HCT and HDT are only weakly related and may measure distinct abilities (Garfinkel et al. 2015; Hickman et al. 2020). One benefit of the HDT is that it can limit expectation effects. However, it may not be a pure measure of interoception (as it requires dual attention to an auditory cue), and it may be confounded by differences in the delay between objective and felt heartbeats. Another more general issue limiting these approaches is that most individuals show relatively poor cardiac interoception at rest (Khalsa and Lapidus 2016). To address this, some studies have utilized pharmacological interventions to increase arousal and enhance the intensity of cardiac sensations (e.g., Khalsa et al. 2009a; Terhaar et al. 2012). One recent example of this work found that administration of a peripheral adrenergic agonist (isoproterenol) during electroencephalography (EEG) led to larger magnitude changes in heartbeat-evoked potentials in individuals with generalized anxiety disorder (GAD) compared to healthy controls (HCs), particularly at low dose levels (Verdonk et al. 2024). Consistent results were also found in women with GAD, where low-dose isoproterenol caused greater cardiorespiratory arousal than in HCs, along with stronger interoceptive sensations and greater anxiety levels (Teed et al. 2022). While such approaches are certainly helpful in testing causal questions and addressing the issue that most people show poor cardiac perception at rest, concerns mentioned about the HCT and HDT measures themselves remain present (i.e., confounds associated with expectation, variable delays in perception, dual attention, test–retest reliability, etc.). These concerns, combined with low convergent validity, maintain the current lack of consensus regarding gold-standard performance metrics (Desmedt et al. 2023). Beyond this, existing measures are purely descriptive and therefore do not afford mechanistic interpretation (e.g., separable contributions of expectation, attention, and learning, among others). Thus, new measurement approaches are also likely needed and may be offered by the tools of computational psychiatry discussed below. Combining perturbations with these computational tools could further allow selective manipulation and assessment of different underlying mechanisms, potentially offering information of unique relevance to the development of novel interventions.

Although the majority of human interoception studies have focused on heartbeat perception, it is worth highlighting that other modalities have also been investigated. One clinically relevant example is respiratory interoception, which has been associated with various cognitive and affective processes (Boyadzhieva and Kayhan 2021; Heck et al. 2022; Varga and Heck 2017; Zelano et al. 2016) and has links to multiple affective disorder symptoms, most notably panic (Gorman et al. 2001; Pohl et al. 1988). Assessment of interoceptive accuracy in the respiratory domain has often made use of loaded breathing paradigms that adjust inspiratory resistance while individuals breathe through a mouthpiece or mask. Participants are asked to detect when a resistance is introduced (Gottfried et al. 1978) or to discriminate between levels of resistance (Webster and Colrain 2000). However, these approaches also have specific limitations (Nikolova et al. 2022). For instance, inspiratory resistance can induce negative affect, which could potentially confound results in some cases (e.g., negative affect or associated arousal could potentially influence breathing or detection thresholds, and affective responses could potentially vary between participants). The necessary timeframes to complete these tests can also be fairly lengthy, and establishing detection thresholds for participants can be somewhat imprecise. This has recently motivated the development of new paradigms with more efficient and controlled resistance delivery systems. For example, Harrison et al. (2021a) recently presented the Filter Detection Task, allowing efficient quantification of various measures of both signal detection and metacognition. Nikolova et al. (2022) also recently proposed the Respiratory Resistance Sensitivity Task, which uses continuous resistance adjustments and might be more sensitive to individual differences. These newer tasks may therefore help to address some of the aforementioned limitations, although others (e.g., negative affect induction) may remain.

One further domain of study within interoception research has been gastrointestinal perception. Until recently, however, available paradigms have been limited by their invasive nature. For example, mechanical distension paradigms involve the insertion and subsequent inflation of a balloon (Aziz et al. 1997; Hobson et al. 2000; Hölzl et al. 1996; Remes-Troche et al. 2012). Less invasive paradigms can also cause distress, such as water ingestion until satiation, tolerance to gas via lipid infusion (van Dyck et al. 2016), or gastric stimulation – chemical (Yang et al. 2006), thermal (Villanova et al. 1997), or electrical (Hobson et al. 2000). To address these limitations, a novel task was recently presented in which individuals swallow a mechanical capsule that is programmed to vibrate at various times while in a participant's stomach (Mayeli et al. 2023; Smith et al. 2021b). This task was developed in tandem with computational modeling, so we return to it further below.

As is apparent in this brief review, interoception research would benefit from more sophisticated paradigms and measurement tools. While perturbation approaches and more recent non-cardiac paradigms have important strengths, some remain limited in the mechanistic insights they can offer. As we will see below, task paradigms and computational models are ideally developed together to answer specific mechanistic questions, and existing paradigms could be modified toward that end. Along these same lines, interoception research could also benefit from a more precise mechanistic theory with which to understand the different processes that may be measured in different task contexts. This could be important, as not all domains of interoception may be of equal relevance to a given psychiatric disorder. It also remains unclear whether interoceptive awareness in one domain (e.g., cardiac) should be expected to correlate with awareness in other domains (e.g., gastrointestinal). At present, some evidence suggests at least partially dissociable neural correlates of information processing for different organ systems (Engelen et al. 2024; Haruki and Ogawa 2023). Studies testing associations between interoception measures across domains have also found mixed results (Crucianelli et al. 2021; Garfinkel et al. 2016; Herbert et al. 2012). However, the relatively greater focus to date on cardiac interoception limits the ability to answer such questions with confidence. As we review below, computational psychiatry has begun to offer resources that may help address several of these issues – most importantly with respect to providing a clear mechanistic framework with which to interpret results, improving specificity and interpretability of outcome measures, and task development/optimization.

3 Computational Psychiatry

In parallel to the recent progression in interoception research described above, the past few decades have also shown a gradual shift in focus toward understanding the brain from a computational perspective. In broad terms, it is assumed that the brain performs specific computations associated with interpreting the meaning of sensory signals, learning the likely outcomes of different action options, and selecting actions based on the expected value of those outcomes, among others. Computational neuroscience seeks to link these functions both to their neurobiological underpinnings and to how they generate observable behavior. One major advantage of cognitive computational models is that, by affording simulation of sequential operations hypothesized to be performed by the brain, they allow unique measures of patterns in perceptual and behavioral dynamics across task trials. Important patterns of this kind are often present due to learning. This contrasts with simpler arithmetic metrics like means and variances across trials, which can average over these dynamics and inadvertently prevent their detection (Teufel and Fletcher 2016). Computational approaches can further help in disentangling the influence of cognitive mechanisms that interact to influence choice in non-linear, high-dimensional ways, such as interactions between expectation and sensory signal strength when one or both of these factors evolve over time (e.g., based on learning or stimulus order). Estimating the values for parameters in a task model for each individual that reflect these complex patterns (e.g., learning rates or prior expectations) therefore offers novel opportunities to collect interpretable mechanistic information and associated individual differences.

This research program has also led to the emerging discipline of computational psychiatry, which aims to evaluate the clinical utility of neurocomputational measures, such as their potential ability to act as severity markers or predictors of treatment outcome, represent mechanisms of change in treatment, and/or improve psychiatric nosology. With respect to nosology, one major goal has been to generate computational phenotypes – that is, multidimensional characterizations of computational mechanisms at the individual level. By deriving these computational measures within a single mechanistic framework, this approach may help to address some of the concerns discussed above with respect to limitations in previous tasks and linked difficulties in the interpretability of inconsistent results. Here, the hope has been that, by removing some of the associated noise and interpretability issues, such computational phenotypes might better account for the heterogeneity within, and co-morbidity between, existing diagnostic categories (Patzelt et al. 2018). This is further motivated by previous concerns that current diagnostic approaches offer limited information, as common diagnostic labels can be assigned to a wide array of distinct symptom profiles (Fried and Nesse 2015), exacerbated by the high co-morbidity between most disorders (Watson 2005). At a broader level, these issues also motivated the creation of the Research Domain Criteria (RDoC) initiative by the National Institute of Mental Health (NIMH; (Cuthbert and Insel 2013)). In place of a categorical system, RDoC aimed to characterize patients along a large number of continuous dimensions. Computational psychiatry therefore dovetails with this approach and can be seen, in part, as a means of carrying out a subset of its broader aims.

Two major threads in computational psychiatry have focused on probabilistic (Bayesian) inference (for book-length and less technical introductions, respectively, see Ma et al. 2022; Sprevak and Smith 2023) and reinforcement learning (RL; for book-length and less technical introductions, respectively, see Sutton and Barto 2018; Wilson and Collins 2019). With respect to the former, it is assumed that, to some degree, the brain approximates optimal inference when determining the meaning of the sensory input it receives. This is because, in general, multiple states of the world can generate the same sensory signals. Thus, the brain must infer the most likely causes of those signals. This in turn requires a model of how sensory signals are generated by different states of the world as well as prior beliefs about which states are most likely a priori. The optimal solution for probabilistic inference in this context is provided by Bayes' theorem:

$$p(s|o,m) = \frac{p(o|s,m)p(s|m)}{p(o|m)} \tag{1}$$

Here, o corresponds to direct observations (sensory input), s corresponds to possible states of the world (events outside of the brain that may have caused that sensory input), and m corresponds to a model of how states and observations are related. For example, an observed facial expression might be used to infer someone's emotional state, based on a model of how emotions generate specific facial expressions. Here, Bayes' theorem formally describes the relationship between specific conditional probabilities entailed by the model. So-called "priors," $p(s|m)$, refer to the initial probability of each possible state of the world within the model. The term $p(o|s, m)$ is often called the "likelihood," which indicates the probability of making an observation if some state of the world is present. Finally, the marginal likelihood, $p(o|m)$, refers to the probability of an observation in a model overall (i.e., across all possible states). Together, these probabilities can be used to calculate the "posterior" belief, $p(s|o, m)$, which represents the updated probability of each possible state of the world after receiving the new observation. Beliefs can also evolve over time, such that a posterior probability at one timepoint can act as a prior belief for the next time point. Crucially, Bayes' theorem is often intractable to calculate exactly. Therefore, it is assumed that the brain approximates this inference process in some manner.

The other major thread in computational psychiatry mentioned above is RL. This approach models how individuals learn to make choices that maximize reward (on average and in the long run). "Model-based" RL algorithms assume that the brain uses a model of how states of the world (probabilistically) change over time given different actions, as well as the reward each action in each state (i.e., each state-action pair) is expected to provide. In contrast, "model-free" RL algorithms do not include explicit beliefs about how states change over time; instead, the reward associated with each state-action pair is learned slowly through trial and error, and the action with the highest expected reward within each state is chosen (i.e., without "thinking ahead" about future states/actions). The RL framework further assumes learning proceeds by computing the difference between expected and observed rewards (a reward prediction error; RPE) and then adjusting expected rewards to better match observed rewards (where the magnitude of this adjustment is controlled by a learning rate).

Bayesian inference and reward-based decision models can also be combined. One example framework for doing so is *active inference*, which models decision-making in contexts where the state of the world is uncertain (for a detailed tutorial, see Smith et al. 2022). In short, this framework assumes that individuals use a model of the world to infer (using approximate Bayesian inference) which actions are most likely to lead to states with the highest probability of generating rewarding observations. Unexpected observations are then used to update beliefs about both states of the world and the observations they are expected to generate. In this context, individuals are also assumed to possess exploration drives that motivate information-seeking behavior aimed at reducing uncertainty about states or other probabilistic beliefs about the world. Aside from active inference, there are also a range of other models within the RL framework that have been proposed to solve similar tasks involving decision-making under uncertainty in various ways.

Independent of the specific framework in question, links recently highlighted between interoception and reward stress the importance of making this connection. For example, while computational frameworks often treat reward as an abstract signal, it has become increasingly clear that reward and associated reinforcement processes are conveyed by afferent interoceptive channels. These channels can signal movement toward or away from homeostatic set points and act as a basis for both positive and negative rewards (Keramati and Gutkin 2011, 2014; Weber et al. 2024). We return to this point further below.

4 Interoceptive Computation

The computational frameworks described above have straightforward applications to interoception and allostatic regulation. As interoception is a perceptual modality, Bayesian inference has been a natural starting point. In this context, the brain can be modeled as inferring interoceptive variables such as body temperature, heart rate, respiration, and gastrointestinal states, among others, using the same types of conditional probabilities described above (Petzschner et al. 2021). Subsequent actions can then be understood as either unconscious changes in visceral regulatory signals or skeletomotor behaviors aiming to indirectly adjust interoceptive states. One formal proposal, for example, is that viable physiological ranges can be modeled as target (prior probability) distributions over observations that an individual aims to produce (Stephan et al. 2016). These observations can then be generated through actions that are themselves inferred in a Bayesian manner (i.e., making choices expected to generate observations consistent with those target distributions). For example, a model may assign a high prior probability to glucose levels within a healthy range of values and then choose to eat food or adjust insulin levels to maintain observed (i.e., sensed) glucose at levels consistent with that prior distribution. However, sensory signals arising from the body can be noisy and ambiguous. Thus, both perception and action are carried out under uncertainty.

Several theoretical models have been proposed with respect to interoception and computation. One example is the embodied predictive interoception coding (EPIC) model (Barrett and Simmons 2015), which considers how variation in the columnar structure of cortical regions may offer clues to computational function. In particular, the authors suggest that granular cortices, which contain a clear layer of granule cells (layer 4), are organized in a manner that allows computation of prediction errors, which could then be passed to other regions to drive belief updating (via superficial layers 1–3). In contrast, dysgranular and agranular regions, which do not have a clear granular layer, are suggested to be better positioned to convey prediction signals (via deep layers 5 and 6; i.e., which are continually updated to minimize prediction error). This setup can implement approximate Bayesian inference using a well-known algorithm called predictive coding (Friston 2005), which more broadly suggests that the brain contains distinct populations of neurons that represent predictions and prediction errors, and where bidirectional interactions between these populations converge on predictions that minimize these errors. The EPIC model can here be seen as an elaboration of earlier predictive accounts of interoception, such as that proposed by Seth (2013). These earlier accounts offered broader conceptual descriptions of how the brain might use predictive coding as an algorithm to accomplish interoception. However, they did not incorporate the detailed links to allostasis or the specific neural model of how this algorithm might be implemented that was proposed by the EPIC model.

With respect to larger-scale anatomy, the first cortical region known to process afferent interoceptive information – primary interoceptive cortex – is the (granular) posterior insula (Craig 2002). As described above, models such as the EPIC model suggest prediction error signals are computed in this region with respect to predictions received from the (dysgranular/agranular) anterior insula and cingulate cortices implicated in visceromotor control (among many other functions). On the one hand, prediction error signals are assumed to propagate back to these visceromotor regions to update beliefs about bodily (e.g., autonomic, hormonal, and immunological) variables. On the other hand, it is also suggested that some prediction signals act as visceromotor commands and initiate changes in bodily states to match those predictions. Although not fully distinguished in the EPIC model, action selection in this setup therefore requires at least two different types of predictions: those that reflect beliefs regarding current bodily states and those that reflect "desired" bodily states (i.e., predicted in the sense that they are consistent with survival). As we will see below, some precise formal models drawing on active inference provide resources to simulate the distinct roles of these different types of predictions.

More broadly, active inference models capture belief-like and desire-like predictions using a quantity called expected free energy, which decision-making aims to minimize. The expected free energy of an action sequence or policy (π) is defined as follows:

$$G(\pi) = -E_{q(o,s|\pi)}[\ln q(s|o,\pi) - \ln q(s|\pi)] - E_{q(o|\pi)}[\ln p(o|C)]$$

The first term on the right motivates information-seeking to reduce uncertainty. Formally, this is the difference between approximate prior and posterior beliefs about the expected states if one chose a specific policy and observed the expected outcomes (q denotes an approximate "best guess" distribution). Maximizing this difference minimizes expected free energy and leads to greater information gain. In the context of interoception, these prior/posterior beliefs would aim to approximate the true state of the body. The second term on the right reflects the probability of desired outcomes (*C*), in this case, corresponding to the target internal states of the body. Outside the context of interoception, it can be seen more generally as driving decisions expected to maximize rewarding/preferred outcomes.

Of course, beliefs about the state of the body can be inaccurate, which can in turn interfere with identifying decisions that will keep the body within preferred states. This has been the basis of several proposals focused on how interoceptive inference might become maladaptive in psychiatric disorders. Many of these have focused on over-weighting prior beliefs or under-weighting the influence of sensory signals (in both cases, preventing effective belief updating), although several other mechanisms are possible (e.g., see Barrett et al. 2016; Clark et al. 2018; Paulus et al. 2019; Petzschner et al. 2021; Petzschner et al. 2017; Stephan et al. 2016; Unal et al. 2021) and are based on extrapolation from evidence for predictive processing in other sensory and cognitive domains (recently reviewed by Hodson et al. 2024). While considerable focus has been placed on the insula, it is unlikely that this structure alone implements the computation of interoceptive prediction errors. Neural responses in a range of other areas have also recently been observed, depending on the interoceptive domain in question (as recently reviewed by Engelen et al. 2023b). For example, heartbeat-evoked responses have been found in the insula (Babo-Rebelo et al. 2016; Engelen et al. 2023a), but also in the somatosensory cortex (Al et al. 2021; Azzalini et al. 2021; Azzalini et al. 2019) and ventromedial prefrontal cortex (VMPFC; Azzalini et al. 2021). Specific responses associated with respiration have also been observed in the somatosensory cortex (Herrero et al. 2018; Perl et al. 2019), cingulate cortex (Kluger and Gross 2021; Perl et al. 2019), and hippocampus (Zelano et al. 2016). In the gastric system, neuroimaging studies have similarly detected responses in the insula and somatosensory cortex, along with cingulate cortices, hippocampus, and vmPFC (Wang et al. 2008). More recent work has further highlighted posterior occipital, parietal, and precuneus regions in processing afferent gastrointestinal signals (Haruki and Ogawa 2023; Mayeli et al. 2023; Smith et al. 2021b). Overall, these studies therefore indicate both overlapping (e.g., insula) and distinct activation patterns between interoceptive modalities, suggesting the flow of interoceptive information within the cortex is distributed. Thus, many brain regions could play a role in representing maladaptive beliefs about the body in psychiatric disorders. However, it is important to consider that some brain regions highlighted in these studies are likely better understood as implementing integrated processing at higher (multimodal) levels in a hierarchy (e.g., posterior cingulate, precuneus, and hippocampus), as opposed to simply low-level sensory prediction errors (Feldman et al. 2024). This implies a gradual integration of visceral information both across modalities and with exteroceptive information, all of which could be used to guide allostasis. Below, we introduce detailed computational models of interoception and allostasis and then highlight their potential implications for psychiatry.

5 Formal Models

In a recent paper, Tschantz et al. (2022) present several simulations using predictive coding and active inference to demonstrate how approximate Bayesian inference can be used to model interoception and allostasis. Their first simulation used a unimodal model of reactive homeostatic control. The model used a fixed prior to represent the acceptable (homeostatic) range of an example interoceptive variable (body temperature). The simulated "agent" observed varying temperatures and generated prediction errors by comparing them to its fixed prior. In place of belief updating, prediction error minimization was achieved by implementing autonomic (vasodilation) action to bring the observed temperature closer to the acceptable range. In their second simulation, they used a multimodal model that incorporated an exteroceptive modality, which could update predictions about future interoceptive states. For example, when external sensory information indicated an imminent run, this led the agent to anticipate increased body temperature. In response, autonomic action was generated to transiently reduce the target body temperature (i.e., by adjusting the fixed prior over the interoceptive state) to counteract this expected increase. This successfully prevented body temperature from moving outside the acceptable range (for another set of similar simulations, see Unal et al. (2021)).

In a third set of simulations, the authors showed how manipulating the precision (inverse variance) of both the fixed prior (essentially, the width of the acceptable value range) and the prediction error (i.e., the noise in the interoceptive signal) could influence model performance. When these precisions were too high or too low, autonomic regulation became less effective, allowing simulation of possible mechanisms of dysregulation in clinical contexts. In a final set of simulations, they modeled how goal-directed actions could be selected to regulate interoceptive variables when autonomic mechanisms could not do so alone (e.g., going to the fridge to get water or turning on the air conditioning to reduce body temperature). This required a hierarchical structure and computing the expected free energy of different policies (plans), as described above. Here, the lower level of the model still used a predictive coding-like approach to process continuous sensory data. The higher-level selected policies were based on more distal expected future states and actions (e.g., how long they would run and how temperature would change over time). Policy selection adjusted priors in the lower-level model as well as controlled discrete actions, such as deciding to bring a water bottle. It therefore allowed explicit simulation of mechanisms underlying allostasis-directed skeletomotor behavior.

Some of the aforementioned simulations were also built on similar simulations and a resulting theory of how fatigue and depressive symptoms could result from sustained dysregulation of homeostasis control (Stephan et al. 2016). In short, they proposed that the brain may also encode beliefs about the degree to which allostatic actions are successful. With repeated failures, expected success in regulation may become low, in which case optimal regulation strategies may lead to fatigue (and perhaps a sense of hopelessness and lack of control).

Other formal models of how interoception guides allostatic control have also been proposed along similar lines. For example, Sennesh et al. (2022) recently described an Allostatic Path-Integrated Control (APIC) model within the framework of control theory. The APIC model considers viscerosensory variables in terms of the control-theoretic concepts of "regulated resources" and "controlled processes." Regulated resources are quantities that are stable over time and cannot be changed instantaneously. Examples of regulated resources include body temperature, blood oxygen, glucose, and carbon dioxide levels. These variables do not have a single fixed target value and can optimally exist over a range of values, where range limits are assumed to be determined by genetic, environmental, and behavioral factors. Controlled processes can be construed as actions that modulate regulated resources (i.e., implementing allostasis). For example, sweating and shivering are examples of controlled processes that regulate body temperature. Allostatic action (the controlled process) is then formulated as a function of the associated (interoceptively detected) regulated resource. Based on this framework, the authors demonstrate how stochastic optimal control theory allows mathematic modeling of interoception-guided allostasis (in particular, the APIC model utilizes a path-integral control approach; for details, see Sennesh et al. (2022)). This bears similarity to the active inference approach described by Tschantz et al. (2022), although couched within a different modeling tradition. However, the authors acknowledge that it could be understood as an active inference model in a broad sense (Sennesh et al. 2022).

As one further example of current simulation work (Allen et al. 2022a), an active inference model was also recently used to help account for ways in which interoceptive and exteroceptive perception may interact. Specifically, this model simulated fast versus slow transitions between cardiac (diastole and systole) states in combination with the perception of neutral or threatening external stimuli. By building on the prior belief that systoles would more often be observed in combination with external threat (e.g., consistent with elevated heart rate), this carried the interesting consequence that the threatening stimulus was also perceived more clearly (i.e., with a more confident posterior belief) when presented in synchrony with heartbeats, consistent with previous literature (Garfinkel and Critchley 2016; Garfinkel et al. 2014). This simple model also allowed them to simulate expected changes in heart rate variability as a function of stimulus presentation and how the relationship between interoception, exteroception, and cardiac cycle was altered when removing ("lesioning") different connections within the model. This therefore illustrates another way in which formal models can simulate and make potentially novel predictions to guide future research.

Building on these simulation studies, other computational models have also been used in conjunction with interoception tasks to measure individual differences based on empirical data. We turn to these studies in the following section.

6 Empirical Modeling

In one recent empirical modeling study (Smith et al. 2020a), an HTT (in which individuals press a key each time they feel their heartbeat) was used in a transdiagnostic population consisting of individuals diagnosed with anxiety, depression, substance use, and eating disorders, as well as healthy volunteers. The task entailed 60-s trials in each of three conditions: guessing, no-guessing, and breath-hold. In the guessing condition, participants were allowed to guess even if they were not confident that they felt something. In the no-guessing condition, they were instructed to press the key only when they were confident that they in fact felt a heartbeat (probing influences of less precise prior beliefs). The same no-guessing instructions were repeated in the breath-hold condition, but this trial type was conducted while participants held their breath as a means of amplifying the afferent cardiac signal (i.e., increasing signal precision). A Bayesian computational model afforded estimation of two parameters in each participant: precision of prior beliefs (pHB) and afferent interoceptive signal precision (IP). As expected, pHB was reduced within the no-guessing condition relative to the guessing condition. Also as expected, IP values were amplified by the breath-hold, but only in healthy participants. This suggested a transdiagnostic reduction in the ability to update beliefs about the precision (i.e., noisiness/unreliability) of bodily signals. These effects were also replicated in a second healthy comparison sample (Smith et al. 2020b) and in a second transdiagnostic clinical sample (Lavalley et al. 2024).

A recent "heart rate discrimination" task has also been used to estimate similar metrics of bias and precision (Legrand et al. 2022). Here, participants were asked to attend to their heartbeats and then judge whether the perceived rate was faster or slower than a set of auditory tones. Fitted psychometric functions showed a negative bias (underestimation of heart rate) and lower precision when compared to an exteroceptive control task. Metacognitive measures also revealed less confidence for the interoceptive versus exteroceptive condition.

A cardiac interoception training study (Suksasilp et al. 2024) also recently combined the HDT with an extended hierarchical version of the Bayesian model developed by Smith et al. (2020a). This study replicated the training approach taken by Quadt et al. (2021) in a community sample and showed that improvements in task performance were associated with increases in IP across training sessions, moderated by a Bayesian learning rate parameter reflecting noise in the afferent signal. Importantly, the training also reduced anxiety scores over time, and greater decreases in anxiety were predicted by several computational parameters (i.e., greater increases in IP, less biased initial priors, and less signal noise [i.e., greater learning rate]), but not by traditional accuracy scores. This study therefore offers a promising initial example of how computational modeling might be uniquely useful in explaining and predicting experimentally induced changes in both interoception and affective outcomes. It will be important for future studies to extend this line of work to other training approaches, interoception tasks, and clinical populations.

In the respiratory domain, one recent study combined the filter detection task with neuroimaging and a computational model (Harrison et al. 2021b). Here, neuroimaging revealed that both predictions and prediction errors in the model were linked to activity patterns within the anterior insula and that these activations were sensitive to anxiety levels. More recently, Allen et al. (2022b) also proposed a theoretical account of how respiratory interoception may be enacted in the brain. These advances open avenues for a more thorough empirical investigation of disruptions in respiratory interoception across psychiatric disorders. However, computational approaches for studying respiratory interoception remain in the early stages.

In the gastrointestinal domain, Smith et al. (2021b) recently combined a Bayesian learning model with a novel gastrointestinal interoception task in which individuals press a key each time they feel a vibration within their stomach. To elicit these vibrations, the authors used a pre-programmed mechanical capsule that, after being swallowed, can deliver precisely timed vibration stimuli of varying strengths (while EEG, heart rate, and skin conductance data were simultaneously collected). Behavior was best explained by a five-parameter model, including: baseline interoceptive signal precision, change in signal precision as a function of vibration strength, an initial prior bias toward feeling a vibration, learning rate (updating prior beliefs) following perception of a vibration, and a decay rate (decreasing expectations to feel a vibration over time) in the absence of vibration. The central result was that event-related potentials (ERPs; as examined by EEG) were associated with several computational parameters in theoretically predicted ways. Specifically, interoceptive signal precision was associated with stronger ERPs in parieto-occipital regions linked to early stimulus processing (suggesting stronger belief updating). Additionally, weaker ERPs in frontal regions were associated with learning rates, decay rates, and initial prior expectations, all in directions suggesting that frontal responses were suppressed in the context of stronger prior biases. These results were jointly consistent with what would be expected under hierarchical predictive coding.

In summary, there are exciting new avenues emerging for studying interoception with computational models. However, across interoceptive domains, relatively little empirical work has been done.

7 Molecular Basis

The molecular mechanisms underlying interoception, and interoceptive computation, are only beginning to be explored and involve a variety of receptor systems and pathways that convey signals from the body to the central nervous system. For instance, the glossopharyngeal and vagus nerves, which innervate the gut, lungs, aortic body, aortic sinus, and core vasculature, are pivotal mediators of interoception. Rajendran et al. (2023) have recently highlighted the intricate roles of the vagus nerve in modulating various cardiovascular functions, emphasizing the need to decode the molecular and genetic diversity of vagal neurons to potentially pinpoint specific processes that comprise cardiovascular interoception. Additional interoceptive signals are transmitted by the lamina I of the spinal cord via the spinothalamic pathway and through the solitary nucleus projections. As underscored by Prescott and Liberles (2022), these diverse vagal sensory neuron types play crucial roles in a multitude of physiological functions. The signals are then processed and integrated within different brain regions, including the brainstem, thalamus, insula, somatosensory cortex, and anterior cingulate cortex. Others (Su et al. 2022) have focused on the nervous system's interaction with the lungs to better understand the neurobiology of respiratory interoception and found that Trpv1+ versus peptidergic Tac1+ vagal neurons show a complex terminal morphology, which targets both airway smooth muscles and vasculature (including lymphatics, and alveoli). These systems are important for differential afferents and create a molecular signature of lung-innervating sensory neurons.

Within this context, understanding the computational mechanisms of interoception at a molecular level is a new frontier. Abstractly, for example, one might aim to characterize gene expression levels as encoding and maintaining innate and learned prior expectations (and their precisions), at least in part via maintenance of changes in the synaptic connection efficacies associated with Hebbian learning and allostasis. More concretely, investigations to date have instead focused on mapping and classification based on intrinsic molecular features. For example, recent studies have aimed to differentiate neurons in the parabrachial nucleus (PB), a region associated with various homeostatic functions, to better understand their specific roles in neural circuit functions (Karthik et al. 2022). This will be a crucial step before characterizing the computations carried out by those circuits. These authors presented a molecular framework that organizes patterns of adult gene expression and predicts output projections to the amygdala, hypothalamus, and hindbrain to organize, interpret, and communicate PB-related information.

A growing body of work also exists examining bi-directional communication between the brain and peripheral immune system (Koren and Rolls 2022). The associated regulation and maintenance of the immunological state of the body has been referred to as "immunoception," representing yet another interoceptive modality of interest. For example, one recent study identified neurons in the insula that are activated during peripheral inflammation (Koren et al. 2021). These same neurons could also reinstate similar patterns of inflammation when subsequently stimulated. Of clinical relevance, it is also known that a persistent state of low-level peripheral inflammation and increased immune reactivity is associated with depression and other co-morbid psychopathology (Beckett and Niklison-Chirou 2022; Costi et al. 2021; Strawbridge et al. 2019; Yuan et al. 2019). Beyond the immune system, it is worth noting that interoceptive processing may benefit from other non-neuronal systems as well. In particular, some have proposed that "Glioception" – the glial sensing and transducing of signals (including osmotic, chemical, and mechanical status of extracellular milieu) – may play an important role in interoceptive processing. Specific glial cells can also trigger visceral regulatory responses by modulating activity within interoception-related neuronal networks (Fabbri et al. 2023).

Taken together, our understanding of the molecular systems that affect interoceptive processing in general, and interoceptive computation in particular, is still in its infancy. Understanding the complexities of the molecular biology and computational processes underlying interoception could provide important insights into human physiology, cognition, and various physical and mental health conditions. Future research will undoubtedly unveil more details about these intricate systems, offering more opportunities for interventions and therapies targeting interoceptive dysfunction. Moreover, we will need to conduct animal studies with sophisticated molecular and computational tools to unravel these relationships.

Along these lines, one recent study provided initial evidence for an interoceptive coding scheme used within parallel afferent vagal pathways (Zhao et al. 2022). Based on a range of molecular and physiological characterization methods, they specifically demonstrate how sensory vagal neurons are organized into functional units and convey multiple dimensions of information. These dimensions reflect organ-specific information and differentiate the tissue layers involved as well as the stimulus type. This multidimensional coding scheme could therefore naturally combine with the hierarchical and integrative computational architectures discussed above in extracting more abstract, multimodal information. This will be an important direction to test in future work.

The genetic variability of interoceptive systems in humans is largely unexplored, but associations have been identified between genetic polymorphisms and phenotypes related to the glossopharyngeal and vagus nerves, and the solitary nucleus. Li et al. (2022) examined alterations in the post-transcriptional regulation in the insula, aiming to determine how pharmacological interventions, such as cocaine use, may permanently affect interoceptive neural circuits. A deeper understanding of the molecular biology and genetics of interoception could potentially fundamentally alter our understanding of the computational processes underlying it. The correlations between genetic polymorphisms and functional variations in interoceptive pathways suggest that our genetic makeup could significantly influence how we process and respond to internal bodily signals. Understanding these genetic influences could therefore allow us to create more accurate computational models that account for individual variations in interoceptive processes. For instance, if certain genetic variants were found to modulate the sensitivity of interoceptive receptors or the efficacy of signal transmission along interoceptive pathways, these factors could be incorporated into computational models to predict interoceptive responses in different individuals or populations. One promising example along these lines is a study that examined how an adenosine 2A receptor genotype variant modulated resting-state functional connectivity between the insula and other brain regions (Geiger et al. 2016). Here this genotype was found to specifically modulate connectivity within a fronto-insular network that was also associated with interoceptive accuracy on the HCT. In computational terms, this result could therefore be taken to suggest a genetic modulation of the baseline precision weight assigned to interoceptive signals within the insula, thereby influencing the psychological impact of those signals. It will be important for future research to test whether, by incorporating this type of information, future computational models might better help to predict an individual's susceptibility to disorders associated with interoceptive dysfunction, such as anxiety disorders, depression, eating disorders, and somatic symptom disorders. In principle, this genetic information could be gathered at very early stages in development and thus contribute to early intervention and prevention efforts. Additionally, by studying the molecular mechanisms that underlie these genetic effects, we may be able to identify new targets for pharmacological interventions or biofeedback techniques that can enhance or modulate interoceptive function. Thus, advances in the molecular biology and genetics of interoception could not only enrich our theoretical understanding of interoception but also pave the way for personalized medicine approaches to mental health and well-being.

8 Future Directions: Computational Phenotyping and Nosology

It is clear in the literature reviewed above that computational approaches for studying interoception, its clinical relevance, and its biological underpinnings remain in very early stages. Thus, it is important to consider possible future directions that could extend this important effort. Here, we suggest several potential lines of research that could move the field forward.

In general, a major aim for future work pertains to computational phenotyping (Patzelt et al. 2018; Schwartenbeck and Friston 2016), which has the potential to improve current psychiatric nosology. As is exemplified by existing research, fitting models to data offers a way of measuring individual differences in computational mechanisms (e.g., prior beliefs, precisions, learning rates, etc.). By providing a phenotypic description of individuals in terms of computational measures, we may be able to differentiate subgroups of individuals within diagnostic categories whose symptoms arise via distinct pathways and who may respond to distinct treatments. We can also characterize individuals along multidimensional continua to complement diagnostic categories. The overarching goal then is to identify interoceptive computational mechanisms, their biological underpinnings, and their psychosocial consequences – with the aim to offer clinicians additional information regarding factors such as severity, prognosis, and treatment. To do so will require future work in several specific directions.

One important direction will be to expand the space of computational models considered. While Bayesian models of perception and decision-making have received the most attention in current interoception research, there are multiple widely used computational models that have not yet been used in this context. One example is drift–diffusion models (DDMs). DDMs are fit to reaction-time and accuracy data within two-choice decision tasks, most often in the context of perceptual judgments in response to noisy/imprecise stimuli (Bitzer et al. 2014). In their simplest form, these models assume that reaction times reflect different rates of perceptual evidence accumulation toward a decision bound. The characteristic rate of this evidence accumulation, starting bias, and height of the decision bounds for each individual can be estimated from task behavior. While widely used with exteroceptive stimuli, we are not aware of work using DDMs with interoception tasks. Yet, for many of the tasks above, reaction times could be calculated with respect to interoceptive stimuli (response time after heartbeats, capsule vibrations, changes in respiratory resistance, etc.). This approach could provide novel insights into how these mechanisms differ from exteroceptive signal processing and whether they are affected in various clinical conditions.

In recent years, the RL framework has also been used to describe homeostatic control processes (Keramati and Gutkin 2011, 2014). Specifically, so-called "homeostatic RL" has proposed that movement (or perhaps expected movement) of physiological variables toward viable ranges is the underlying basis of reward. This transforms reward from an abstract signal into something of direct biological relevance that could be mediated by interoception. More recently, Weber et al. (2024) have expanded on this by reviewing a larger body of work connecting specific interoceptive signals (e.g., reflecting changes in hydration, temperature, and blood glucose levels) to dissociable reward signals and distinct reward functions, in part through interactions with the dopamine system. One future research direction could therefore be to combine RL-based decision tasks with interoception-based rewards and punishments (e.g., probabilistic, choice-dependent changes in breathing resistance) and test for differences in psychiatric populations.

Existing computational approaches for analyzing task data could also be extended in important ways. For example, research to date has predominantly focused on single domains of interoception in isolation. Another important direction might therefore be to study relationships between interoceptive channels and how they are integrated. For example, it is currently unclear whether performance on cardiac perception tasks would predict performance on gastrointestinal or respiratory perception tasks. It is also unclear whether interoceptive signals in one channel modulate perception in other channels. Given the tight links between organ systems, and ways in which the nervous system coordinates their function, one might expect such influences (Calderon et al. 2017; Ford et al. 1995). In addition to these possible cross-modal effects, one might also consider multimodal integration (e.g., perception of high-arousal states evidenced by both elevated heart rate and increased breathing rate). Hierarchical computational models are well-suited to capture such processes (Smith et al. 2019; Smith et al. 2017).

Another important direction will be to identify the effects of medication on interoceptive computation. To the best of our knowledge, only one of the empirical modeling studies mentioned above examined the potential influences of medication status on cardiac interoception, reporting no effects on either prior biases or signal precision estimates (Smith et al. 2020a). More studies are therefore needed, in both the cardiac domain and other interoceptive domains. Future work using the other modeling paradigms mentioned above in psychiatric populations will provide straightforward opportunities for doing so (Harrison et al. 2021b; Smith et al. 2021b). This could also be informed by more general proposals regarding possible computational roles of neuromodulators that are targeted by common psychiatric medications (Browning and Lan 2022).

Unraveling the intricacies of the molecular biology and genetics of computational processes underlying interoception will require a multi-disciplinary approach that combines the strengths of molecular biology, neuroscience, genetics, computational modeling, and clinical research. In molecular biology, cutting-edge techniques, such as single-cell RNA sequencing, could be employed to decipher the specific gene expression patterns of individual neurons within interoceptive pathways, thereby revealing the molecular constituents of these neurons at an unprecedented level of detail. This could be applied using a number of feasible approaches, such as blood, tissue, or cerebral spinal fluid sampling, or through other measures such as extra-cellular vesicle tracking. Using such techniques, one recent study identified that a particular micro-RNA (miR93) known to modulate peripheral inflammatory responses was differentially expressed in depressed individuals (Burrows et al. 2024). Non-depressed individuals with higher miR93 expression also showed stronger interoception-related responses within the insula. Future work could build on such approaches to characterize the specific influence of miR93 on interoceptive computation and perhaps work toward testing means of modulating its expression to evaluate possible changes in symptoms and/or interoceptive function, and this is just one example of how such molecular approaches could be taken more generally going forward. Advanced imaging techniques like multiphoton microscopy could also be applied in parallel to help visualize the real-time functioning of these molecular pathways at a cellular level.

Future work could also take advantage of large-scale genome-wide association studies (GWAS) to identify genetic variants associated with interoceptive traits or disorders. For instance, one recent study identified candidate loci for distinct phenotypes of pain sensitivity that could have an interoceptive component (Zhang et al. 2023). Another GWAS study also identified loci associated with anorexia nervosa that overlap with previous studies of autoimmune disorders, which might have links to the immunoception processes mentioned above (Duncan et al. 2017). This approach could be extended to study other modalities of interoception going forward. In tandem with GWAS approaches, functional genomics and gene editing technologies, such as CRISPR-Cas9, could also be used to elucidate the functional effects of identified genetic variants. In computational neuroscience, novel models could be developed that incorporate these molecular and genetic data to simulate the computational processes underlying interoception. Finally, these models could be validated and refined using behavioral and physiological data from human subjects or animal models, collected through experimental paradigms that engage interoceptive processes or challenge interoceptive functions. It will also be essential to develop novel tools to measure interoceptive function at multiple levels, from the cellular to the behavioral, and to create databases that integrate these multi-scale data. These integrated efforts will allow for a comprehensive understanding of the molecular biology and genetics underlying the computational processes of interoception.

9 Conclusion

In this chapter, we have reviewed current work applying computational models to study interoception in both healthy and clinical populations. We have also considered important future directions. In doing so, it is clear that this field is in its very early stages. However, new models and paradigms offer promising approaches that may be crucial for answering outstanding mechanistic questions and point the way toward novel treatment targets. We look forward to future applications of these approaches and the insights they may provide.

References

Adams KL, Edwards A, Peart C, Ellett L, Mendes I, Bird G, Murphy J (2022) The association between anxiety and cardiac interoceptive accuracy: A systematic review and meta-analysis. Neurosci Biobehav Rev 140:104754. https://doi.org/10.1016/j.neubiorev.2022.104754

Al E, Iliopoulos F, Nikulin VV, Villringer A (2021) Heartbeat and somatosensory perception. NeuroImage 238:118247. https://doi.org/10.1016/j.neuroimage.2021.118247

Allen M, Levy A, Parr T, Friston KJ (2022a) In the body's eye: the computational anatomy of interoceptive inference. PLoS Comput Biol 18(9):e1010490. https://doi.org/10.1371/journal.pcbi.1010490

Allen M, Varga S, Heck DH (2022b) Respiratory rhythms of the predictive mind. Psychol Rev 130: 1066. https://doi.org/10.1037/rev0000391

Aziz Q, Andersson JL, Valind S, Sundin A, Hamdy S, Jones AK, Foster ER, Långström B, Thompson DG (1997) Identification of human brain loci processing esophageal sensation using positron emission tomography. Gastroenterology 113(1):50–59. https://doi.org/10.1016/s0016-5085(97)70079-9

Azzalini D, Rebollo I, Tallon-Baudry C (2019) Visceral signals shape brain dynamics and cognition. Trends Cogn Sci 23(6):488–509. https://doi.org/10.1016/j.tics.2019.03.007

Azzalini D, Buot A, Palminteri S, Tallon-Baudry C (2021) Responses to heartbeats in ventromedial prefrontal cortex contribute to subjective preference-based decisions. J Neurosci 41(23): 5102–5114. https://doi.org/10.1523/jneurosci.1932-20.2021

Babo-Rebelo M, Wolpert N, Adam C, Hasboun D, Tallon-Baudry C (2016) Is the cardiac monitoring function related to the self in both the default network and right anterior insula? Philos Trans R Soc B Biological Sci 371(1708):20160004. https://doi.org/10.1098/rstb.2016.0004

Barrett LF, Simmons WK (2015) Interoceptive predictions in the brain. Nat Rev Neurosci 16(7): 419–429. https://doi.org/10.1038/nrn3950

Barrett LF, Quigley KS, Hamilton P (2016) An active inference theory of allostasis and interoception in depression. Philos Trans R Soc Lond Ser B Biol Sci 371(1708):20160011. https://doi.org/10.1098/rstb.2016.0011

Beckett CW, Niklison-Chirou MV (2022) The role of immunomodulators in treatment-resistant depression: case studies. Cell Death Discov 8(1):367. https://doi.org/10.1038/s41420-022-01147-6

Beckie TM (2012) A systematic review of allostatic load, health, and health disparities. Biol Res Nurs 14(4):311–346

Bitzer S, Park H, Blankenburg F, Kiebel SJ (2014) Perceptual decision making: drift-diffusion model is equivalent to a Bayesian model. Front Hum Neurosci 8:102. https://doi.org/10.3389/fnhum.2014.00102

Boyadzhieva A, Kayhan E (2021) Keeping the breath in mind: respiration, neural oscillations, and the free energy principle [review]. Front Neurosci 15. https://doi.org/10.3389/fnins.2021.647579

Browning M, Lan D (2022) What can reinforcement learning models of dopamine and serotonin tell us about the action of antidepressants? Computational Psychiatry 6(1):166–188

Burrows K, Figueroa-Hall LK, Stewart JL, Alarbi AM, Kuplicki R, Hannafon BN, Tan C, Risbrough VB, McKinney BA, Ramesh R, Victor TA, Aupperle R, Savitz J, Teague TK, Khalsa SS, Paulus MP (2024) Exploring the role of neuronal-enriched extracellular vesicle miR-93 and interoception in major depressive disorder. Transl Psychiatry 14(1):199. https://doi.org/10.1038/s41398-024-02907-x

Calderon PGB, Habib M, Kappel F, de Los Reyes AA, t. (2017) Control aspects of the human cardiovascular-respiratory system under a nonconstant workload. Math Biosci 289:142–152. https://doi.org/10.1016/j.mbs.2017.05.008

Choquette EM, Flux MC, Moseman SE, Chappelle S, Naegele J, Upshaw V, Morton A, Paulus MP, Feinstein JS, Khalsa SS (2023) The impact of floatation therapy on body image and anxiety in anorexia nervosa: a randomised clinical efficacy trial. EClinicalMedicine 64:102173. https://doi.org/10.1016/j.eclinm.2023.102173

Clark JE, Watson S, Friston KJ (2018) What is mood? A computational perspective. Psychol Med 48(14):2277–2284. https://doi.org/10.1017/S0033291718000430

Corneille O, Desmedt O, Zamariola G, Luminet O, Maurage P (2020) A heartfelt response to Zimprich et al. (2020), and Ainley et al. (2020)'s commentaries: acknowledging issues with the HCT would benefit interoception research. Biol Psychol 152:107869. https://doi.org/10.1016/j.biopsycho.2020.107869

Costi S, Morris LS, Collins A, Fernandez NF, Patel M, Xie H, Kim-Schulze S, Stern ER, Collins KA, Cathomas F, Parides MK, Whitton AE, Pizzagalli DA, Russo SJ, Murrough JW (2021) Peripheral immune cell reactivity and neural response to reward in patients with depression and anhedonia. Transl Psychiatry 11(1):565. https://doi.org/10.1038/s41398-021-01668-1

Craig AD (2002) How do you feel? Interoception: the sense of the physiological condition of the body. Nat Rev Neurosci 3(8):655–666. https://doi.org/10.1038/nrn894

Critchley HD, Garfinkel SN (2017) Interoception and emotion. Curr Opin Psychol 17:7–14. https://doi.org/10.1016/j.copsyc.2017.04.020

Crucianelli L, Enmalm A, Ehrsson HH (2021) Probing interoception via thermosensation: no specific relationships across multiple interoceptive sub-modalities. bioRxiv. https://doi.org/10.1101/2021.03.04.433866

Cuthbert BN, Insel TR (2013) Toward the future of psychiatric diagnosis: the seven pillars of RDoC. BMC Med 11(1):126. https://doi.org/10.1186/1741-7015-11-126

Dale A, Anderson D (1978) Information variables in voluntary control and classical conditioning of heart rate: field dependence and heart-rate perception. Percept Mot Skills 47(1):79–85. https://doi.org/10.2466/pms.1978.47.1.79

Desmedt O, Luminet O, Corneille O (2018) The heartbeat counting task largely involves non-interoceptive processes: evidence from both the original and an adapted counting task. Biol Psychol 138:185–188. https://doi.org/10.1016/j.biopsycho.2018.09.004

Desmedt O, Luminet O, Walentynowicz M, Corneille O (2023) The new measures of interoceptive accuracy: A systematic review and assessment. Neurosci Biobehav Rev 153:105388. https://doi.org/10.1016/j.neubiorev.2023.105388

Duncan L, Yilmaz Z, Gaspar H, Walters R, Goldstein J, Anttila V, Bulik-Sullivan B, Ripke S, Consortium EDWGOTPG, Thornton L, Hinney A, Daly M, Sullivan PF, Zeggini E, Breen G, Bulik CM, Duncan L, Yilmaz Z, Gaspar H et al (2017) Significant locus and metabolic genetic correlations revealed in genome-wide association study of anorexia nervosa. Am J Psychiatry 174(9):850–858. https://doi.org/10.1176/appi.ajp.2017.16121402

Dunn BD, Galton HC, Morgan R, Evans D, Oliver C, Meyer M, Cusack R, Lawrence AD, Dalgleish T (2010a) Listening to your heart. How interoception shapes emotion experience and intuitive decision making. Psychol Sci 21(12):1835–1844. https://doi.org/10.1177/0956797610389191

Dunn BD, Stefanovitch I, Evans D, Oliver C, Hawkins A, Dalgleish T (2010b) Can you feel the beat? Interoceptive awareness is an interactive function of anxiety- and depression-specific symptom dimensions. Behav Res Ther 48(11):1133–1138. https://doi.org/10.1016/j.brat.2010.07.006

Eggart M, Lange A, Binser MJ, Queri S, Müller-Oerlinghausen B (2019) Major depressive disorder is associated with impaired interoceptive accuracy: a systematic review. Brain Sci 9(6):131. https://doi.org/10.3390/brainsci9060131

Engelen T, Buot A, Grèzes J, Tallon-Baudry C (2023a) Whose emotion is it? Perspective matters to understand brain-body interactions in emotions. NeuroImage 268:119867. https://doi.org/10.1016/j.neuroimage.2023.119867

Engelen T, Solcà M, Tallon-Baudry C (2023b) Interoceptive rhythms in the brain. Nat Neurosci 26(10):1670–1684. https://doi.org/10.1038/s41593-023-01425-1

Engelen T, Schuhmann T, Sack AT, Tallon-Baudry C (2024) The cardiac, respiratory and gastric rhythms independently modulate corticospinal excitability. bioRxiv, 2024.2009.2010.612221. https://doi.org/10.1101/2024.09.10.612221

Eshkevari E, Rieger E, Musiat P, Treasure J (2014) An investigation of interoceptive sensitivity in eating disorders using a heartbeat detection task and a self-report measure. Eur Eat Disord Rev 22(5):383–388. https://doi.org/10.1002/erv.2305

Fabbri R, Spennato D, Conte G, Konstantoulaki A, Lazzarini C, Saracino E, Nicchia GP, Frigeri A, Zamboni R, Spray DC, Benfenati V (2023) The emerging science of Glioception: contribution of glia in sensing, transduction, circuit integration of interoception. Pharmacol Ther 245:108403. https://doi.org/10.1016/j.pharmthera.2023.108403

Feldman MJ, Bliss-Moreau E, Lindquist KA (2024) The neurobiology of interoception and affect. Trends Cogn Sci 28(7):643–661. https://doi.org/10.1016/j.tics.2024.01.009

Flux MC, Fine TH, Poplin T, Al Zoubi O, Schoenhals WA, Schettler J, Refai HH, Naegele J, Wohlrab C, Yeh HW, Lowry CA, Levine JC, Smith R, Khalsa SS, Feinstein JS (2022) Exploring the acute cardiovascular effects of floatation-REST. Front Neurosci 16:995594. https://doi.org/10.3389/fnins.2022.995594

Ford MJ, Camilleri MJ, Hanson RB, Wiste JA, Joyner MJ (1995) Hyperventilation, central autonomic control, and colonic tone in humans. Gut 37(4):499–504. https://doi.org/10.1136/gut.37.4.499

Fried EI, Nesse RM (2015) Depression is not a consistent syndrome: an investigation of unique symptom patterns in the STAR* D study. J Affect Disord 172:96–102

Friston K (2005) A theory of cortical responses. Philos Trans R Soc Lond Ser B Biol Sci 360(1456):815–836. https://doi.org/10.1098/rstb.2005.1622

Garfinkel SN, Critchley HD (2016) Threat and the body: how the heart supports fear processing. Trends Cogn Sci 20(1):34–46. https://doi.org/10.1016/j.tics.2015.10.005

Garfinkel SN, Minati L, Gray MA, Seth AK, Dolan RJ, Critchley HD (2014) Fear from the heart: sensitivity to fear stimuli depends on individual heartbeats. J Neurosci 34(19):6573–6582. https://doi.org/10.1523/jneurosci.3507-13.2014

Garfinkel SN, Seth AK, Barrett AB, Suzuki K, Critchley HD (2015) Knowing your own heart: distinguishing interoceptive accuracy from interoceptive awareness. Biol Psychol 104:65–74. https://doi.org/10.1016/j.biopsycho.2014.11.004

Garfinkel SN, Manassei MF, Hamilton-Fletcher G, In den Bosch Y, Critchley HD, Engels M (2016) Interoceptive dimensions across cardiac and respiratory axes. Philos Trans R Soc Lond Ser B Biol Sci 371(1708):20160014. https://doi.org/10.1098/rstb.2016.0014

Geiger MJ, Domschke K, Homola GA, Schulz SM, Nowak J, Akhrif A, Pauli P, Deckert J, Neufang S (2016) ADORA2A genotype modulates interoceptive and exteroceptive processing in a fronto-insular network. Eur Neuropsychopharmacol 26(8):1274–1285. https://doi.org/10.1016/j.euroneuro.2016.05.007

Gorman JM, Kent J, Martinez J, Browne S, Coplan J, Papp LA (2001) Physiological changes during carbon dioxide inhalation in patients with panic disorder, major depression, and premenstrual dysphoric disorder: evidence for a central fear mechanism. Arch Gen Psychiatry 58(2):125–131. https://doi.org/10.1001/archpsyc.58.2.125

Gottfried SB, Altose MD, Kelsen SG, Fogarty CM, Cherniack NS (1978) The perception of changes in airflow resistance in normal subjects and patients with chronic airways obstruction. Chest 73(2 Suppl):286–288. https://doi.org/10.1378/chest.73.2_supplement.286

Greenwood BM, Garfinkel SN (2024) Interoceptive mechanisms and emotional processing. Annu Rev Psychol. https://doi.org/10.1146/annurev-psych-020924-125202

Guidi J, Lucente M, Sonino N, Fava GA (2021) Allostatic load and its impact on health: a systematic review. Psychother Psychosom 90(1):11–27. https://doi.org/10.1159/000510696

Harrison OK, Garfinkel SN, Marlow L, Finnegan SL, Marino S, Köchli L, Allen M, Finnemann J, Keur-Huizinga L, Harrison SJ, Stephan KE, Pattinson KTS, Fleming SM (2021a) The filter detection task for measurement of breathing-related interoception and metacognition. Biol Psychol 165:108185. https://doi.org/10.1016/j.biopsycho.2021.108185

Harrison OK, Köchli L, Marino S, Luechinger R, Hennel F, Brand K, Hess AJ, Frässle S, Iglesias S, Vinckier F, Petzschner FH, Harrison SJ, Stephan KE (2021b) Interoception of breathing and its relationship with anxiety. Neuron 109(24):4080–4093.e4088. https://doi.org/10.1016/j.neuron.2021.09.045

Haruki Y, Ogawa K (2023) Cardiac and gastric interoceptive awareness have distinct neural substrates. eneuro 10(1). https://doi.org/10.1523/eneuro.0157-22.2023

Heck DH, Correia BL, Fox MB, Liu Y, Allen M, Varga S (2022) Recent insights into respiratory modulation of brain activity offer new perspectives on cognition and emotion. Biol Psychol 170:108316. https://doi.org/10.1016/j.biopsycho.2022.108316

Heim N, Bobou M, Tanzer M, Jenkinson PM, Steinert C, Fotopoulou A (2023) Psychological interventions for interoception in mental health disorders: A systematic review of randomized-controlled trials. Psychiatry Clin Neurosci 77(10):530–540. https://doi.org/10.1111/pcn.13576

Herbert BM, Muth ER, Pollatos O, Herbert C (2012) Interoception across modalities: on the relationship between cardiac awareness and the sensitivity for gastric functions. PLoS One 7(5):e36646. https://doi.org/10.1371/journal.pone.0036646

Herrero JL, Khuvis S, Yeagle E, Cerf M, Mehta AD (2018) Breathing above the brain stem: volitional control and attentional modulation in humans. J Neurophysiol 119(1):145–159. https://doi.org/10.1152/jn.00551.2017

Hickman L, Seyedsalehi A, Cook JL, Bird G, Murphy J (2020) The relationship between heartbeat counting and heartbeat discrimination: A meta-analysis. Biol Psychol 156:107949. https://doi.org/10.1016/j.biopsycho.2020.107949

Hobson AR, Sarkar S, Furlong PL, Thompson DG, Aziz Q (2000) A cortical evoked potential study of afferents mediating human esophageal sensation. Am J Physiol Gastrointest Liver Physiol 279(1):G139–G147. https://doi.org/10.1152/ajpgi.2000.279.1.G139

Hodson R, Mehta M, Smith R (2024) The empirical status of predictive coding and active inference. Neurosci Biobehav Rev 157:105473. https://doi.org/10.1016/j.neubiorev.2023.105473

Hölzl R, Erasmus LP, Möltner A (1996) Detection, discrimination and sensation of visceral stimuli. Biol Psychol 42(1–2):199–214. https://doi.org/10.1016/0301-0511(95)05155-4

Huys QJ, Maia TV, Frank MJ (2016) Computational psychiatry as a bridge from neuroscience to clinical applications. Nat Neurosci 19(3):404–413. https://doi.org/10.1038/nn.4238

Karthik S, Huang D, Delgado Y, Laing JJ, Peltekian L, Iverson GN, Grady F, Miller RL, McCann CM, Fritzsch B, Iskusnykh IY, Chizhikov VV, Geerling JC (2022) Molecular ontology of the parabrachial nucleus. J Comp Neurol 530(10):1658–1699. https://doi.org/10.1002/cne.25307

Keramati M, Gutkin B (2011) A reinforcement learning theory for homeostatic regulation. Adv Neural Inf Proces Syst 24

Keramati M, Gutkin B (2014) Homeostatic reinforcement learning for integrating reward collection and physiological stability. eLife 3:e04811. https://doi.org/10.7554/eLife.04811

Khalsa SS, Lapidus RC (2016) Can interoception improve the pragmatic search for biomarkers in psychiatry? Front Psych 7:121. https://doi.org/10.3389/fpsyt.2016.00121

Khalsa SS, Rudrauf D, Sandesara C, Olshansky B, Tranel D (2009a) Bolus isoproterenol infusions provide a reliable method for assessing interoceptive awareness. Int J Psychophysiol 72(1): 34–45. https://doi.org/10.1016/j.ijpsycho.2008.08.010

Khalsa SS, Rudrauf D, Tranel D (2009b) Interoceptive awareness declines with age. Psychophysiology 46(6):1130–1136. https://doi.org/10.1111/j.1469-8986.2009.00859.x

Khalsa SS, Craske MG, Li W, Vangala S, Strober M, Feusner JD (2015) Altered interoceptive awareness in anorexia nervosa: effects of meal anticipation, consumption and bodily arousal. Int J Eat Disord 48(7):889–897

Khalsa SS, Adolphs R, Cameron OG, Critchley HD, Davenport PW, Feinstein JS, Feusner JD, Garfinkel SN, Lane RD, Mehling WE, Meuret AE, Nemeroff CB, Oppenheimer S, Petzschner FH, Pollatos O, Rhudy JL, Schramm LP, Simmons WK, Stein MB et al (2018) Interoception and mental health: a roadmap. Biol Psychiatry Cogn Neurosci Neuroimaging 3(6):501–513. https://doi.org/10.1016/j.bpsc.2017.12.004

Kluger DS, Gross J (2021) Respiration modulates oscillatory neural network activity at rest. PLoS Biol 19(11):e3001457. https://doi.org/10.1371/journal.pbio.3001457

Koren T, Rolls A (2022) Immunoception: defining brain-regulated immunity. Neuron 110(21): 3425–3428. https://doi.org/10.1016/j.neuron.2022.10.016

Koren T, Yifa RE, Amer M, Krot M, Boshnak N, Ben-Shaanan TL, Azulay-Debby H, Zalayat I, Avishai E, Hajjo H, Schiller M, Haykin H, Korin B, Farfara D, Hakim F, Kobiler O, Rosenblum K, Rolls A (2021) Insular cortex neurons encode and retrieve specific immune responses. Cell 184(24):5902–5915.e 5917. https://doi.org/10.1016/j.cell.2021.10.013

Lavalley CA, Hakimi N, Taylor S, Kuplicki R, Forthman KL, Stewart JL, Paulus MP, Khalsa SS, Smith R (2024) Transdiagnostic failure to adapt interoceptive precision estimates across affective, substance use, and eating disorders: A replication and extension of previous results. Biol Psychol 191:108825. https://doi.org/10.1016/j.biopsycho.2024.108825

Legrand N, Nikolova N, Correa C, Brændholt M, Stuckert A, Kildahl N, Vejlø M, Fardo F, Allen M (2022) The heart rate discrimination task: A psychophysical method to estimate the accuracy and precision of interoceptive beliefs. Biol Psychol 168:108239. https://doi.org/10.1016/j.biopsycho.2021.108239

Li W, Zhang C, Wang YY, Xiao L, Feng Y, Huo X, Wang C, Sun Y, Wang F, Sun T (2022) Alterations of RNAs in the insula related to cocaine-induced condition place preference in adolescent mice. Biochem Biophys Res Commun 621:109–115. https://doi.org/10.1016/j.bbrc.2022.06.080

Ma WJ, Kording KP, Goldreich D (2022) Bayesian models of perception and action. MIT Press, Cambridge, MA

Mayeli A, Al Zoubi O, White EJ, Chappelle S, Kuplicki R, Morton A, Bruce J, Smith R, Feinstein JS, Bodurka J, Paulus MP, Khalsa SS (2023) Parieto-occipital ERP indicators of gut mechanosensation in humans. Nat Commun 14(1):3398. https://doi.org/10.1038/s41467-023-39058-4

McEwen BS (1998) Stress, adaptation, and disease. Allostasis and allostatic load. Ann NY Acad Sci 840:33–44. PM: 9629234 (NOT IN FILE)

McFarland RA (1975) Heart rate perception and heart rate control. Psychophysiology 12(4): 402–405. https://doi.org/10.1111/j.1469-8986.1975.tb00011.x

Murphy J, Catmur C, Bird G (2019) Classifying individual differences in interoception: implications for the measurement of interoceptive awareness. Psychon Bull Rev 26(5):1467–1471. https://doi.org/10.3758/s13423-019-01632-7

Nikolova N, Harrison O, Toohey S, Brændholt M, Legrand N, Correa C, Vejlø M, Jensen MS, Fardo F, Allen M (2022) The respiratory resistance sensitivity task: an automated method for quantifying respiratory interoception and metacognition. Biol Psychol 170:108325. https://doi.org/10.1016/j.biopsycho.2022.108325

Patzelt EH, Hartley CA, Gershman SJ (2018) Computational phenotyping: using models to understand individual differences in personality, development, and mental illness. Personal Neurosci 1:e18. https://doi.org/10.1017/pen.2018.14

Paulus MP, Stein MB (2010) Interoception in anxiety and depression. Brain Struct Funct 214:451–463. https://doi.org/10.1007/s00429-010-0258-9

Paulus MP, Feinstein JS, Khalsa SS (2019) An active inference approach to interoceptive psychopathology. Annu Rev Clin Psychol 15:97–122. https://doi.org/10.1146/annurev-clinpsy-050718-095617

Perl O, Ravia A, Rubinson M, Eisen A, Soroka T, Mor N, Secundo L, Sobel N (2019) Human non-olfactory cognition phase-locked with inhalation. Nat Hum Behav 3(5):501–512. https://doi.org/10.1038/s41562-019-0556-z

Petzschner FH, Weber LAE, Gard T, Stephan KE (2017) Computational psychosomatics and computational psychiatry: toward a joint framework for differential diagnosis. Biol Psychiatry 82(6):421–430. https://doi.org/10.1016/j.biopsych.2017.05.012

Petzschner FH, Garfinkel SN, Paulus MP, Koch C, Khalsa SS (2021) Computational models of Interoception and body regulation. Trends Neurosci 44(1):63–76. https://doi.org/10.1016/j.tins.2020.09.012

Pohl R, Yeragani VK, Balon R, Rainey JM, Lycaki H, Ortiz A, Berchou R, Weinberg P (1988) Isoproterenol-induced panic attacks. Biol Psychiatry 24:891–902. https://doi.org/10.1016/0006-3223(88)90224-7

Pollatos O, Traut-Mattausch E, Schroeder H, Schandry R (2007) Interoceptive awareness mediates the relationship between anxiety and the intensity of unpleasant feelings. J Anxiety Disord 21(7):931–943. https://doi.org/10.1016/j.janxdis.2006.12.004

Pollatos O, Kurz AL, Albrecht J, Schreder T, Kleemann AM, Schöpf V, Kopietz R, Wiesmann M, Schandry R (2008) Reduced perception of bodily signals in anorexia nervosa. Eat Behav 9(4):381–388. https://doi.org/10.1016/j.eatbeh.2008.02.001

Pollatos O, Traut-Mattausch E, Schandry R (2009) Differential effects of anxiety and depression on interoceptive accuracy. Depress Anxiety 26(2):167–173. https://doi.org/10.1002/da.20504

Prescott SL, Liberles SD (2022) Internal senses of the vagus nerve. Neuron 110(4):579–599. https://doi.org/10.1016/j.neuron.2021.12.020

Quadt L, Garfinkel SN, Mulcahy JS, Larsson DEO, Silva M, Jones A-M, Strauss C, Critchley HD (2021) Interoceptive training to target anxiety in autistic adults (ADIE): A single-center, superiority randomized controlled trial. eClinicalMedicine:39, 101042. https://doi.org/10.1016/j.eclinm.2021.101042

Rajendran PS, Hadaya J, Khalsa SS, Yu C, Chang R, Shivkumar K (2023) The vagus nerve in cardiovascular physiology and pathophysiology: from evolutionary insights to clinical medicine. Semin Cell Dev Biol 156:190. https://doi.org/10.1016/j.semcdb.2023.01.001

Remes-Troche JM, Attaluri A, Chahal P, Rao SS (2012) Barostat or dynamic balloon distention test: which technique is best suited for esophageal sensory testing? Dis Esophagus 25(7):584–589. https://doi.org/10.1111/j.1442-2050.2011.01294.x

Ritz T, Meuret AE, Corner S (2024) Interoceptive interventions. In: Interoception: a comprehensive guide. Springer International Publishing, Cham, pp 405–433

Schwartenbeck P, Friston K (2016) Computational phenotyping in psychiatry: a worked example. eneuro 3(4). https://doi.org/10.1523/ENEURO.0049-16.2016

Sennesh E, Theriault J, Brooks D, van de Meent J-W, Barrett LF, Quigley KS (2022) Interoception as modeling, allostasis as control. Biol Psychol 167:108242. https://doi.org/10.1016/j.biopsycho.2021.108242

Seth AK (2013) Interoceptive inference, emotion, and the embodied self. Trends Cogn Sci 17(11):565–573. https://doi.org/10.1016/j.tics.2013.09.007

Smith R, Thayer JF, Khalsa SS, Lane RD (2017) The hierarchical basis of neurovisceral integration. Neurosci Biobehav Rev 75:274–296. https://doi.org/10.1016/j.neubiorev.2017.02.003

Smith R, Lane RD, Parr T, Friston KJ (2019) Neurocomputational mechanisms underlying emotional awareness: insights afforded by deep active inference and their potential clinical relevance. Neurosci Biobehav Rev 107:473–491. https://doi.org/10.1016/j.neubiorev.2019.09.002

Smith R, Kuplicki R, Feinstein J, Forthman KL, Stewart JL, Paulus MP, Tulsa I, Khalsa SS (2020a) A Bayesian computational model reveals a failure to adapt interoceptive precision estimates across depression, anxiety, eating, and substance use disorders. PLoS Comput Biol 16(12): e1008484. https://doi.org/10.1371/journal.pcbi.1008484

Smith R, Kuplicki R, Teed A, Upshaw V, Khalsa SS (2020b) Confirmatory evidence that healthy individuals can adaptively adjust prior expectations and interoceptive precision estimates. In: Verbelen T, Lanillos P, Buckley C, De Boom C (eds) Active inference, Communications in computer and information science, vol 1326. Springer, Cham, pp 156–164. https://doi.org/10.1007/978-3-030-64919-7_16

Smith R, Badcock P, Friston KJ (2021a) Recent advances in the application of predictive coding and active inference models within clinical neuroscience. Psychiatry Clin Neurosci 75(1):3–13. https://doi.org/10.1111/pcn.13138

Smith R, Mayeli A, Taylor S, Al Zoubi O, Naegele J, Khalsa SS (2021b) Gut inference: A computational modelling approach. Biol Psychol 164:108152. https://doi.org/10.1016/j.biopsycho.2021.108152

Smith R, Friston KJ, Whyte CJ (2022) A step-by-step tutorial on active inference and its application to empirical data. J Math Psychol 107:102632. https://doi.org/10.1016/j.jmp.2021.102632

Sprevak M, Smith R (2023) An introduction to predictive processing models of perception and decision-making. Top Cogn Sci. https://doi.org/10.1111/tops.12704

Stephan KE, Manjaly ZM, Mathys CD, Weber LA, Paliwal S, Gard T, Tittgemeyer M, Fleming SM, Haker H, Seth AK, Petzschner FH (2016) Allostatic self-efficacy: A metacognitive theory of Dyshomeostasis-induced fatigue and depression. Front Hum Neurosci 10:550. https://doi.org/10.3389/fnhum.2016.00550

Stewart JL, Khalsa SS, Kuplicki R, Puhl M, Investigators T, Paulus MP (2020) Interoceptive attention in opioid and stimulant use disorder. Addict Biol 25(6):e12831

Strawbridge R, Hodsoll J, Powell TR, Hotopf M, Hatch SL, Breen G, Cleare AJ (2019) Inflammatory profiles of severe treatment-resistant depression. J Affect Disord 246:42–51. https://doi.org/10.1016/j.jad.2018.12.037

Su Y, Barr J, Jaquish A, Xu J, Verheyden JM, Sun X (2022) Identification of lung innervating sensory neurons and their target specificity. Am J Physiol Lung Cell Mol Physiol 322(1):L50–L63. https://doi.org/10.1152/ajplung.00376.2021

Sugawara A, Terasawa Y, Katsunuma R, Sekiguchi A (2020) Effects of interoceptive training on decision making, anxiety, and somatic symptoms. Biopsychosoc Med 14:7. https://doi.org/10.1186/s13030-020-00179-7

Suksasilp C, McLanachan A, Quadt L, Boulton B, Mulcahy J, Critchley HD, Smith R, Garfinkel SN (2024) A hierarchical Bayesian model reveals increased precision weighting for afferent cardiac signals, and reduced anxiety, as a function of interoceptive training. bioRxiv

Sutton RS, Barto AG (2018) Reinforcement learning: an introduction. The MIT Press, Cambridge

Teed AR, Feinstein JS, Puhl M, Lapidus RC, Upshaw V, Kuplicki RT, Bodurka J, Ajijola OA, Kaye WH, Thompson WK, Paulus MP, Khalsa SS (2022) Association of generalized anxiety disorder with autonomic hypersensitivity and blunted ventromedial prefrontal cortex activity during peripheral adrenergic stimulation: a randomized clinical trial. JAMA Psychiatry 79(4): 323–332. https://doi.org/10.1001/jamapsychiatry.2021.4225

Terhaar J, Viola FC, Bär K-J, Debener S (2012) Heartbeat evoked potentials mirror altered body perception in depressed patients. Clin Neurophysiol 123(10):1950–1957. https://doi.org/10.1016/j.clinph.2012.02.086

Teufel C, Fletcher PC (2016) The promises and pitfalls of applying computational models to neurological and psychiatric disorders. Brain 139(10):2600–2608. https://doi.org/10.1093/brain/aww209

Tschantz A, Barca L, Maisto D, Buckley CL, Seth AK, Pezzulo G (2022) Simulating homeostatic, allostatic and goal-directed forms of interoceptive control using active inference. Biol Psychol 169:108266. https://doi.org/10.1016/j.biopsycho.2022.108266

Unal O, Eren OC, Alkan G, Petzschner FH, Yao Y, Stephan KE (2021) Inference on homeostatic belief precision. Biol Psychol 165:108190. https://doi.org/10.1016/j.biopsycho.2021.108190

van Dyck Z, Vögele C, Blechert J, Lutz AP, Schulz A, Herbert BM (2016) The water load test as a measure of gastric interoception: development of a two-stage protocol and application to a healthy female population. PLoS One 11(9):e0163574. https://doi.org/10.1371/journal.pone.0163574

Varga S, Heck DH (2017) Rhythms of the body, rhythms of the brain: respiration, neural oscillations, and embodied cognition. Conscious Cogn 56:77–90. https://doi.org/10.1016/j.concog.2017.09.008

Verdonk C, Teed AR, White EJ, Ren X, Stewart JL, Paulus MP, Khalsa SS (2024) Heartbeat-evoked neural response abnormalities in generalized anxiety disorder during peripheral adrenergic stimulation. Neuropsychopharmacology 49(8):1246–1254. https://doi.org/10.1038/s41386-024-01806-5

Villanova N, Azpiroz F, Malagelada JR (1997) Perception and gut reflexes induced by stimulation of gastrointestinal thermoreceptors in humans. J Physiol 502(1):215–222

Wang G-J, Tomasi D, Backus W, Wang R, Telang F, Geliebter A, Korner J, Bauman A, Fowler JS, Thanos PK, Volkow ND (2008) Gastric distention activates satiety circuitry in the human brain. NeuroImage 39(4):1824–1831. https://doi.org/10.1016/j.neuroimage.2007.11.008

Wareing L, Readman MR, Longo MR, Linkenauger SA, Crawford TJ (2024) The utility of heartrate and heartrate variability biofeedback for the improvement of interoception across behavioural, physiological and neural outcome measures: a systematic review. Brain Sci 14(6):579. https://www.mdpi.com/2076-3425/14/6/579

Watson D (2005) Rethinking the mood and anxiety disorders: a quantitative hierarchical model for DSM-V. J Abnorm Psychol 114(4):522–536. https://doi.org/10.1037/0021-843x.114.4.522

Weber L, Yee D, Small D, Petzschner FH (2024) Rethinking reinforcement learning: the interoceptive origin of reward. University of Oxford, Oxford. https://doi.org/10.31234/osf.io/be6nv

Webster KE, Colrain IM (2000) The relationship between respiratory-related evoked potentials and the perception of inspiratory resistive loads. Psychophysiology 37(6):831–841

Whitehead WE, Drescher VM, Heiman P, Blackwell B (1977) Realtion of heart rate control to heartbeat perception. Biofeedback Self Regul 2(4):317–392

Wilson RC, Collins AG (2019) Ten simple rules for the computational modeling of behavioral data. elife 8. https://doi.org/10.7554/eLife.49547

Yang M, Li ZS, Xu XR, Fang DC, Zou DW, Xu GM, Sun ZX, Tu ZX (2006) Characterization of cortical potentials evoked by oesophageal balloon distention and acid perfusion in patients with functional heartburn. Neurogastroenterol Motil 18(4):292–299. https://doi.org/10.1111/j.1365-2982.2006.00761.x

Yuan N, Chen Y, Xia Y, Dai J, Liu C (2019) Inflammation-related biomarkers in major psychiatric disorders: a cross-disorder assessment of reproducibility and specificity in 43 meta-analyses. Transl Psychiatry 9(1):233. https://doi.org/10.1038/s41398-019-0570-y

Zelano C, Jiang H, Zhou G, Arora N, Schuele S, Rosenow J, Gottfried JA (2016) Nasal respiration entrains human limbic oscillations and modulates cognitive function. J Neurosci 36(49):12448–12467. https://doi.org/10.1523/jneurosci.2586-16.2016

Zhang L, Pan Y, Huang G, Liang Z, Li L, Zhang M, Zhang Z (2023) A brain-wide genome-wide association study of candidate quantitative trait loci associated with structural and functional phenotypes of pain sensitivity. Cereb Cortex 33(11):7297–7309. https://doi.org/10.1093/cercor/bhad039

Zhao Q, Yu CD, Wang R, Xu QJ, Dai Pra R, Zhang L, Chang RB (2022) A multidimensional coding architecture of the vagal interoceptive system. Nature 603(7903):878–884. https://doi.org/10.1038/s41586-022-04515-5

Perceptual Dysfunction in Eating Disorders

Erin E. Reilly, Tiffany A. Brown, and Guido K. W. Frank

Contents

Abstract Eating disorders (EDs) are characterized by abnormal responses to food and weight-related stimuli and are associated with significant distress, impairment, and poor outcomes. Because many of the cardinal symptoms of EDs involve disturbances in perception of one's body or abnormal affective or cognitive reactions to food intake and how that affects one's size, there has been longstanding interest in characterizing alterations in sensory perception among differing ED diagnostic groups. Within the current review, we aimed to critically assess the existing research on exteroceptive and interoceptive perception and how sensory perception may influence ED behavior. Overall, existing research is most consistent regarding

Erin E. Reilly and Tiffany A. Brown contributed equally to this work.

E. E. Reilly
Department of Psychiatry and Behavioral Science, University of California, San Francisco, San Francisco, CA, USA

T. A. Brown
Department of Psychology, Auburn University, Auburn, AL, USA

G. K. W. Frank (✉)
Department of Psychiatry, University of California, San Diego, CA, USA
e-mail: gfrank@health.ucsd.edu

Curr Topics Behav Neurosci (2025) 74: 157–190
https://doi.org/10.1007/7854_2024_470
Published Online: 11 May 2024

alterations in taste, visual, tactile, and gastric-specific interoceptive processing in EDs, with emerging work indicating elevated respiratory and cardiovascular sensitivity. However, this work is far from conclusive, with most studies unable to speak to the precise etiology of observed perceptual differences in these domains and disentangle these effects from affective and cognitive processes observed within EDs. Further, existing knowledge regarding perceptual disturbances in EDs is limited by heterogeneity in methodology, lack of multimodal assessment protocols, and inconsistent attention to different ED diagnoses. We propose several new avenues for improving neurobiology-informed research on sensory processing to generate actionable knowledge that can inform the development of innovative interventions for these serious disorders.

Keywords Eating disorders · Exteroception · Interoception · Sensory perception · Sensory processing

1 Introduction

Eating disorders (EDs), including anorexia nervosa (AN), bulimia nervosa (BN), binge eating disorder (BED), and avoidant/restrictive food intake disorder (ARFID), are characterized by marked dysregulation in eating behaviors with some experiencing altered subjective experience of body weight, shape (American Psychiatric Association 2013). Across diagnostic presentations, EDs are associated with significant medical consequences, functional impairment, and cost to society (Jenkins et al. 2011; Peebles and Sieke 2019; Santomauro et al. 2021; Streatfeild et al. 2021). Factors contributing to the etiology and maintenance of ED symptoms remain undercharacterized (Culbert et al. 2015; Solmi et al. 2020; Stern and Bulik 2020). Meta-analytic data from treatment outcomes research suggests that many individuals experience persistence of symptoms or relapse over time (Berends et al. 2018; Linardon and Wade 2018; Vocks et al. 2010), highlighting a critical need to better understand these pernicious illnesses to inform the design of more effective treatments.

EDs are often characterized by symptoms that suggest alterations in the perceptual experience of one's body and internal sensations relating to food intake (American Psychiatric Association 2013); accordingly, there is a growing body of research that has explored contributors to these perceptual differences, as well as how they may contribute to behavioral and cognitive symptoms of EDs. Within the current review, we provide a focused narrative review of empirical work on sensory processing disturbances in EDs, with the larger goal of informing novel research in this domain that can inform innovative prevention and treatment efforts. To do so, our review will (1) examine relevant definitions and common methods of assessing different facets of sensory perception, (2) outline the existing literature regarding perceptual alterations in EDs organized across forms of sensory perception, and

(3) detail important areas for future study. Across all sections, we will include studies conducted across all ED diagnostic groups as they were available. Further, while much of the existing work employs behavioral assessments, we will include emphasis on neurobiological measurement whenever possible.

2 Definitions and Assessment

Sensory perception is a multifaceted process that involves detection, integration, and interpretation of sensory information across a range of external and internal inputs (Coren 2013; Mather 2016; Pleger and Villringer 2013). Because of the complexity of this process, dysfunction in sensory perception can be variable in its etiology; individuals may present with abnormalities in their ability to sense, integrate, interpret, regulate, or react to sensory signals, all resulting in perceptual abnormalities. Further, deficits or alterations in perception can be concentrated within one specific sensory modality and due to peripheral dysfunction or can be broad and represent alterations in other related domains, such as cognition, language, memory, affective processing, and attention, among others (Wolff et al. 2021; Pecher and Zwaan 2005; Azzalini et al. 2019; Critchley and Garfinkel 2017).

Sensory perception can be further broken down into subprocesses of *exteroception* (i.e., perception of stimuli originating outside of the body) and *interoception* (i.e., the perception of stimuli originating within the body) (Khalsa et al. 2018a). While certain types of sensory input or stimuli may fall clearly into the exteroceptive or interoceptive domains (e.g., touch as exteroception), it is also important to acknowledge that there are sensory domains for which the distinction may be less clear, such as gustation, taste, and proprioception (Chen et al. 2021). Although a thorough review of the biological processes that govern interoceptive and exteroceptive processing is beyond the scope of the current review, it is important to underscore the complexity of these systems, which involve integration of afferent and efferent signaling from the central and peripheral nervous systems, executed by a range of receptors (for reviews on related topics, see Pleger and Villringer 2013; Berntson and Khalsa 2021; Grunwald et al. 2003; Pollatos and Herbert 2018).

Sensory perception can be assessed at many levels of analysis, ranging from self-report measurements to neuroimaging (for reviews, see Khalsa et al. 2018a; Proff et al. 2022; Suksasilp and Garfinkel 2022). Because exteroceptive processing involves input from external cues via the visual, auditory, tactile, and olfactory systems, assessment of facets of exteroception generally involve a range of sensory tasks, such as the ability to detect, distinguish between, and identify different scents (i.e., "sniffin' sticks" method (Kobal et al. 1996)), or ability to detect standardized touch or other tactile stimuli (Essick et al. 2010). Behavioral assessments of interoceptive processing historically used in the human literature involve testing the awareness of and response to internal bodily functions, such as awareness of one's heartbeat (Schandry 1981), or breathing during inspiratory restriction loading

(Lopata et al. 1977), among others, although there has been recent work further expanding assessments across domains and units of analysis (Garfinkel et al. 2022). As noted above, in addition to responses to objective sensory stimuli, an individual's awareness of their general ability to perceive sensations (e.g., patient-reported trait-level pain tolerance) and/or their beliefs about and interpretations of specific sensory sensations (e.g., whether or not they believe themselves more sensitive to sensations) can also be measured via self-report.

3 Existing Research on Sensory Perception in Eating Disorders

Within the current literature review, we outline existing studies probing differences across sensory modalities often equated with "exteroception," including touch, olfaction, gustation, vision, and broad work on self-reported "sensory sensitivity." We then move into a review of research on interoceptive processing, including work on perception of gastrointestinal sensations, respiration, cardiovascular activity, and self-reported interoceptive awareness. Across domains, we discuss exteroception and interoception related to both disorder-specific and more general cues. Because the cardinal symptoms of EDs often involve abnormal affective and behavioral responses to food and body-related stimuli, understanding the extent to which sensory processing and perception of disorder-specific cues (e.g., food) is altered *independent of* cognitive and affective responses to food is challenging – and in some cases, impossible – to ascertain; thus, we prioritize review of research that can speak to this distinction and do not review work primarily focused on affective or cognitive processing in isolation.

4 Exteroceptive Processing

General Sensory Sensitivity Research on self-reported sensory sensitivity in AN suggests that individuals with AN, regardless of weight status, self-report enhanced sensitivity to and overresponsiveness to sensory experiences across domains (e.g., taste, touch) and motivated avoidance of sensory experiences compared to control women (Brand-Gothelf et al. 2016; Zucker et al. 2013; Merwin et al. 2013), which correlates with more severe symptom profiles (Zucker et al. 2013). However, one recent study conducted in recent-onset AN indicated similar sensory sensitivity to controls, suggesting that these patterns may reflect a scar of the illness, rather than a predisposing factor (Kitajima et al. 2022). Overall, there is limited work on self-reported sensitivity in BN, with the majority of studies finding no differences between BN groups and controls on self-reported sensory sensitivity (Brand-Gothelf et al. 2016).

Emerging work also suggests sensory sensitivity may be altered in ARFID, particularly among specific subgroups or presentations of patients (Dovey et al. 2019; Thomas et al. 2017a); however, as is the case across sensory domains, most of this research has been almost fully comprised of community or other clinical samples endorsing food selectivity, such as ASD and/or anxiety, rather than individuals diagnosed with ARIFD (Proff et al. 2022; Cermak et al. 2010; Wildes et al. 2012).

Touch To date, studies focused on perception of tactile stimuli in EDs have explored either (a) affective, soft touch (i.e., pleasurable touch), or (b) pain perception.

Affective Touch Research on affective touch has been conducted almost exclusively within samples of individuals with AN, both actively ill and recovered; findings from this work suggest altered neural activation associated with the anticipation and experience of soft touch (gentle strokes with a soft brush administered to the forearm or palm) in a sample of women recovered from AN (Bischoff-Grethe et al. 2018), as well as lower ratings of pleasantness of these sensations (Bischoff-Grethe et al. 2018; Crucianelli et al. 2016; Davidovic et al. 2018). Follow-up research exploring whether these differences in the experience of affective touch are generating at the central or peripheral level has been mixed in its findings, with some studies indicating that differences in experience of affective touch may be related to altered "bottom-up" afferent processing (Crucianelli et al. 2016), with others yielding findings consistent with altered "top-down" efferent processing (Crucianelli et al. 2021). Only one study to date has explored neural and behavioral responses to affective touch in women recovered from BN; findings suggested that the BN group demonstrated increased neural responses to the anticipation of affective touch (Wierenga et al. 2020), which represented the opposite pattern that has been observed in the same paradigm in women recovered from AN (Bischoff-Grethe et al. 2018). Altogether, work on neural and behavioral responses to pleasant, tactile stimuli is most developed in AN and suggests altered neural response and decreased pleasantness associated with these sensations, although the etiology of these effects remains unclear and could be secondary to observed alterations in interpersonal/social and affective processes in these samples (Christensen and Haynos 2020; Haynos and Fruzzetti 2011; Lavender et al. 2015; Mason et al. 2021; Schmidt and Treasure 2006).

Pain While a range of differing definitions of pain exist (Raja et al. 2020), recent consensus defines pain as an unpleasant sensory and emotional experience associated with, or resembling that associated with, actual or potential tissue damage (Raja et al. 2020; IASP 2020). Of note, the experience of pain often involves, but does not necessitate, nociception, which is defined as the neural encoding of any noxious stimuli. Research on potential alterations in pain perception in EDs were historically motivated by both empirical observations of altered opioid system activity in EDs (Kaye et al. 1987) and theoretical hypotheses that alterations in the experience of pain may contribute to altered body image or facilitate engagement in ED symptoms despite experience of aversive sensations (Faris et al. 1992; Raymond et al. 1999a).

There has been considerable work on alterations in pain perception across different ED diagnoses, although the specific facets of pain processing and stimuli used to invoke pain via nociception in the laboratory are varied. For instance, studies have explored group differences in pain thresholds (i.e., the intensity at which a noxious stimulus is perceived as painful, sometimes called pain sensitivity), pain tolerance (i.e., the amount of time a participant can tolerate exposure to noxious probes), self-reported affective responses to pain probes, and neural or physiological responses to these same probes. Further, nociceptive stimuli used in these paradigms have varied, with most studies using painful heat probes (Lautenbacher et al. 1991; Lautenbacher et al. 1990), cold pressor challenges (Lapidus et al. 2020), or "mechanical" pain via ischemic pressure (Stein et al. 2003; De Zwaan et al. 1996a; De Zwaan et al. 1996b).

Results from this work broadly suggest that individuals with AN, BN, and BED may demonstrate elevated pain thresholds (Faris et al. 1992; Raymond et al. 1999a; Lautenbacher et al. 1991; Bär et al. 2013; Papežová et al. 2005; Raymond et al. 1995) when exposed to mechanical and heat-based pain probes and altered neural responses to painful stimuli in the insula, prefrontal cortex, and pons in response to heat (Bär et al. 2013; Strigo et al. 2013). Of note, findings are most consistent in EDs where binge eating and purging behaviors are present (e.g., BN, AN binge/purge subtype) and in ill samples. Studies on pain processing in samples recovered from EDs have been mixed in their findings, with some work suggesting that increased tolerance of heat stimuli may be related to acute restriction and be state-based in nature (Bär et al. 2006; Krieg et al. 1993) and other studies supporting persistent, elevated thresholds for ischemic and thermal pain (Stein et al. 2003) and continued alterations in neural response to heat-based pain post-recovery (Strigo et al. 2013). More specifically, work in recovered AN samples using heat probes suggested increased activity in the dorsolateral prefrontal cortex (dlPFC), right anterior insula, and cingulate during pain anticipation and decreased activation in the insula and increased activity in the dlPFC during pain experience (Strigo et al. 2013). However, very few studies to date have explored responses to pain using neuroimaging in EDs other than AN, and existing studies have been limited to consideration of thermal pain.

More recent research on pain tolerance in EDs has focused on how elevated pain tolerance may confer risk for dual engagement in ED behaviors and self-harm; specifically, drawing from theoretical work in suicidal behavior (Van Orden et al. 2010), researchers suggested that increased pain tolerance over time in ED samples may result in increased capacity for non-suicidal self-injury and suicide risk (Claes et al. 2012; Smith et al. 2020a; Witte et al. 2016), although it is important to note that this work has remained primarily theoretical in nature or has used self-reported pain tolerance or engagement in NSSI as a proxy for pain tolerance.

Studies exploring the contributors to observed elevated pain thresholds are limited in number and challenging to summarize, with existing work finding associations between tolerance of thermal and pressure-based pain and a range of different indicators, including body temperature, psychological and affective constructs, blood pressure, cortisol, and autonomic function (Lautenbacher et al. 1990;

Girdler et al. 1998; Raymond et al. 1999b; Yamamotova et al. 2017). In sum, work on pain thresholds does support alterations in this process across EDs and most consistently in binge-spectrum EDs. Emerging work using neuroimaging also supports that alterations in thermal pain perception are associated with differential activation in the insula, dlPFC, and pons and may persist following recovery in AN. However, the extent to which these observed differences in pain threshold and experience have an etiology driven by top-down or bottom-up biological processes, their link with illness state, and their specificity to different types of nociceptive stimuli remains unclear.

Taste Because EDs are characterized by altered consumption of and responses to food stimuli, researchers have considered the possibility that those with EDs demonstrate alterations in their taste perception that may contribute to these symptoms' onset or persistence. Several recent reviews have provided an overview of gustation in patients with EDs (Chao et al. 2020; Kinnaird et al. 2018; Kot et al. 2020; Leland et al. 2022). Results indicated that data from psychophysical assessments consistently supports altered gustatory function in ill AN, such that patients with AN often demonstrate impairments in identification and a higher threshold for detecting different tastes (Aschenbrenner et al. 2008; Casper et al. 1980; Eiber et al. 2002; Fernández-Aranda et al. 2016; Goldzak-Kunik et al. 2012; Jirik-Babb and Katz 1988; Lacey et al. 1977; Nakai et al. 1987; Masuda et al. 1996; Frank et al. 2016a). Limited work exploring neural correlates of taste classification, for instance, suggested reduced classification accuracy in the insula for ill AN and obese subjects, with no differences in recovered AN and BN groups compared to healthy controls (Frank et al. 2016a). Ratings for pleasantness of sweet tastes have been mixed in AN, with some studies indicating lower preference for sweet taste probes and higher aversion to fat tastes (Drewnowski et al. 1988; Sunday and Halmi 1990; Szalay et al. 2010) and others suggesting no difference in rated pleasantness of sucrose compared to control groups (Goldzak-Kunik et al. 2012). Notably, data on changes in gustation throughout treatment suggests an influence of illness state, such that patients with AN demonstrate improvements in taste identification and normalization of thresholds over the course of treatment, while in some studies, remaining lower in function than controls (Aschenbrenner et al. 2008; Casper et al. 1980; Nakai et al. 1987; Masuda et al. 1996; Frank et al. 2022). Psychophysical findings in BN tend to be more conflicting (Aschenbrenner et al. 2008; Eiber et al. 2002; Nakai et al. 1987; Frank et al. 2016a; Blazer et al. 2008; Dazzi et al. 2013; Weiland et al. 2011), and one study exploring gustation differences concentrated in the palate alone versus the whole mouth suggested that any observed gustation differences could be due to purging-related palate damage (Rodin et al. 1990). Work on ratings of pleasantness broadly suggests higher preference for sweet tastes in BN compared to AN and controls (Rodin et al. 1990; Drewnowski et al. 1987; Klein et al. 2009), but similar fat-related aversion to AN samples (Drewnowski et al. 1988; Sunday and Halmi 1990; Drewnowski et al. 1987). Of note, research on taste in EDs has generally not incorporated updated research characterizing distinct cells and receptors for differing types of taste (Chandrashekar et al. 2006; Chaudhari and Roper 2021), idiosyncratic distribution of taste receptors across the mouth (Spence 2022), or interactions between taste and other sensory systems peripherally and centrally (Finger and Kinnamon 2011; Avery et al. 2020).

Conceptual work has outlined a potential for alterations in taste in avoidant/restrictive food intake disorder (ARFID), such that a subset of patients with ARFID may represent "supertasters" and have altered gustation and/or oral texture sensitivity that contributes to food sensitivity and avoidance (Thomas et al. 2017b). However, empirical work supporting this supposition has been conducted in samples of selective eaters drawn from the community (Nagao-Sato et al. 2023; Sorokowska et al. 2022) or other clinical samples (e.g., autism spectrum disorders (Zickgraf et al. 2022)) rather than clinical samples of individuals diagnosed with ARFID.

In addition to paradigms probing self-report and behavioral indicators of gustatory function, there is also growing work testing the neural substrates of taste processing in EDs (for related reviews, see Chao et al. 2020; Kot et al. 2020; Leland et al. 2022; Steward et al. 2018). These studies have primarily evaluated neural structural and functional responses to sweet taste or palatable foods (e.g., milkshake) (Chao et al. 2020). In both ill and recovered AN, work using structural neuroimaging suggests lower grey matter volume in areas implicated in hedonic and motivational facets of taste processing (Frank et al. 2013a; Frank et al. 2013b). Analyses from DTI exploring white matter integrity and connectivity further support these findings, such that women currently underweight with AN demonstrate alterations in connectivity among regions implicated in homeostasis, emotion, and food-reward processing that may correlate with ratings of subjective taste (Kot et al. 2020; Frank et al. 2013a; Frank et al. 2016b).

Results from functional neuroimaging studies generally suggest that predictable administration of sweet taste stimuli is often related to reduced activation in sensory and reward circuitry in both ill and recovered AN (Oberndorfer et al. 2013a; Wagner et al. 2008; Frank et al. 2013b; Wagner et al. 2015) and less sensitivity to hunger state compared to controls (Kaye et al. 2020). Contrastingly, some work suggests that individuals ill with and recovered from BN and BED demonstrate increased activation from these regions (Wagner et al. 2015; Oberndorfer et al. 2013b), with some exceptions indicating decreased responsivity in the insula for receipt of milkshake (Bohon and Stice 2011). Of note, existing findings suggest that the predictability of a given taste stimulus influences brain responses across diagnoses. More specifically, research exploring administration of taste stimuli within a classical conditioning task suggest that patterns in brain response reverse for *unexpected* receipt or omission of taste stimuli (i.e., "prediction errors"), such that individuals with current or a history of AN generally show increased brain activity in salient stimulus-related regions and individuals with current or past BN tend to have decreased neural responses to unexpected receipt or omission (Frank et al. 2011; Frank et al. 2021). One recent longitudinal, multimodal study indicated that changes in taste perception (i.e., decreased sweetness perception) over time moderated changes in neural prediction error responses to taste, suggesting an association between changes in these two processes *in addition to* observed changes in BMI (Frank et al. 2022).

Altogether, data support altered psychophysical, self-report, and neural responses to taste stimuli in individuals with EDs, although the extent to which psychophysical and self-reported taste aberrations normalize with treatment is less clear. Further, gustatory processes have well-characterized interactions with and contributions from olfaction (Sinding et al. 2023; Chen et al. 2023); the extent to which current paradigms engage both processes simultaneously or how observed findings may be related to taste versus olfaction remain unanswered by current data. Finally, because normative taste processing involves enactment of reward and motivational circuitry to promote adaptive behavior (Small 2006), it is and will remain challenging to discern the extent to which altered responses to taste in EDs are secondary to affective symptoms (e.g., fear of weight gain).

Olfaction In a similar manner to taste research, initial research on olfaction within EDs was prompted by altered responses to food within these populations. Studies assessing olfaction in EDs have focused primarily on behavioral indicators of olfaction, including scent discrimination, olfactory threshold, identification, and overall olfactory processing. Most existing work has been completed in AN, with fewer studies conducted in BN, binge eating disorder, ARFID, and other specified EDs (Leland et al. 2022). Overall, findings regarding whether olfaction is altered in AN and BN, and, if so, the nature of this difference, are often conflicting (for reviews, see Leland et al. 2022; Islam et al. 2015; Mai et al. 2020). One meta-analysis suggested that altered olfaction may represent a marker of severity in AN, rather than provide diagnostic value (Mai et al. 2020); however, somewhat inconsistent with this possibility, studies probing changes in olfaction associated with treatment have yielded mostly null results (Aschenbrenner et al. 2008; Roessner et al. 2005; Stein et al. 2012). One study to date has explored olfaction in ARFID, finding that youth with greater avoidant/restrictive eating and food fussiness demonstrated poorer performance on scent discrimination and higher fasting levels of serum PYY, compared to controls (Stull et al. 2021).

Beyond baseline comparisons with healthy controls, some recent studies have instead explored the relationship between olfaction and neurocognitive processes relevant to EDs, such as set shifting and alternation learning (Stein et al. 2012; Karavia et al. 2022), which were mixed in their findings. Specifically, a study in adult patients with AN suggested that poorer olfactory identification related to greater challenges in set shifting (Karavia et al. 2022), whereas another study in a mixed diagnostic sample found no association between olfactory discrimination and threshold and alternation learning performance (Stein et al. 2012).

Finally, research using neurobiological assessments to probe response to smell stimuli remains limited and primarily focuses on responsivity of mesolimbic circuitry to food-based odors in patient populations. One recent investigation in individuals with recurrent binge eating using electroencephalography (EEG) indicated greater N2 amplitudes for pictures of chocolate primed with a chocolate scent compared to neutral scents and pictures, which was correlated to increased craving and not observed in healthy controls (Wolz et al. 2017). Another study exploring neural responses to food and nonfood odors in AN and BN suggested lower activation in the ventral tegmental area in AN compared to BN, lower activation in the caudate in BN compared to HC when hungry, and decreased activity in the ventral pallidum and insula in response to odors of highly palatable foods in both groups compared to control subjects (Jiang et al. 2019). Importantly, while these studies suggest group differences in neural activity to food-related smells, it remains unclear whether altered responses to odors of palatable foods represent alterations in the sensory experience of this stimuli or differences in the affective or value-driven response to this stimuli (i.e., a rewarding versus a fearful response). Further, as noted above, work on olfaction in EDs has yet to incorporate considerations of interactions between smell and other related systems, such as taste (Sinding et al. 2023; Chen et al. 2023).

Visual System EDs are commonly characterized by disturbances in body image, which may include perceptual (accuracy of one's judgment of their size, shape, and weight), affective (body dissatisfaction, anxiety), and cognitive (beliefs about body weight or shape, mental body representations) components (Cash and Deagle III 1997). Given that body image disturbance (BID) in EDs is often extreme, to the point of delusion, researchers have been interested in studying how the visual system may play a role in BID. We first review the literature characterizing BID in EDs, discuss the relative contribution of visuospatial deficits versus cognitive-affective disturbances in BID, and review the literature on integrative visual system difficulties in EDs.

Body Size and Weight Overestimation Overall, research supports that adolescents and adults with EDs, particularly AN and BN, overestimate their body size compared to controls (Gardner and Brown 2014; Kinner et al. 2021; Mölbert et al. 2017). Research on individuals with BED suggests that patients with BED report higher degrees of weight and shape concerns than individuals without BED, but there is limited evidence regarding overestimation of body size (Ahrberg et al. 2011). A recent meta-analysis supported moderate effect size differences between individuals with AN or BN and controls (Hedge's $g = 0.63$), with effect sizes being larger for BN versus AN (Mölbert et al. 2017). Additionally, while individuals with EDs tend to overestimate whole body size, there is some evidence to support that certain body parts may have a greater likelihood of being overestimated. For example, body parts that are more subject to fluctuations in weight or body size (e.g., waist) are more likely to be overestimated in eating disorder patients compared to controls than body parts that are less influenced by weight fluctuations (e.g., shoulders (Mölbert et al. 2017)). Individuals with AN also tend to persist in overestimation even after being provided with feedback and tend to update their body size estimates at a slower rate than healthy controls (Saramandi et al. 2022). Critically, body size overestimation has been associated with increased negative affect, anxiety, body dissatisfaction, drive for thinness, eating disorder symptom severity, poor outcome, and relapse (Mölbert et al. 2017; Brown et al. 2021a; Hagman et al. 2015; Keel et al. 2005; Kinnear et al. 2021). In addition to visual overestimation of body size, a small body of research supports that individuals with AN are also more likely to overestimate the size of high-energy food objects (e.g., a cream bun) versus equally sized nonfood objects (e.g., a jewelry box) compared to healthy controls (Yellowlees et al. 2019). Similarly, individuals with AN estimated the volume of small- and medium-sized meals as being significantly larger than control women (Milos et al. 2013). However, a recent study found that individuals with AN did not overestimate the size of "unsafe" or high-calorie foods, compared to "safe" or lower-calorie foods compared to controls (Brown et al. 2021a). Given the relevance of body size overestimation across ED diagnoses, research has also explored the relative contribution of perceptual, cognitive, and attitudinal mechanisms of BID.

4.1 Multisensory Integration

Visuoperceptual Disturbances Versus Cognitive-Affective Factors Prior neurobiological research has implicated that alterations in the precuneus (which is involved in self-referential processing, imagery, and memory) and parietal lobes (which have roles in spatial and body representations, body ownership, and visuospatial processing) may play a role in altered bodily perception in EDs, largely based on research in AN (Gaudio and Quattrocchi 2012; Preston and Ehrsson 2016). As such, some research has speculated that perceptual disturbances may drive BID in AN (Alexi et al. 2019; Dakanalis et al. 2016). However, other researchers suggest that cognitive and affective features (e.g., drive for thinness, fear of or anxiety about weight gain) of BID may play a larger role in driving body image distortions than perceptual disturbances, citing the lack of research supporting that visual body perception is altered outside of cognitive and affective influences (Mölbert et al. 2017; Hagman et al. 2015; Frank and Treasure 2016). Recent research has suggested that visuoperceptual and attitudinal BID mechanisms may be independent of one another, rather than interactive (Cornelissen et al. 2019). A review by Gaudio and Quattrocchi (2012) summarized neurobiological research findings across perceptual and affective aspects of BID in AN and found that while perceptual disturbances were largely related to alterations in the precuneus and inferior parietal lobe, the affective component of BID was more likely to be assorted with alterations in the prefrontal cortex, insula, and amygdala, areas of the brain noted in cognition, emotional processing, motivation, and decision-making. Thus, perceptual and cognitive and affective factors likely both play a role in BID, likely via multiple neural processes and potentially difficulty integrating different sensory systems. However, that review was based on studies over a decade ago, and more novel and technically advanced studies are lacking.

More recent work supports dysfunction within the visual system in individuals with AN and has linked this dysfunction to symptom severity. In a combined neuroimaging and psychophysiological study, individuals with AN demonstrated hypoactivity in visual areas when viewing both facial and non-facial (houses) stimuli, which was particularly prominent for low spatial frequency images (Li et al. 2015). This suggests that individuals with AN may have deficits in the holistic appraisal of visual stimuli, suggesting that visual system dysfunction may contribute to perceptual distortions in AN. In contrast, a similar study exploring low spatial frequency images of faces found that individuals with AN exhibited low spatial frequency processing-stream abnormalities in higher-order systems, but connectivity within occipitotemporal face processing networks was unaffected in AN (Moody et al. 2015). The authors of that study suggested that this may have been due to the relative lack of salience of facial images in AN compared to body-related features (like thighs or hips). Moody and colleagues later expanded these stimuli to full bodies (Moody et al. 2021) and found that in individuals with AN, activity and connectivity in the dorsal visual network were associated with AN symptoms and subjective ratings of the appearance of others' bodies. Another study explored possible alterations in the default mode network, which plays a key role in self-related information (Via et al. 2018). Individuals with AN showed a hyperactivation of the dorsal posterior cingulate cortex when viewing video clips of their own body and decreased resting-state connectivity between regions involved in the default mode network, supporting task-specific and task-independent alterations in self-related processing in individuals with AN. Overall, these results suggest that AN may have visual sensory processing deficits that may contribute to BID.

Visuo-Tactile Integration Indeed, the case has also been made for individuals with AN having difficulty in visual-tactile integration (Case et al. 2012; Zopf et al. 2016). Much of this research has been based on the rubber hand illusion, during which individuals experience ownership over a fake rubber hand when it is stroked synchronously with their own hand, hidden out of view. Prior research supports that individuals with EDs have demonstrated a greater sense of ownership over the fake hand compared to controls, with susceptibility to the illusion being associated with the severity of ED psychopathology (Eshkevari et al. 2012). Another study explored body-location perception in AN using an action-oriented hand-reaching task in combination with the rubber hand illusion (Zopf et al. 2016). Results supported that hand movement trajectories in individuals with AN were more shifted compared to controls and demonstrated larger errors, suggesting that hand location estimates were more influenced by external visual hand information and less influenced by proprioceptive information. Taken together, results suggest that individuals with EDs may overly rely on visual body information, which may lead to distorted body perception. Further, recent research in this area has demonstrated that while individuals with EDs did not demonstrate differential ownership over the fake hand compared to controls, they did display a significant reduction in hand size estimation following the illusion, which did not occur in controls, which the authors suggested may indicate that bodily experience may be more malleable or unstable in individuals with EDs compared to controls (Carey and Preston 2019).

Allocentric Lock Hypothesis Given the multiple sensory systems implicated in BID in EDs, the allocentric lock hypothesis (Riva and Gaudio 2012; Riva and Gaudio 2018) suggests that BID in AN is the result of multisensory processing deficits in integrating expected versus actual bodily experiences. This hypothesis notes that individuals with AN have impairments in the ability (1) to link internal body signals to their potential consequences and (2) of updating offline knowledge, beliefs, and attitudes about one's body (allocentric) with real-time perception-driven inputs (egocentric). Overall, this hypothesis is in line with research supporting the relevance of both possible visuo-perceptual deficits and cognitive processes in individuals with AN. Consistent with this theory, during a proof of concept study in community women, body avoidance and uneasiness predicted perceptual body image disturbance when individuals were recreating their perceived body using virtual reality in a third-person (allocentric), but not a first-person (egocentric), perspective (Corno et al. 2018). Extending this to clinical samples, individuals with AN were more likely than healthy controls to rotate their shoulders to fit through a door frame, supporting body size overestimation. Further, negative body attitudes were associated with the magnitude of overestimation, suggesting altered body schema in AN (Beckmann et al. 2021). This hypothesis has suggested that the use of virtual reality may be a useful technique for helping to reintegrate cognitive and perceptual deficits in individuals with AN (Riva and Gaudio 2018).

5 Interoception

General Self-Report Literature It is important to distinguish between terminology around interoception within the self-report literature. While interoception refers to the overall process of how the nervous system senses, integrates, stores, and represents information about the state of the inner body at conscious and unconscious levels (Khalsa et al. 2018a), most of the self-report literature has used the term interoceptive awareness (IA). IA has had a number of conceptualizations within the literature (Khalsa et al. 2018a). IA was originally coined by Garner and colleagues (Garner et al. 1983) in reference to their developed subscale of the Eating Disorder Inventory (EDI). However, most of the items in the EDI-IA scale reference emotional awareness or alexithymia, rather than interoception-related symptoms. Later versions of the EDI (EDI-3) changed this subscale name to interoceptive deficits, although the conceptualization remained the same. Thus, this early conceptualization of IA differs from how this term is used currently, which considers IA to be far more comprehensive and includes multiple dimensions (e.g., attention, detection, discrimination, accuracy, insight, sensibility), with self-report measures representing one level of IA (Khalsa et al. 2018a).

Prior research supports that individuals with EDs exhibit self-reported deficits in IA. Indeed, a meta-analysis of studies examining the EDI-IA subscale demonstrated a large effect size difference between individuals with EDs and non-ED controls (effect size = 1.62 (Jenkinson et al. 2018)). Regarding diagnostic differences, patients with AN and BN both reported significantly greater IA deficits compared to patients with BED, but patients with AN and BN did not significantly differ from one another. Moderator analyses supported that IA deficits were greater in younger patients overall, but among AN and BN diagnoses, IA deficits were larger in older patients. IA deficits were also greater in samples with lower BMIs (Jenkinson et al. 2018). Notably, the EDI assesses IA using a unidimensional structure and also focuses more strongly on alexithymia; however, more recent research supports the importance of exploring self-reported IA across multiple dimensions. One recent popular assessment is the Multidimensional Assessment of Interoceptive Awareness (MAIA) (Mehling et al. 2012) and the more recent MAIA-2 (Mehling et al. 2018). The factor structure of this measure has been replicated in eating disorder samples, with individuals who distracted from uncomfortable body sensations (MAIA Not Distracting), had a low ability to regulate distress by attending to their bodily signals (MAIA Self-Regulation), and trusted their bodies the least (MAIA Trusting) having the most severe ED symptoms (Brown et al. 2017). Another self-report measure that has been less used in the ED literature, but could prove useful as a multidimensional self-report of IA, is the Interoceptive Accuracy Scale (IAS (Murphy et al. 2018)), which assesses an individual's self-reported accuracy in perceiving bodily sensations across a variety of interoceptive domains. Hybrid approaches, such as those outlined in Edwards-Duric and colleagues (Edwards-Duric et al. 2020), which combine self-report and sensory-specific manipulations may also be a helpful methodology for assessing self-reported IA in EDs.

Gastrointestinal (GI) Interoception In addition to self-reported interoceptive deficits, individuals with EDs have also demonstrated deficits in gastrointestinal interoception. Research supports that individuals with EDs may be hyper- or hyposensitive to signals of hunger and fullness and that prolonged restriction or binge eating/purging can interfere with gastrointestinal functioning (for review, see Khalsa et al. 2022). Indeed, both upper and lower GI symptoms, including nausea, gastric pain, early satiety, postprandial fullness bloating, constipation, and diarrhea are common across individuals with EDs and further interfere with food intake (Khalsa et al. 2022; Zucker and Bulik 2020). As such, frameworks have underscored the importance and relevance of developmental aversive visceral conditioning in EDs and specifically AN and suggested that early experiences with GI pain or discomfort may increase GI hypervigilance and preoccupation, which in turn can increase risk for food avoidance (Zucker and Bulik 2020). Other factors include disturbed hormone regulation that could interfere with satiety and hunger signaling and be related to altered internal GI sensation (Eddy et al. 2015). In AN, refeeding further exacerbates GI discomfort and anxiety which further reinforces food avoidance, maintaining ED symptoms. Given this, research has proposed that interoceptive exposure-based interventions that help recontextualize GI sensations may be promising to help address GI-specific interoceptive concerns in EDs (Zucker and Bulik 2020).

GI-specific interoceptive deficits have been measured through a variety of modalities including self-report (anxiety and hypersensitivity/hypervigilance for GI sensations (Brown et al. 2021b; Kerr et al. 2016)), behavioral (e.g., water load tests (Brown et al. 2022; Kerr et al. 2022; van Dyck et al. 2021)), physiological (EGG (Kerr et al. 2022; van Dyck et al. 2021)), and neuroimaging tasks (Kerr et al. 2016).

Self-report, anxiety, and hypervigilance about GI sensations have been associated across ED diagnoses with elevated body dissatisfaction, cognitive restraint, purging, restricting, and excessive exercise (Brown et al. 2021b), consistent with the hypotheses in Zucker and Bulik (2020) that GI distress may exacerbate or maintain body image concerns, food restriction, and compensatory behaviors in EDs.

Behavioral tasks to date have largely focused on water load tests and test meals to assess willingness to approach gastric sensations and sensations of fullness. During a water load test, individuals with BN or BED consumed significantly more water until reporting satiation compared to BMI-matched controls (van Dyck et al. 2021). Individuals with BN or BED were also more likely to have slowed electrical activity of the stomach compared to controls. Further, the percentage of normal gastric electrical activity was significantly smaller in BN and BED and was negatively related to the number of objective binge-eating episodes per week (van Dyck et al. 2021). In contrast, individuals with AN consume significantly less water during a water load task, but reported greater post-task fullness and negative affect, compared to controls, supporting that stomach sensations and fullness are perceived as aversive in individuals with AN (Brown et al. 2022). Interestingly, individuals with AN exhibit decreased rates of normal gastric electrical activity post-water load task when weight restored compared to underweight (Kerr et al. 2022). This was contrary to the authors' hypotheses that weight restoration would improve gastric motility and instead suggests that individuals with AN experience aberrant gastric physiology when weight restored, but not when they are acutely ill. Although results are in need of replication, this is in line with prior literature supporting aversive GI symptoms in individuals with AN during the refeeding process and acute weight restoration (e.g., delayed gastric emptying and increased fullness/early satiety (Zucker and Bulik 2020). Future research could benefit from also exploring individuals recovered from AN to understand whether these decreases in normal gastric activity persist after weight restoration and normal eating are maintained longer-term, which supports theories and research suggesting that reductions in GI symptoms in AN may reinforce restriction. Regarding test meal studies, for individuals in a nonclinical sample with lower levels of eating restraint, fullness ratings during a test meal were associated with greater caloric estimates; however, for individuals with higher levels of eating restraint, fullness and calorific estimates were not significantly related, reinforcing that restrictive eating may have direct consequences on interoceptive capabilities (Datta et al. 2021). Most of the behavioral tasks to date have focused on endogenous gastric state perturbations (e.g., consumption of water, food). However, exogenous forms of perturbation, in which signals are modulated and the person's reaction is assessed across subjective perception and objective physiology/circuits (Khalsa et al. 2018a), also have relevance for understanding of interoceptive dysfunction in EDs and is an area for further research.

Regarding neuroimaging tasks, individuals with AN demonstrated decreased activity in the dorsal mid-insula while attending to stomach sensations, and this lower activity was associated with greater scores on measures of anxiety and psychopathology, supporting associations between gastric interoception and anxiety in EDs (Kerr et al. 2016). Neuroimaging work also supports altered brain response in individuals with AN in response to food pictures, with AN participants who reported high-intensity stomach sensations exhibiting elevated activity in the amygdala and subgenual anterior cingulate cortex in response to highly palatable foods (Kerr et al.

2017). Similarly, neuroimaging research supports that reduced functional connectivity between central autonomic network regions and motor, premotor, frontal, parietal, and visual brain regions in AN in response to an interoceptive perturbation (isoproterenol infusion) is associated with elevated trait anxiety (De la Cruz et al. 2023). These results suggest that gastric sensations may contribute to a strong negative emotional response to high energy-dense food in individuals with AN, which may help explain food avoidance (Kerr et al. 2017).

Respiration Individuals with EDs have also demonstrated alterations in response to respiratory sensations across behavioral and neuroimaging tasks. During an experimental task pre- and post-meal, AN participants were more likely than controls to report detection of interoceptive respiratory sensations driven by false discriminations at low arousal levels suggesting possible hypersensitivity to respiratory sensations (Khalsa et al. 2015). Individuals with AN also endorsed greater pre-meal dyspnea compared to healthy controls. Taken together, these results suggest the presence of violations of expected respiratory interoceptive sensations ("prediction errors") in individuals with AN, particularly during meal anticipation, which may maintain (or be a result of) anxiety around eating (Khalsa et al. 2015). Further enforcing the relevance of respiratory sensations for EDs beyond AN, during a breath-holding task, individuals with EDs reported significantly more stress, feelings of suffocation, and suffocation fear than healthy controls, even though there were no group differences in breath-hold duration or physiological responses (Lapidus et al. 2020). Further, anxiety sensitivity (hyperreactivity to anxiety-based cues) was related to suffocation fear in the ED group (Lapidus et al. 2020).

Neuroimaging studies have also identified differences in respiratory interoception during an aversive breathing load task in individuals recovered from AN and BN compared to healthy controls (Berner et al. 2018; Berner et al. 2019). For individuals recovered from AN, during *anticipation* of restricted breathing, the recovered AN group had reduced activation in the right mid-insula camper to controls; however, during *receipt* of the aversive breathing load, the recovered AN group shows increased neural activation compared to the control group (Berner et al. 2018). Further, post-task, individuals recovered from AN demonstrated increased neural activation, with this hyperactivation being associated with severity of their past ED symptoms. The authors suggested that this pattern of deactivation in anticipation of the averse interoceptive stimuli and subsequent overactivation in response to receipt of the stimuli suggests that restriction may be one method of avoiding uncomfortable and unpredictable interoceptive changes in individuals with AN (Berner et al. 2018). In contrast to participants with AN, those recovered from BN showed increased activation in the mid-insula, superior frontal gyrus, putamen, dorsal anterior cingulate, posterior cingulate, and amygdala during anticipation of the aversive breathing load compared to controls; however, during receipt of the stimuli, neural activation in individuals recovered from BN did not decline to the degree that it did in controls. The authors suggested that the observed neural hyperactivity during anticipation and subsequent blunting of reduced activity during the task support that homeostatic instability may play a role in BN (Berner et al. 2019). Overall, results support that while overlooked until recent years, respiratory interoception is altered in EDs and warrants further investigation. The data suggest that altered respiratory interoception in EDs reflects broader interoceptive deficits outside of the GI system that may either be a cause or consequence of anxiety and ED symptoms.

Cardiovascular Patients with AN have displayed significantly poorer cardiac interoceptive sensitivity (sensitivity to cardiac signals) compared to controls during a heartbeat perception task (Pollatos et al. 2008). However, other studies have found no differences between individuals with AN and controls in cardiac interoceptive accuracy (Khalsa et al. 2015; Lutz et al. 2019). Of note, during an isoproterenol challenge, individuals with AN were more likely to report the detection of cardiac sensations than controls, driven by hypersensitivity to cardiac sensations at low levels of arousal in anticipation of a meal (Khalsa et al. 2015). Consistent with this, in a similar study, individuals with AN reported experiencing heartbeat sensations during meal-related saline infusions (e.g., when no cardiorespiratory modulation occurred). Individuals with AN also demonstrated a trend towards higher levels of panic during meal anticipation, suggesting increased cardiac hypersensitivity in AN. Consistent with this research, EEG methodology exploring the physiology of cardiac interoception has found that individuals with AN demonstrated heightened cortical processing of cardiac sensations (indicated by significantly larger heartbeat-evoked potentials) compared to controls (Lutz et al. 2019). This heightened sensitivity to cardiac sensations may either be the result of or contribute to increased anxiety observed in individuals with AN. Notably, another study found that individuals with AN demonstrated greater interoceptive accuracy on a heartbeat perception task when they were focused on other people's faces (decreasing self-focused attention) versus their own (increasing self-focused attention), which was the opposite pattern found in control (Pollatos et al. 2016). Further, greater body dissatisfaction in patients with AN was associated with poorer cardiac interoceptive accuracy when focused on their own faces (but not others' faces), suggesting that body image concerns and self-focused attention may serve to increase body avoidance in AN. While most of the cardiac interoception literature has focused on individuals with AN, individuals recovered from BN have also demonstrated lower interoceptive accuracy on the heartbeat perception task compared to controls after accounting for age (Klabunde et al. 2013). However, another study found that individuals currently ill with BN did not differ from controls on interoceptive accuracy after controlling for BMI, depression, and anxiety (Pollatos and Georgiou 2016). Thus, the literature on cardiac interoception may be less conclusive in BN (Klabunde et al. 2017). Notably in a mixed ED sample, Eshkevari and colleagues (Eshkevari et al. 2014) found no significant differences between individuals with an ED and healthy controls on the heartbeat detection task and performance did not differ from that expected by chance. Thus, the authors suggested the need for additional measures of interoceptive sensitivity in EDs. Interestingly, a more recent study which coupled the heartbeat detection task with an inspiratory breath-hold found that including the breath-hold perturbation increased precision estimates for cardiac signals in healthy individuals, but not for individuals with EDs, suggesting reduced interoceptive precision in EDs. Taken together, results also support the use of perturbation approaches (e.g., isoproterenol infusion, inspiratory breath-hold) in addition to heartbeat detection paradigms in assessing cardiac interoception in EDs (Khalsa et al. 2018a; Khalsa et al. 2015; Smith et al. 2020b; Khalsa et al. 2018b). Overall, results support that context (mealtime, focus of attention), and potentially diagnosis, may play a role in whether individuals with EDs are hyper- or hyposensitive to cardiac interoceptive signals.

6 Summary of Extero- and Interoceptive Processing in Eating Disorders

There exist a large number of studies focused on facets of interoceptive and exteroceptive sensory perceptual processing in EDs, as well as a range of reviews that provide summaries of this research (Kot et al. 2020; Leland et al. 2022; Mai et al. 2020; Mai et al. 2020; Mölbert et al. 2017; Mölbert et al. 2017; Jenkinson et al. 2018; Klabunde et al. 2017). In reviewing this work, several notable trends emerge, as well as several main gaps. First, findings regarding perceptual alterations to date appear most consistent for processing of disorder-specific stimuli (e.g., taste processing of food probes, visual responses to one's weight and shape, gastric interoception associated with digestion). While this trend may simply reflect the fact that the preponderance of research has been concentrated in disorder-specific domains, it seems likely that deficits or dysfunctions in sensory processing are heterogeneous within the larger ED population. More specifically, dysfunctional sensory perception of body and food-related stimuli may be present in the majority of patients, whereas other forms of altered perceptual processing (e.g., emerging research in areas such as respiration (Lapidus et al. 2020; Berner et al. 2018; Berner et al. 2019)) are likely altered in subgroups of this heterogeneous population and could represent differing ED phenotypes. Of note, despite a large and growing number of studies probing these processes, there remain a paucity of well-powered investigations; this limitation is particularly limiting for the field's ability to evaluate potential phenotypes in existing work.

Second, there remain significant knowledge gaps regarding how observed perceptual disturbances in EDs may contribute to the onset of symptoms versus represent a secondary effect of symptom engagement. We incorporate several possibilities into a framework for understanding sensory perception in EDs to guide future research in Fig. 1. More specifically, the extent to which affective and cognitive processes (e.g., learned fear responses to high-fat foods) drive neural, behavioral, and self-reported differences in sensory experience remains under-characterized. Additionally, while increasing work has begun to explore neurobiological correlates of sensory processing, these studies have remained primarily descriptive and cross-sectional, with few cohesive theoretical accounts and tests of how these processes may drive symptom engagement or top-down versus bottom-up

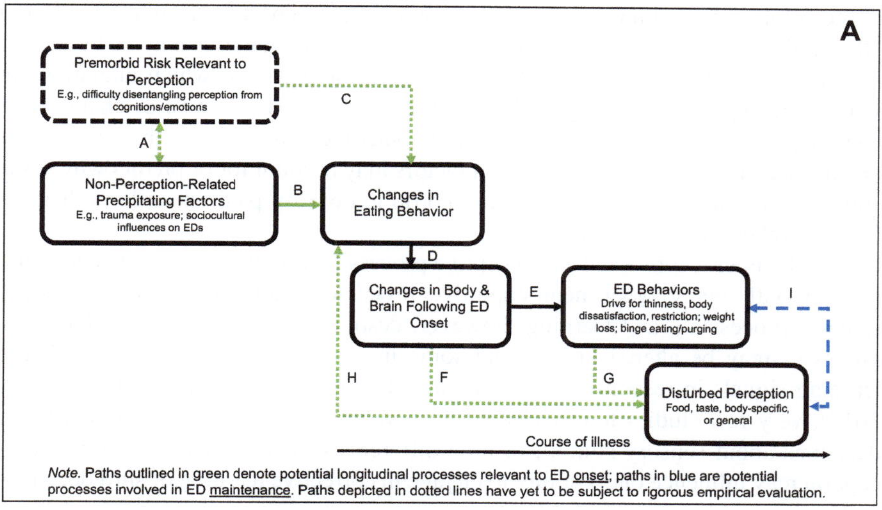

B

Verbal Description of Model	Questions for Future Research
• **ED Symptom Etiology.** Within this model, pre-morbid risk factors relevant to perception (e.g., challenges distinguishing between body sensations and emotions) may interact with precipitating risk factors or events (e.g., stress, trauma, desire for weight loss) to prompt initial changes in eating behavior (e.g., going on a diet) (Paths A, B, and C). These changes in eating behavior then result in changes to brain functioning (neurotransmitter levels or receptor densities) or body processes (e.g., weight, hormone levels) (Path D), which influence emergence of consistent ED symptoms (Path E) and perhaps the emergence of disturbed perception (Path F). Within this model, disturbances in perception may arise as a result of changes in brain or body and associated with eating behaviors (Path F), engagement in ED behaviors themselves (Path G). However, it is also possible that disturbances in perception may influence initial changes in eating behaviors (Path H), thus representing a process relevant to how EDs emerge.	• **ED Symptom Etiology.** ➢ Are there abnormalities in premorbid perceptual processes (biological or affective) that may interact with precipitating events to increase risk for initial onset of ED behaviors? (Paths A, B, and C) ➢ When does disturbed perception emerge in the course of illness, and how does this relate to the timeline for changes in eating behavior and body/brain (Paths F and G)? Could disturbed perception, intero- or exteroception, contribute to these changes? (Path H)
• **ED Symptom Maintenance**. Once the ED behaviors have emerged, it is possible that relationships between ED behaviors and disturbances in perception are reciprocal (Path I), such that engagement in ED behaviors may represent maladaptive responses to sensory or perceptual cues and ED behaviors may perpetuate differences in somatosensory experience.	• **ED Symptom Maintenance.** ➢ How do interactions between perceptual experiences and ED behaviors play out in real time? (Path I)

Fig. 1 Theoretical model of associations between ED symptoms and perceptual disturbances

contributions to these observed effects. For instance, accumulating data using neuroimaging techniques suggests altered neural responses in the insula, prefrontal cortex, and striatal regions to taste stimuli (Chao et al. 2020; Frank et al. 2016a; Frank et al. 2013a; Oberndorfer et al. 2013b). However, these findings only represent a first step towards characterizing the etiology and maintenance of EDs; there is a need to clarify (a) when these altered neural responses emerge within a course of illness; (b) the extent to which observed neurobiological processing of stimuli may be primary or secondary to behavioral, cognitive, and affective symptoms; and (c) how changes in these altered signals influence subjective experience and symptom maintenance. Ongoing, large-scale, publicly available longitudinal studies such as the Adolescent Brain Cognitive Development Study (ABCD (Karcher and Barch 2021)) offer one method through which some of these hypotheses may be tested; for instance, these datasets may be able to speak to the extent to which alterations in sensory perception may precede illness onset or diagnosis, as well as characterize factors that may account for multi-finality in outcomes associated with interoceptive or exteroceptive dysfunction (i.e., what factors may account for or predict why some children with interoceptive dysfunction go on to develop panic disorder, others an ED, and others no psychopathology?).

Finally, interoceptive and exteroceptive processing do not occur in a vacuum and interact with many other neurocognitive processes, including decision-making, salience processing, and learning; however, despite research suggesting that these processes may be altered in EDs and some initial research demonstrating altered learning signals in response to taste stimuli (Frank et al. 2016c; DeGuzman et al. 2017), very few studies have explored how individuals with EDs learn and make decisions about sensory stimuli, or how differences in sensory processing or perception may influence one's ability to engage in corrective learning or shift decision-making.

7 Future Directions in Research on Sensory Perception Dysfunction in Eating Disorders

Expanding Beyond AN and Considering Transdiagnostic Approaches Depending on the domain assessed, research on perceptual disturbances in EDs has predominantly focused on AN and, to a lesser extent, BN. Research on BED and ARFID, as well as other specified EDs, such as atypical AN, remains extremely limited, despite theoretical and empirical reasons to be believed that these patient populations may also experience alterations in sensory processing and perception (Brown et al. 2017; Khalsa et al. 2022; van Dyck et al. 2021; Nicholas et al. 2021). Accordingly, it is critical that future research expand to consider the ways in which sensory perception may be similar or diverse across ED diagnostic presentations through a wider range of sampling.

In addition to expanding samples to include other diagnostic categories, it is also recommended that future research consider studying perceptual processing in EDs within a transdiagnostic framework (for an example of transdiagnostic framework that could be tested, see Fig. 1). Particularly in light of data suggesting high levels of diagnostic crossover within EDs throughout the lifespan (Breithaupt et al. 2022; Eddy et al. 2008), as well as documented heterogeneity within diagnostic groups (Beltz et al. 2018), it may be the case that mixed findings discussed above represent the presence of multiple subgroups of patients who may share a diagnosis but demonstrate different patterns of perceptual disturbances with distinct neurobiological substrates. Research in EDs can draw from other work in psychiatry leveraging novel methods from data science and computational neuroscience to better characterize heterogeneity within, as well as similarities between, distinct diagnostic categories (Feczko et al. 2019; Williams 2016).

Multimodal, Dynamic, Assessment of Sensory Perception As knowledge regarding the neurobiological processes that govern exteroceptive and interoceptive processing evolves (Berntson and Khalsa 2021; Allen 2020) alongside innovations in assessment (Suksasilp and Garfinkel 2022), there emerge promising opportunities to expand the type and complexity of assessments probing sensory processing in EDs (Cusack et al. 2022). Suksasilp and Garfinkel et al. (2022) outline a framework for characterizing comprehensive measurement of interoception, which includes neural representation of afferent signals, strength of afferent signals, preconscious impact of afferent signals, interoceptive accuracy, self-report and beliefs, interoceptive insight, attention to interoceptive sensations, and attribution of interoceptive sensations, noting that the majority of research in interoception focuses on one of these areas, rather than multiple areas simultaneously (Suksasilp and Garfinkel 2022). Khalsa and Lapidus (2016) also highlight that including both direct perturbation of the system being studied and subjective reports of symptoms experience, autonomic activity, and behavior allows assessment to be more naturalistic, avoid floor effects, and allow greater standardization of the context and expectancies (Khalsa and Lapidus 2016). Research on sensory perception in EDs should heed these calls and expand assessment across different levels of analysis. Exploring the ways in subprocesses of sensory perception interaction may be particularly useful, as recent studies suggest that discrepancies in or dissociations between these measurements may be meaningful in conceptualizing psychopathology and more precisely isolating the etiology of observed effects (Garfinkel et al. 2015; Dang et al. 2020; Ferentzi et al. 2018).

In addition to considering multimodal assessments for sensory perception, we recommend adoption of novel methodologies that facilitate (a) broadening of the contexts in which sensory data is collected outside of the laboratory and (b) increases in the volume of data collected per participant. Of note, the concept of "context" relevant to sensory perception is multifaceted and includes a range of factors, including differing dimensions of time and space (e.g., where the participant is located, time of year, time of day), affective state, or physiological state (e.g., menstrual status). Related to place and time, while experimental, lab-based paradigms can probe many of these contextual factors through thoughtful research design, lab-based paradigms may be limited in their ability to speak to dynamic fluctuations in sensory experience through personalized probes that may drive engagement in ED behaviors. Lab-based data also is limited in the amount of intraindividual data it can provide. Innovations in wearable, passive assessment tools and increased accessibility of technology for ecological momentary assessments of relevant sensory symptoms (Ralph-Nearman et al. 2019) offer the potential to gauge fluctuations in sensory perception in patients with EDs in the natural environment and increase the amount of data gathered from each participant (Velkoff and Smith 2023; Höller et al. 2021). Despite an inherent decrease in experimental control associated with ecological data collection, these methods can provide large amounts of data for each participant in the natural environment and thus may provide auxiliary information that can be integrated with findings from in-lab studies to facilitate greater understanding of individual differences in these processes (also see discussion of heterogeneity above) (Cusack et al. 2022).

Adoption of Computational Models There is increased interest in the potential for methods from computational neuroscience to better characterize alterations in interception and bodily processing observed in psychiatric disorders (Khalsa et al. 2018a; Garfinkel et al. 2022; Owens et al. 2018; Petzschner et al. 2021). More specifically, mathematical models that seek to quantify the processes through which organisms encode and incorporate different types of information within a changing environment to drive effective behavior are commonly used to advance knowledge regarding cognitive and motor processes, with good reason to believe that they can be adapted for use across other domains, including interoception (Petzschner et al. 2021). Indeed, development of computational models of homeostatic processes offers the long-term potential to evaluate more precisely the places at which individuals with EDs may have deficits, as well as a comprehensive integration of interoceptive and exteroceptive systems across multiple units of analysis (Khalsa et al. 2022; Petzschner et al. 2021).

There are a number of existing computational frameworks that have been proposed as having direct relevance for characterizing sensory processing. For instance, growing work supports the promise of the active inference computational framework for capturing alterations in how individuals with dysfunctional interoceptive processing may predict, process, and update expectations related to sensory processes (Paulus et al. 2019; Smith et al. 2021), as well as other alternative frameworks, such as homoeostatic reinforcement learning (Keramati et al. 2017; Keramati and Gutkin 2014). However, few studies to date have applied these methods to the study of EDs (Khalsa et al. 2022; Smith et al. 2020b).

Changes in Sensory Perception Throughout Treatment Although some initial research described above suggests that interoceptive and exteroceptive processing may be malleable in response to intervention (Kinnaird et al. 2018), additional longitudinal work is required to both explore whether alterations in sensory perception change throughout existing treatments for EDs and, if so, what elements of these interventions may account for observed changes. For example, transdiagnostic cognitive-behavioral therapy approaches often include targeting of maladaptive beliefs about or interpretations of physiological sensations (e.g., mislabeling of changes in physiological state or affect as "feelings of fatness") (Anderson et al. 2022; Morales et al. 2022; Mehak and Racine 2020). It could be the case that subjective experience of interoceptive and exteroceptive stimuli change indirectly through existing cognitive and behavioral therapeutic exercises (Fairburn 2008); however, no work has explored this possibility. Further, some emerging research supports a role for interventions that focus on sensory and perceptual processing; for instance, interoceptive exposure (Boswell et al. 2019; Zucker et al. 2019) refers to a set of treatment techniques wherein patients are asked to purposefully elicit aversive sensory sensations with the goal of increasing tolerance of or shifting interpretations of these sensations (Boswell et al. 2013); this approach has received broad empirical support in other clinical populations endorsing alterations in sensory processing, such as panic disorder and irritable bowel syndrome (Pompoli et al. 2018; Craske et al. 2011), with some increasing work supporting use in EDs (Boswell et al. 2019; Hildebrandt et al. 2021). However, future research must determine more precisely what – if any – facets of perceptual processing are targeted by these interventions (e.g., does this intervention change one's detection or interpretation of specific bodily signals?).

8 Conclusion

In the current chapter, we provided a broad overview of research on disturbances in sensory perception that have been observed across ED diagnoses. Altogether, findings consistently indicate alterations in facets of sensory processing in EDs and the critical need for ongoing work in this domain. However, investigations that adopt new methods of assessment and analysis, expand sample characteristics, and better capture dynamic changes in sensory processing over time are necessary to inform more precise etiological models of EDs, disentangle somatic versus cognitive-emotional influences on perceptual disturbances, and improve neurobiological and psychological or behavioral treatment offerings.

References

Ahrberg M, Trojca D, Nasrawi N, Vocks S (2011) Body image disturbance in binge eating disorder: a review: body image disturbance. Eur Eat Disord Rev 19(5):375–381. https://doi.org/10.1002/erv.1100

Alexi J, Palermo R, Rieger E, Bell J (2019) Evidence for a perceptual mechanism relating body size misperception and eating disorder symptoms. Eat Weight Disord Stud Anorex Bulim Obes. 24(4):615–621. https://doi.org/10.1007/s40519-019-00653-4

Allen M (2020) Unravelling the neurobiology of interoceptive inference. Trends Cogn Sci 24(4): 265–266

American Psychiatric Association (2013) Diagnostic and statistical manual of mental disorders, 5th edn. American Psychiatric Association. https://doi.org/10.1176/appi.books.9780890425596

Anderson LM, Hall LM, Crosby RD et al (2022) "Feeling fat," disgust, guilt, and shame: Preliminary evaluation of a mediation model of binge-eating in adults with higher-weight bodies. Body Image 42:32–42

Aschenbrenner K, Scholze N, Joraschky P, Hummel T (2008) Gustatory and olfactory sensitivity in patients with anorexia and bulimia in the course of treatment. J Psychiatr Res 43(2):129–137. https://doi.org/10.1016/j.jpsychires.2008.03.003

Avery JA, Liu AG, Ingeholm JE, Riddell CD, Gotts SJ, Martin A (2020) Taste quality representation in the human brain. J Neurosci 40(5):1042–1052

Azzalini D, Rebollo I, Tallon-Baudry C (2019) Visceral signals shape brain dynamics and cognition. Trends Cogn Sci 23(6):488–509

Bär KJ, Boettger S, Wagner G et al (2006) Changes of pain perception, autonomic function, and endocrine parameters during treatment of anorectic adolescents. J Am Acad Child Adolesc Psychiatry 45(9):1068–1076. https://doi.org/10.1097/01.chi.0000227876.19909.48
Bär KJ, Berger S, Schwier C, Wutzler U, Beissner F (2013) Insular dysfunction and descending pain inhibition in anorexia nervosa: pain and anorexia nervosa. Acta Psychiatr Scand 127(4): 269–278. https://doi.org/10.1111/j.1600-0447.2012.01896.x
Beckmann N, Baumann P, Herpertz S, Trojan J, Diers M (2021) How the unconscious mind controls body movements: body schema distortion in anorexia nervosa. Int J Eat Disord 54(4):578–586
Beltz AM, Moser JS, Zhu DC, Burt SA, Klump KL (2018) Using person-specific neural networks to characterize heterogeneity in eating disorders: illustrative links between emotional eating and ovarian hormones. Int J Eat Disord 51(7):730–740
Berends T, Boonstra N, Van Elburg A (2018) Relapse in anorexia nervosa: a systematic review and meta-analysis. Curr Opin Psychiatry 31(6):445–455
Berner LA, Simmons AN, Wierenga CE et al (2018) Altered interoceptive activation before, during, and after aversive breathing load in women remitted from anorexia nervosa. Psychol Med 48(1): 142–154. https://doi.org/10.1017/S0033291717001635
Berner LA, Simmons AN, Wierenga CE et al (2019) Altered anticipation and processing of aversive interoceptive experience among women remitted from bulimia nervosa. Neuropsychopharmacology 44:1265–1273. https://doi.org/10.1038/s41386-019-0361-4
Berntson GG, Khalsa SS (2021) Neural circuits of interoception. Trends Neurosci 44(1):17–28
Bischoff-Grethe A, Wierenga CE, Berner LA et al (2018) Neural hypersensitivity to pleasant touch in women remitted from anorexia nervosa. Transl Psychiatry 8(1):161
Blazer T, Latzer Y, Nagler RM (2008) Salivary and gustatory alterations among bulimia nervosa patients. Eur J Clin Nutr 62(7):916–922. https://doi.org/10.1038/sj.ejcn.1602801
Bohon C, Stice E (2011) Reward abnormalities among women with full and subthreshold bulimia nervosa: a functional magnetic resonance imaging study. Int J Eat Disord 44(7):585–595. https://doi.org/10.1002/eat.20869
Boswell JF, Farchione TJ, Sauer-Zavala S, Murray HW, Fortune MR, Barlow DH (2013) Anxiety sensitivity and interoceptive exposure: a transdiagnostic construct and change strategy. Behav Ther 44(3):417–431
Boswell JF, Anderson LM, Oswald JM, Reilly EE, Gorrell S, Anderson DA (2019) A preliminary naturalistic clinical case series study of the feasibility and impact of interoceptive exposure for eating disorders. Behav Res Ther 117:54–64
Brand-Gothelf A, Parush S, Eitan Y, Admoni S, Gur E, Stein D (2016) Sensory modulation disorder symptoms in anorexia nervosa and bulimia nervosa: a pilot study. Int J Eat Disord 49(1):59–68
Breithaupt L, Kahn DL, Slattery M et al (2022) Eighteen-month course and outcome of adolescent restrictive eating disorders: persistence, crossover, and recovery. J Clin Child Adolesc Psychol 51(5):715–725
Brown TA, Berner LA, Jones MD et al (2017) Psychometric evaluation and norms for the multidimensional assessment of interoceptive awareness (MAIA) in a clinical eating disorders sample. Eur Eat Disord Rev 25(5):411–416. https://doi.org/10.1002/erv.2532
Brown TA, Shott ME, Frank GK (2021a) Body size overestimation in anorexia nervosa: contributions of cognitive, affective, tactile and visual information. Psychiatry Res 297:113705
Brown TA, Reilly EE, Murray HB, Perry TR, Kaye WH, Wierenga CE (2021b) Validating the visceral sensitivity index in an eating disorder sample. Int J Eat Disord 54(6):986–994. https://doi.org/10.1002/eat.23471
Brown TA, Perry TR, Kaye WH, Wierenga CE (2022) Pilot study of a water load test as a measure of gastric interoception in anorexia nervosa. Eat Weight Disord Stud Anorex Bulim Obes. 27(6): 2223–2228. https://doi.org/10.1007/s40519-022-01376-9
Carey M, Preston C (2019) Investigating the components of body image disturbance within eating disorders. Front Psych 10:635

Case LK, Wilson RC, Ramachandran VS (2012) Diminished size–weight illusion in anorexia nervosa: evidence for visuo-proprioceptive integration deficit. Exp Brain Res 217:79–87

Cash TF, Deagle EA III (1997) The nature and extent of body-image disturbances in anorexia nervosa and bulimia nervosa: a meta-analysis. Int J Eat Disord 22(2):107–126

Casper RC, Kirschner B, Sandstead HH, Jacob RA, Davis JM (1980) An evaluation of trace metals, vitamins, and taste function in anorexia nervosa. Am J Clin Nutr 33(8):1801–1808. https://doi.org/10.1093/ajcn/33.8.1801

Cermak SA, Curtin C, Bandini LG (2010) Food selectivity and sensory sensitivity in children with autism spectrum disorders. J Am Diet Assoc 110(2):238–246

Chandrashekar J, Hoon MA, Ryba NJP, Zuker CS (2006) The receptors and cells for mammalian taste. Nature 444(7117):288–294. https://doi.org/10.1038/nature05401

Chao AM, Roy A, Franks AT, Joseph PV (2020) A systematic review of taste differences among people with eating disorders. Biol Res Nurs 22(1):82–91

Chaudhari N, Roper SD (2021) Taste: from peripheral receptors to perception. Curr Opin Physio 21:44–47. https://doi.org/10.1016/j.cophys.2021.04.001

Chen WG, Schloesser D, Arensdorf AM et al (2021) The emerging science of interoception: sensing, integrating, interpreting, and regulating signals within the self. Trends Neurosci 44(1):3–16

Chen YP, Ding Z, Yu Y et al (2023) Recent advances in investigating odor-taste interactions: psychophysics, neuroscience, and microfluidic techniques. Trends Food Sci Technol 138:500–510. https://doi.org/10.1016/j.tifs.2023.06.019

Christensen KA, Haynos AF (2020) A theoretical review of interpersonal emotion regulation in eating disorders: enhancing knowledge by bridging interpersonal and affective dysfunction. J Eat Disord 8(1):1–10

Claes L, Soenens B, Vansteenkiste M, Vandereycken W (2012) The scars of the inner critic: perfectionism and nonsuicidal self-injury in eating disorders. Eur Eat Disord Rev 20(3): 196–202. https://doi.org/10.1002/erv.1158

Coren S. Sensation and perception. Published online 2013

Cornelissen KK, Widdrington H, McCarty K, Pollet TV, Tovée MJ, Cornelissen PL (2019) Are attitudinal and perceptual body image the same or different? Evidence from high-level adaptation. Body Image 31:35–47

Corno G, Serino S, Cipresso P, Baños RM, Riva G (2018) Assessing the relationship between attitudinal and perceptual component of body image disturbance using virtual reality. Cyberpsychology Behav Soc Netw 21(11):679–686

Craske MG, Wolitzky-Taylor KB, Labus J et al (2011) A cognitive-behavioral treatment for irritable bowel syndrome using interoceptive exposure to visceral sensations. Behav Res Ther 49(6–7):413–421

Critchley HD, Garfinkel SN (2017) Interoception and emotion. Curr Opin Psychol 17:7–14

Crucianelli L, Cardi V, Treasure J, Jenkinson PM, Fotopoulou A (2016) The perception of affective touch in anorexia nervosa. Psychiatry Res 239:72–78

Crucianelli L, Demartini B, Goeta D et al (2021) The anticipation and perception of affective touch in women with and recovered from anorexia nervosa. Neuroscience 464:143–155

Culbert KM, Racine SE, Klump KL (2015) Research review: what we have learned about the causes of eating disorders – a synthesis of sociocultural, psychological, and biological research. J Child Psychol Psychiatry 56(11):1141–1164. https://doi.org/10.1111/jcpp.12441

Cusack CE, Ralph-Nearman C, Nicholas JK, Levinson CA (2022) New directions in research on somatic concerns in individuals with eating disorders. New Ideas Psychol 66:100937. https://doi.org/10.1016/j.newideapsych.2022.100937

Dakanalis A, Gaudio S, Serino S, Clerici M, Carrà G, Riva G (2016) Body-image distortion in anorexia nervosa. Nat Rev Dis Primer. 2(1):1–2

Dang J, King KM, Inzlicht M (2020) Why are self-report and behavioral measures weakly correlated? Trends Cogn Sci 24(4):267–269

Datta N, Bidopia T, Datta S et al (2021) Internal states and interoception along a spectrum of eating disorder symptomology. Physiol Behav 230:113307. https://doi.org/10.1016/j.physbeh.2020.113307

Davidovic M, Karjalainen L, Starck G, Wentz E, Björnsdotter M, Olausson H (2018) Abnormal brain processing of gentle touch in anorexia nervosa. Psychiatry Res Neuroimaging 281:53–60

Dazzi F, Nitto SD, Zambetti G, Loriedo C, Ciofalo A (2013) Alterations of the olfactory-gustatory functions in patients with eating disorders: olfactory-gustatory alterations in ED patients. Eur Eat Disord Rev 21(5):382–385. https://doi.org/10.1002/erv.2238

De la Cruz F, Teed AR, Lapidus RC et al (2023) Central autonomic network alterations in anorexia nervosa following peripheral adrenergic stimulation. Biol Psychiatry Cogn Neurosci Neuroimaging 8(7):720–730

De Zwaan M, Biener D, Schneider C, Stacher G (1996a) Relationship between thresholds to thermally and to mechanically induced pain in patients with eating disorders and healthy subjects. Pain 67(2):511–512. https://doi.org/10.1016/0304-3959(96)03143-0

De Zwaan M, Biener D, Bach M, Wiesnagrotzki S, Stacher G (1996b) Pain sensitivity, alexithymia, and depression in patients with eating disorders: are they related? J Psychosom Res 41(1): 65–70. https://doi.org/10.1016/0022-3999(96)00088-8

DeGuzman M, Shott ME, Yang TT, Riederer J, Frank GKW (2017) Association of Elevated Reward Prediction Error Response with Weight Gain in adolescent anorexia nervosa. Am J Psychiatry 174(6):557–565. https://doi.org/10.1176/appi.ajp.2016.16060671

Dovey TM, Kumari V, Blissett J (2019) Eating behaviour, behavioural problems and sensory profiles of children with avoidant/restrictive food intake disorder (ARFID), autistic spectrum disorders or picky eating: same or different? Eur Psychiatry 61:56–62

Drewnowski A, Bellisle F, Aimez P, Remy B (1987) Taste and bulimia. Physiol Behav 41(6): 621–626

Drewnowski A, Pierce B, Halmi KA (1988) Fat aversion in eating disorders. Appetite 10(2): 119–131

Eddy KT, Dorer DJ, Franko DL, Tahilani K, Thompson-Brenner H, Herzog DB (2008) Diagnostic crossover in anorexia nervosa and bulimia nervosa: implications for DSM-V. Am J Psychiatry 165(2):245–250

Eddy KT, Lawson EA, Meade C et al (2015) Appetite regulatory hormones in women with anorexia nervosa: binge-eating/purging versus restricting type. J Clin Psychiatry 76(01):19–24. https://doi.org/10.4088/JCP.13m08753

Edwards-Duric J, Stevenson RJ, Francis HM (2020) The congruence of interoceptive predictions and hippocampal-related memory. Biol Psychol 155:107929

Eiber R, Berlin I, de Brettes B, Foulon C, Guelfi JD (2002) Hedonic response to sucrose solutions and the fear of weight gain in patients with eating disorders. Psychiatry Res 113(1–2):173–180. https://doi.org/10.1016/S0165-1781(02)00232-9

Eshkevari E, Rieger E, Longo MR, Haggard P, Treasure J (2012) Increased plasticity of the bodily self in eating disorders. Psychol Med 42(4):819–828

Eshkevari E, Rieger E, Musiat P, Treasure J (2014) An investigation of interoceptive sensitivity in eating disorders using a heartbeat detection task and a self-report measure: interoceptive sensitivity in eating disorders. Eur Eat Disord Rev 22(5):383–388. https://doi.org/10.1002/erv.2125

Essick GK, McGlone F, Dancer C et al (2010) Quantitative assessment of pleasant touch. Neurosci Biobehav Rev 34(2):192–203

Fairburn CG (2008) Cognitive behavior therapy and eating disorders. Guilford Press

Faris PL, Raymond NC, De Zwaan M, Howard LA, Eckert ED, Mitchell JE (1992) Nociceptive, but not tactile, thresholds are elevated in bulimia nervosa. Biol Psychiatry 32(5):462–466. https://doi.org/10.1016/0006-3223(92)90134-L

Feczko E, Miranda-Dominguez O, Marr M, Graham AM, Nigg JT, Fair DA (2019) The heterogeneity problem: approaches to identify psychiatric subtypes. Trends Cogn Sci 23(7):584–601

Ferentzi E, Bogdány T, Szabolcs Z, Csala B, Horváth Á, Köteles F (2018) Multichannel investigation of interoception: sensitivity is not a generalizable feature. Front Hum Neurosci 12:223. https://doi.org/10.3389/fnhum.2018.00223

Fernández-Aranda F, Agüera Z, Fernández-García JC et al (2016) Smell–taste dysfunctions in extreme weight/eating conditions: analysis of hormonal and psychological interactions. Endocrine 51(2):256–267. https://doi.org/10.1007/s12020-015-0684-9

Finger TE, Kinnamon SC (2011) Taste isn't just for taste buds anymore. F1000 Biol Rep 3:20. https://doi.org/10.3410/B3-20

Frank GK, Treasure J (2016) Authors' reply: cognitive and emotional factors are involved in body-image distortion. Nat Rev Dis Primer 2(1):1–2

Frank GK, Reynolds JR, Shott ME, O'Reilly RC (2011) Altered temporal difference learning in bulimia nervosa. Biol Psychiatry 70(8):728–735

Frank GK, Shott ME, Hagman JO, Mittal VA (2013a) Taste reward circuitry related brain structures characterize ill and recovered anorexia nervosa and bulimia nervosa. Am J Psychiatry 170(10): 1152. https://doi.org/10.1176/appi.ajp.2013.12101294

Frank GK, Shott ME, Hagman JO, Yang TT (2013b) Localized brain volume and white matter integrity alterations in adolescent anorexia nervosa. J Am Acad Child Adolesc Psychiatry 52(10):1066–1075. e5

Frank GKW, Shott ME, Keffler C, Cornier M (2016a) Extremes of eating are associated with reduced neural taste discrimination. Int J Eat Disord 49(6):603–612. https://doi.org/10.1002/eat.22538

Frank GKW, Shott ME, Riederer J, Pryor TL (2016b) Altered structural and effective connectivity in anorexia and bulimia nervosa in circuits that regulate energy and reward homeostasis. Transl Psychiatry 6(11):e932–e932

Frank GK, Collier S, Shott ME, O'Reilly RC (2016c) Prediction error and somatosensory insula activation in women recovered from anorexia nervosa. J Psychiatry Neurosci 41(5):304–311

Frank GK, Shott ME, Stoddard J, Swindle S, Pryor TL (2021) Association of brain reward response with body mass index and ventral striatal-hypothalamic circuitry among young women with eating disorders. JAMA Psychiatry 78(10):1123–1133

Frank GKW, Shott ME, Pryor T, Swindle S, Stoddard J (2022) Brain reward response in adolescents and young adults with anorexia nervosa is moderated by changes in body weight and sweetness perception. Int J Eat Disord 55(12):1799–1810. https://doi.org/10.1002/eat.23814

Gardner RM, Brown DL (2014) Body size estimation in anorexia nervosa: a brief review of findings from 2003 through 2013. Psychiatry Res 219(3):407–410. https://doi.org/10.1016/j.psychres.2014.06.029

Garfinkel SN, Seth AK, Barrett AB, Suzuki K, Critchley HD (2015) Knowing your own heart: distinguishing interoceptive accuracy from interoceptive awareness. Biol Psychol 104:65–74

Garfinkel SN, Schulz A, Tsakiris M (2022) Addressing the need for new interoceptive methods. Biol Psychol 170:108322. https://doi.org/10.1016/j.biopsycho.2022.108322

Garner DM, Olmstead MP, Polivy J (1983) Development and validation of a multidimensional eating disorder inventory for anorexia nervosa and bulimia. Int J Eat Disord 2(2):15–34

Gaudio S, Quattrocchi CC (2012) Neural basis of a multidimensional model of body image distortion in anorexia nervosa. Neurosci Biobehav Rev 36(8):1839–1847. https://doi.org/10.1016/j.neubiorev.2012.05.003

Girdler SS, Koo-Loeb J, Pedersen CA, Brown HJ, Maixner W (1998) Blood pressure-related hypoalgesia in bulimia nervosa. Psychosom Med 60(6) https://journals.lww.com/psychosomaticmedicine/Fulltext/1998/11000/Blood_Pressure_Related_Hypoalgesia_in_Bulimia.16.aspx

Goldzak-Kunik G, Friedman R, Spitz M, Sandler L, Leshem M (2012) Intact sensory function in anorexia nervosa. Am J Clin Nutr 95(2):272–282. https://doi.org/10.3945/ajcn.111.020131

Grunwald T, Boutros NN, Pezer N et al (2003) Neuronal substrates of sensory gating within the human brain. Biol Psychiatry 53(6):511–519

Hagman J, Gardner RM, Brown DL, Gralla J, Fier JM, Frank GKW (2015) Body size overestimation and its association with body mass index, body dissatisfaction, and drive for thinness in anorexia nervosa. Eat Weight Disord Stud Anorex Bulim Obes. 20(4):449–455. https://doi.org/10.1007/s40519-015-0193-0

Haynos AF, Fruzzetti AE (2011) Anorexia nervosa as a disorder of emotion dysregulation: evidence and treatment implications. Clin Psychol Sci Pract 18(3):183–202. https://doi.org/10.1111/j.1468-2850.2011.01250.x

Hildebrandt T, Peyser D, Sysko R (2021) Lessons learned developing and testing FAMILY-BASED interoceptive exposure for adolescents with l OW-WEIGHT eating disorders. Int J Eat Disord 54(11):2037–2045. https://doi.org/10.1002/eat.23605

Höller I, Stenzel JS, Rath D, Forkmann T (2021) Listen to your heart–ecological momentary assessment of interoceptive accuracy, awareness and sensibility: a pilot study. Int J Environ Res Public Health 18(9):4893

IASP (2020. https://www.iasp-pain.org/resources/terminology/) IASP taxonomy. IASP

Islam MA, Fagundo AB, Arcelus J et al (2015) Olfaction in eating disorders and abnormal eating behavior: a systematic review. Front Psychol:6. https://doi.org/10.3389/fpsyg.2015.01431

Jenkins PE, Hoste RR, Meyer C, Blissett JM (2011) Eating disorders and quality of life: a review of the literature. Clin Psychol Rev 31(1):113–121. https://doi.org/10.1016/j.cpr.2010.08.003

Jenkinson PM, Taylor L, Laws KR (2018) Self-reported interoceptive deficits in eating disorders: a meta-analysis of studies using the eating disorder inventory. J Psychosom Res 110:38–45. https://doi.org/10.1016/j.jpsychores.2018.04.005

Jiang T, Soussignan R, Carrier E, Royet JP (2019) Dysfunction of the mesolimbic circuit to food Odors in women with anorexia and bulimia nervosa: a fMRI study. Front Hum Neurosci 13:117. https://doi.org/10.3389/fnhum.2019.00117

Jirik-Babb P, Katz JL (1988) Impairment of taste perception in anorexia nervosa and bulimia. Int J Eat Disord 7(3):353–360

Karavia A, Kapsali F, Gonidakis F et al (2022) Olfactory capacity in anorexia nervosa: correlations with set-shifting ability. Eat Weight Disord Stud Anorex Bulim Obes. 27(2):535–542. https://doi.org/10.1007/s40519-021-01188-3

Karcher NR, Barch DM (2021) The ABCD study: understanding the development of risk for mental and physical health outcomes. Neuropsychopharmacology 46(1):131–142

Kaye WH, Berrettini WH, Gwirtsman HE et al (1987) Reduced cerebrospinal fluid levels of immunoreactive pro-opiomelanocortin related peptides (including beta-endorphin) in anorexia nervosa. Life Sci 41(18):2147–2155. https://doi.org/10.1016/0024-3205(87)90533-9

Kaye WH, Wierenga CE, Bischoff-Grethe A et al (2020) Neural insensitivity to the effects of hunger in women remitted from anorexia nervosa. Am J Psychiatry 177(7):601–610

Keel PK, Dorer DJ, Franko DL, Jackson SC, Herzog DB (2005) Postremission predictors of relapse in women with eating disorders. Am J Psychiatry 162(12):2263–2268

Keramati M, Gutkin B (2014) Homeostatic reinforcement learning for integrating reward collection and physiological stability. elife 3:e04811

Keramati M, Durand A, Girardeau P, Gutkin B, Ahmed SH (2017) Cocaine addiction as a homeostatic reinforcement learning disorder. Psychol Rev 124(2):130

Kerr KL, Moseman SE, Avery JA, Bodurka J, Zucker NL, Simmons WK (2016) Altered insula activity during visceral interoception in weight-restored patients with anorexia nervosa. Neuropsychopharmacology 41(2):521–528. https://doi.org/10.1038/npp.2015.174

Kerr KL, Moseman SE, Avery JA, Bodurka J, Simmons WK (2017) Influence of visceral interoceptive experience on the Brain's response to food images in anorexia nervosa. Psychosom Med 79(7):777–784. https://doi.org/10.1097/PSY.0000000000000486

Kerr KL, DeVille DC, Moseman SE, Simmons WK (2022) Aberrant gastric functioning in weight-restored but not acute restricting-type anorexia nervosa: an electrogastrography study. Int J Eat Disord 55(10):1384–1389. https://doi.org/10.1002/eat.23802

Khalsa SS, Lapidus RC (2016) Can interoception improve the pragmatic search for biomarkers in psychiatry? Front Psych 7:121

Khalsa SS, Craske MG, Li W, Vangala S, Strober M, Feusner JD (2015) Altered interoceptive awareness in anorexia nervosa: effects of meal anticipation, consumption and bodily arousal: interoception in anorexia nervosa. Int J Eat Disord 48(7):889–897. https://doi.org/10.1002/eat.22387

Khalsa SS, Adolphs R, Cameron OG et al (2018a) Interoception and mental health: a roadmap. Biol Psychiatry Cogn Neurosci Neuroimaging 3(6):501–513. https://doi.org/10.1016/j.bpsc.2017.12.004

Khalsa SS, Hassanpour MS, Strober M, Craske MG, Arevian AC, Feusner JD (2018b) Interoceptive anxiety and body representation in anorexia nervosa. Front Psych 9:444. https://doi.org/10.3389/fpsyt.2018.00444

Khalsa SS, Berner LA, Anderson LM (2022) Gastrointestinal interoception in eating disorders: charting a new path. Curr Psychiatry Rep 24(1):47–60. https://doi.org/10.1007/s11920-022-01318-3

Kinnaird E, Stewart C, Tchanturia K (2018) Taste sensitivity in anorexia nervosa: a systematic review. Int J Eat Disord 51(8):771–784

Kinnear A, Withnell SJ, Witte TK, Smith AR, Szczyglowski K, Bodell LP (2021) Weight misperception and its associations with eating disorder symptoms over the course of residential eating disorder treatment. Int J Eat Disord 54(10):1810–1818

Kinner A, Rosenthal K, Lütz S (2021) Identification and expression of new unspecific peroxygenases – recent advances, challenges and opportunities. Front Bioeng Biotechnol 9: 705630. https://doi.org/10.3389/fbioe.2021.705630

Kitajima T, Otani R, Inoue T, Matsushima N, Matsubara N, Sakuta R (2022) Sensory processing in children and adolescents shortly after the onset of anorexia nervosa: a pilot study. Biopsychosoc Med 16(1):27

Klabunde M, Acheson DT, Boutelle KN, Matthews SC, Kaye WH (2013) Interoceptive sensitivity deficits in women recovered from bulimia nervosa. Eat Behav 14(4):488–492. https://doi.org/10.1016/j.eatbeh.2013.08.002

Klabunde M, Collado D, Bohon C (2017) An interoceptive model of bulimia nervosa: A neurobiological systematic review. J Psychiatr Res 94:36–46. https://doi.org/10.1016/j.jpsychires.2017.06.009

Klein DA, Schebendach JE, Brown AJ, Smith GP, Walsh BT (2009) Modified sham feeding of sweet solutions in women with and without bulimia nervosa. Physiol Behav 96(1):44–50

Kobal G, Hummel TH, Sekinger B, Barz S, Roscher S, Wolf S (1996) "Sniffin'sticks": screening of olfactory performance. Rhinology 34(4):222–226

Kot E, Kucharska K, Monteleone AM, Monteleone P (2020) Structural and functional brain correlates of altered taste processing in anorexia nervosa: a systematic review. Eur Eat Disord Rev 28(2):122–140

Krieg JC, Roscher S, Strian F, Pirke KM, Lautenbacher S (1993) Pain sensitivity in recovered anorexics, restrained and unrestrained eaters. J Psychosom Res 37(6):595–601. https://doi.org/10.1016/0022-3999(93)90054-J

Lacey JH, Stanley PA, Crutchfield M, Crisp AH (1977) Sucrose sensitivity in anorexia nervosa. J Psychosom Res 21(1):17–21. https://doi.org/10.1016/0022-3999(77)90021-6

Lapidus RC, Puhl M, Kuplicki R et al (2020) Heightened affective response to perturbation of respiratory but not pain signals in eating, mood, and anxiety disorders. Vickers K, ed. PLoS One 15(7):e0235346. https://doi.org/10.1371/journal.pone.0235346

Lautenbacher S, Pauls AM, Strian F, Pirke KM, Krieg JC (1990) Pain perception in patients with eating disorders. Psychosom Med 52(6) https://journals.lww.com/psychosomaticmedicine/Fulltext/1990/11000/Pain_perception_in_patients_with_eating_disorders_.8.aspx

Lautenbacher S, Pauls AM, Strian F, Pirke KM, Krieg JC (1991) Pain sensitivity in anorexia nervosa and bulimia nervosa. Biol Psychiatry 29(11):1073–1078

Lavender JM, Wonderlich SA, Engel SG, Gordon KH, Kaye WH, Mitchell JE (2015) Dimensions of emotion dysregulation in anorexia nervosa and bulimia nervosa: a conceptual review of the empirical literature. Clin Psychol Rev 40:111–122. https://doi.org/10.1016/j.cpr.2015.05.010

Leland EM, Xie DX, Kamath V, Seal SM, Lin SY, Rowan NR (2022) Psychophysical chemosensory dysfunction in eating disorders: a qualitative systematic review. Eat Weight Disord Stud Anorex Bulim Obes 27(2):429–447. https://doi.org/10.1007/s40519-021-01189-2

Li W, Lai TM, Bohon C et al (2015) Anorexia nervosa and body dysmorphic disorder are associated with abnormalities in processing visual information. Psychol Med 45(10):2111–2122. https://doi.org/10.1017/S0033291715000045

Linardon J, Wade TD (2018) How many individuals achieve symptom abstinence following psychological treatments for bulimia nervosa? A meta-analytic review. Int J Eat Disord 51(4): 287–294. https://doi.org/10.1002/eat.22838

Lopata M, La Fata J, Evanich MJ, Lourenço RV (1977) Effects of flow-resistive loading on mouth occlusion pressure during CO2 rebreathing. Am Rev Respir Dis 115(1):73–81. https://doi.org/10.1164/arrd.1977.115.1.73

Lutz APC, Schulz A, Voderholzer U, Koch S, van Dyck Z, Vögele C (2019) Enhanced cortical processing of cardio-afferent signals in anorexia nervosa. Clin Neurophysiol 130(9): 1620–1627. https://doi.org/10.1016/j.clinph.2019.06.009

Mai Y, Zhang X, Li Z et al (2020) Olfaction is a marker of severity but not diagnosis in anorexia nervosa: a systematic review and meta-analysis. Neuropsychol Rev 30(2):251–266. https://doi.org/10.1007/s11065-020-09438-4

Mason TB, Lesser EL, Dolgon-Krutolow AR, Wonderlich SA, Smith KE (2021) An updated transdiagnostic review of social cognition and eating disorder psychopathology. J Psychiatr Res 143:602–627

Nozoe S-I, Masuda A, Naruo T, Soejima Y, Nagai N, Tanaka H (1996) Changes in taste responsiveness in patients with anorexia nervosa during behavior therapy. Physiol Behav 59(3):549–553. https://doi.org/10.1016/0031-9384(95)02105-1

Mather G (2016) Foundations of sensation and perception. Psychology Press

Mehak A, Racine SE (2020) Understanding "feeling fat" and its underlying mechanisms: the importance of multimethod measurement. Int J Eat Disord 53(9):1400–1404

Mehling WE, Price C, Daubenmier JJ, Acree M, Bartmess E, Stewart A (2012) The multidimensional assessment of interoceptive awareness (MAIA). PLoS One 7(11):e48230

Mehling WE, Acree M, Stewart A, Silas J, Jones A (2018) The multidimensional assessment of interoceptive awareness, version 2 (MAIA-2). PLoS One 13(12):e0208034

Merwin RM, Moskovich AA, Wagner HR, Ritschel LA, Craighead LW, Zucker NL (2013) Emotion regulation difficulties in anorexia nervosa: relationship to self-perceived sensory sensitivity. Cogn Emot 27(3):441–452

Milos G, Kuenzli C, Soelch CM, Schumacher S, Moergeli H, Mueller-Pfeiffer C (2013) How much should I eat? Estimation of meal portions in anorexia nervosa. Appetite 63:42–47

Mölbert SC, Klein L, Thaler A et al (2017) Depictive and metric body size estimation in anorexia nervosa and bulimia nervosa: a systematic review and meta-analysis. Clin Psychol Rev 57:21–31. https://doi.org/10.1016/j.cpr.2017.08.005

Moody TD, Sasaki MA, Bohon C et al (2015) Functional connectivity for face processing in individuals with body dysmorphic disorder and anorexia nervosa. Psychol Med 45(16): 3491–3503. https://doi.org/10.1017/S0033291715001397

Moody TD, Morfini F, Cheng G et al (2021) Brain activation and connectivity in anorexia nervosa and body dysmorphic disorder when viewing bodies: relationships to clinical symptoms and perception of appearance. Brain Imaging Behav 15(3):1235–1252. https://doi.org/10.1007/s11682-020-00323-5

Morales C, Dolan SC, Anderson DA, Anderson LM, Reilly EE (2022) Exploring the contributions of affective constructs and interoceptive awareness to feeling fat. Eat Weight Disord-Stud Anorex Bulim Obes 27(8):3533–3541

Murphy J, Brewer R, Plans D, Khalsa SS, Catmur C, Bird G (2018) Testing the Independence of self-reported interoceptive accuracy and attention. PsyArXiv. https://doi.org/10.31234/osf.io/fqgu4

Nagao-Sato S, Kawasaki Y, Akamatsu R et al (2023) Sensory profile of picky eaters among college female students. Appetite 185:106518. https://doi.org/10.1016/j.appet.2023.106518
Nakai Y, Kinoshita F, Koh T, Tsujii S, Tsukada T (1987) Taste function in patients with anorexia nervosa and bulimia nervosa. Int J Eat Disord 6(2):257–265
Nicholas JK, van Tilburg MA, Pilato I et al (2021) The diagnosis of avoidant restrictive food intake disorder in the presence of gastrointestinal disorders: opportunities to define shared mechanisms of symptom expression. Int J Eat Disord 54(6):995–1008
Oberndorfer TA, Frank GKW, Simmons AN et al (2013a) Altered insula response to sweet taste processing after recovery from anorexia and bulimia nervosa. Am J Psychiatry 170(10): 1143–1151. https://doi.org/10.1176/appi.ajp.2013.11111745
Oberndorfer TA, Frank GK, Simmons AN et al (2013b) Altered insula response to sweet taste processing after recovery from anorexia and bulimia nervosa. Am J Psychiatry 170(10): 1143–1151
Owens AP, Allen M, Ondobaka S, Friston KJ (2018) Interoceptive inference: From computational neuroscience to clinic. Neurosci Biobehav Rev 90:174–183. https://doi.org/10.1016/j.neubiorev.2018.04.017
Papežová H, Yamamotova A, Uher R (2005) Elevated pain threshold in eating disorders: physiological and psychological factors. J Psychiatr Res 39(4):431–438
Paulus MP, Feinstein JS, Khalsa SS (2019) An active inference approach to interoceptive psychopathology. Annu Rev Clin Psychol 15:97–122
Pecher D, Zwaan RA (2005) Grounding cognition: the role of perception and action in memory, language, and thinking. Cambridge University Press
Peebles R, Sieke EH (2019) Medical complications of eating disorders in youth. Child Adolesc Psychiatr Clin N Am 28(4):593–615. https://doi.org/10.1016/j.chc.2019.05.009
Petzschner FH, Garfinkel SN, Paulus MP, Koch C, Khalsa SS (2021) Computational models of interoception and body regulation. Trends Neurosci 44(1):63–76
Pleger B, Villringer A (2013) The human somatosensory system: from perception to decision making. Prog Neurobiol 103:76–97
Pollatos O, Georgiou E (2016) Normal interoceptive accuracy in women with bulimia nervosa. Psychiatry Res 240:328–332. https://doi.org/10.1016/j.psychres.2016.04.072
Pollatos O, Herbert BM (2018) Interoception: Definitions, dimensions, neural substrates. Embodiment Psychother Pract Guide. Published online:15–27
Pollatos O, Kurz AL, Albrecht J et al (2008) Reduced perception of bodily signals in anorexia nervosa. Eat Behav 9(4):381–388. https://doi.org/10.1016/j.eatbeh.2008.02.001
Pollatos O, Herbert BM, Berberich G, Zaudig M, Krauseneck T, Tsakiris M (2016) Atypical self-focus effect on interoceptive accuracy in anorexia nervosa. Front Hum Neurosci:10. https://doi.org/10.3389/fnhum.2016.00484
Pompoli A, Furukawa TA, Efthimiou O, Imai H, Tajika A, Salanti G (2018) Dismantling cognitive-behaviour therapy for panic disorder: a systematic review and component network meta-analysis. Psychol Med 48(12):1945–1953
Preston C, Ehrsson HH (2016) Illusory obesity triggers body dissatisfaction responses in the insula and anterior cingulate cortex. Cereb Cortex 26(12):4450–4460. https://doi.org/10.1093/cercor/bhw313
Proff I, Williams GL, Quadt L, Garfinkel SN (2022) Sensory processing in autism across exteroceptive and interoceptive domains. Psychol Neurosci 15(2):105–130. https://doi.org/10.1037/pne0000262
Raja SN, Carr DB, Cohen M et al (2020) The revised IASP definition of pain: concepts, challenges, and compromises. Pain 161(9):1976
Ralph-Nearman C, Arevian AC, Puhl M et al (2019) A novel Mobile tool (Somatomap) to assess body image perception pilot tested with fashion models and nonmodels: cross-sectional study. JMIR Ment Health 6(10):e14115. https://doi.org/10.2196/14115
Raymond NC, de Zwaan M, Faris PL et al (1995) Pain thresholds in obese binge-eating disorder subjects. Biol Psychiatry 37(3):202–204

Raymond NC, Faris PL, Thuras PD et al (1999a) Elevated pain threshold in anorexia nervosa subjects. Biol Psychiatry 45(10):1389–1392

Raymond NC, Eckert ED, Hamalainen M et al (1999b) A preliminary report on pain thresholds in bulimia nervosa during a bulimic episode. Compr Psychiatry 40(3):229–233. https://doi.org/10.1016/S0010-440X(99)90009-3

Riva G, Gaudio S (2012) Allocentric lock in anorexia nervosa: new evidences from neuroimaging studies. Med Hypotheses 79(1):113–117

Riva G, Gaudio S (2018) Locked to a wrong body: eating disorders as the outcome of a primary disturbance in multisensory body integration. Conscious Cogn 59:57–59

Rodin J, Bartoshuk L, Peterson C, Schank D (1990) Bulimia and taste: possible interactions. J Abnorm Psychol 99(1):32–39. https://doi.org/10.1037/0021-843X.99.1.32

Roessner V, Bleich S, Banaschewski T, Rothenberger A (2005) Olfactory deficits in anorexia nervosa. Eur Arch Psychiatry Clin Neurosci 255(1):6–9. https://doi.org/10.1007/s00406-004-0525-y

Santomauro DF, Melen S, Mitchison D, Vos T, Whiteford H, Ferrari AJ (2021) The hidden burden of eating disorders: an extension of estimates from the global burden of disease study 2019. Lancet Psychiatry 8(4):320–328

Saramandi A, Crucianelli L, Koukoutsakis A et al (2022) Belief updating about interoception and body size estimation in anorexia nervosa. PsyArXiv. https://doi.org/10.31234/osf.io/rntsf

Schandry R (1981) Heart beat perception and emotional experience. Psychophysiology 18(4):483–488

Schmidt U, Treasure J (2006) Anorexia nervosa: valued and visible. A cognitive-interpersonal maintenance model and its implications for research and practice. Br J Clin Psychol 45(3):343–366. https://doi.org/10.1348/014466505X53902

Sinding C, Saint-Eve A, Thomas-Danguin T (2023) Multimodal sensory interactions. In: Flavor. Elsevier, pp 205–231. https://doi.org/10.1016/B978-0-323-89903-1.00001-3

Small DM (2006) Central gustatory processing in humans. Taste Smell 63:191–220

Smith AR, Forrest LN, Duffy ME, Jones PJ, Joiner TE, Pisetsky EM (2020a) Identifying bridge pathways between eating disorder symptoms and suicidal ideation across three samples. J Abnorm Psychol 129(7):724–736. https://doi.org/10.1037/abn0000553

Smith R, Kuplicki R, Feinstein J et al (2020b) A Bayesian computational model reveals a failure to adapt interoceptive precision estimates across depression, anxiety, eating, and substance use disorders. Gershman SJ, ed. PLoS Comput Biol 16(12):e1008484. https://doi.org/10.1371/journal.pcbi.1008484

Smith R, Badcock P, Friston KJ (2021) Recent advances in the application of predictive coding and active inference models within clinical neuroscience. Psychiatry Clin Neurosci 75(1):3–13

Solmi M, Radua J, Stubbs B et al (2020) Risk factors for eating disorders: an umbrella review of published meta-analyses. Braz J Psychiatry 43:314–323

Sorokowska A, Chabin D, Hummel T, Karwowski M (2022) Olfactory perception relates to food neophobia in adolescence. Nutrition 98:111618. https://doi.org/10.1016/j.nut.2022.111618

Spence C (2022) The tongue map and the spatial modulation of taste perception. Curr Res Food Sci 5:598–610. https://doi.org/10.1016/j.crfs.2022.02.004

Stein D, Kaye WH, Matsunaga H et al (2003) Pain perception in recovered bulimia nervosa patients. Int J Eat Disord 34(3):331–336. https://doi.org/10.1002/eat.10164

Stein D, Gross-Isseroff R, Besserglick R et al (2012) Olfactory function and alternation learning in eating disorders. Eur Neuropsychopharmacol 22(9):615–624. https://doi.org/10.1016/j.euroneuro.2011.12.006

Stern SA, Bulik CM (2020) Alternative frameworks for advancing the study of eating disorders. Trends Neurosci 43(12):951–959

Steward T, Menchon JM, Jiménez-Murcia S, Soriano-Mas C, Fernandez-Aranda F (2018) Neural network alterations across eating disorders: a narrative review of fMRI studies. Curr Neuropharmacol 16(8):1150–1163

Streatfeild J, Hickson J, Austin SB et al (2021) Social and economic cost of eating disorders in the United States: evidence to inform policy action. Int J Eat Disord 54(5):851–868. https://doi.org/10.1002/eat.23486

Strigo IA, Matthews SC, Simmons AN et al (2013) Altered insula activation during pain anticipation in individuals recovered from anorexia nervosa: evidence of interoceptive dysregulation. Int J Eat Disord 46(1):23–33. https://doi.org/10.1002/eat.22045

Stull MR, Harshman SG, Kuhnle M et al (2021) Olfactory performance in youth with full and subthreshold avoidant/restrictive food intake disorder. J Endocr Soc 5(Supplement_1):A630–A631. https://doi.org/10.1210/jendso/bvab048.1285

Suksasilp C, Garfinkel SN (2022) Towards a comprehensive assessment of interoception in a multi-dimensional framework. Biol Psychol 168:108262. https://doi.org/10.1016/j.biopsycho.2022.108262

Sunday SR, Halmi KA (1990) Taste perceptions and hedonics in eating disorders. Physiol Behav 48(5):587–594

Szalay C, Ábrahám I, Papp S et al (2010) Taste reactivity deficit in anorexia nervosa: taste reactivity and anorexia nervosa. Psychiatry Clin Neurosci 64(4):403–407. https://doi.org/10.1111/j.1440-1819.2010.02106.x

Thomas JJ, Lawson EA, Micali N, Misra M, Deckersbach T, Eddy KT (2017a) Avoidant/restrictive food intake disorder: a three-dimensional model of neurobiology with implications for Etiology and treatment. Curr Psychiatry Rep 19(8):54. https://doi.org/10.1007/s11920-017-0795-5.

Thomas JJ, Lawson EA, Micali N, Misra M, Deckersbach T, Eddy KT (2017b) Avoidant/restrictive food intake disorder: a three-dimensional model of neurobiology with implications for Etiology and treatment. Curr Psychiatry Rep 19(8):54. https://doi.org/10.1007/s11920-017-0795-5

van Dyck Z, Schulz A, Blechert J, Herbert BM, Lutz APC, Vögele C (2021) Gastric interoception and gastric myoelectrical activity in bulimia nervosa and binge-eating disorder. Int J Eat Disord 54(7):1106–1115. https://doi.org/10.1002/eat.23291

Van Orden KA, Witte TK, Cukrowicz KC, Braithwaite SR, Selby EA, Joiner TE Jr (2010) The interpersonal theory of suicide. Psychol Rev 117(2):575

Velkoff EA, Smith AR (2023) Feeling the urge: using ecological momentary assessment to test the relationship of interoception with disordered eating and nonsuicidal self-injury. J Psychopathol Clin Sci 132(1):78

Via E, Goldberg X, Sánchez I et al (2018) Self and other body perception in anorexia nervosa: the role of posterior DMN nodes. World J Biol Psychiatry 19(3):210–224

Vocks S, Tuschen-Caffier B, Pietrowsky R, Rustenbach SJ, Kersting A, Herpertz S (2010) Meta-analysis of the effectiveness of psychological and pharmacological treatments for binge eating disorder. Int J Eat Disord 43(3):205–217

Wagner A, Aizenstein H, Mazurkewicz L et al (2008) Altered insula response to taste stimuli in individuals recovered from restricting-type anorexia nervosa. Neuropsychopharmacology 33(3):513–523

Wagner A, Simmons AN, Oberndorfer TA et al (2015) Altered sensitization patterns to sweet food stimuli in patients recovered from anorexia and bulimia nervosa. Psychiatry Res Neuroimaging 234(3):305–313

Weiland R, Macht M, Ellgring H, Groß-Lesch S, Lesch KP, Pauli P (2011) Olfactory and gustatory sensitivity in adults with attention-deficit/hyperactivity disorder. ADHD Atten Deficit Hyperact Disord 3(1):53–60. https://doi.org/10.1007/s12402-010-0052-9

Wierenga CE, Bischoff-Grethe A, Berner LA et al (2020) Increased anticipatory brain response to pleasant touch in women remitted from bulimia nervosa. Transl Psychiatry 10(1):236

Wildes JE, Zucker NL, Marcus MD (2012) Picky eating in adults: results of a web-based survey. Int J Eat Disord 45(4):575–582

Williams LM (2016) Precision psychiatry: a neural circuit taxonomy for depression and anxiety. Lancet Psychiatry 3(5):472–480. https://doi.org/10.1016/S2215-0366(15)00579-9

Witte TK, Zuromski KL, Gauthier JM et al (2016) Restrictive eating: associated with suicide attempts, but not acquired capability in residential patients with eating disorders. Psychiatry Res 235:90–96. https://doi.org/10.1016/j.psychres.2015.11.043

Wolff M, Morceau S, Folkard R, Martin-Cortecero J, Groh A (2021) A thalamic bridge from sensory perception to cognition. Neurosci Biobehav Rev 120:222–235

Wolz I, Sauvaget A, Granero R et al (2017) Subjective craving and event-related brain response to olfactory and visual chocolate cues in binge-eating and healthy individuals. Sci Rep:7. https://doi.org/10.1038/srep41736

Yamamotova A, Bulant J, Bocek V, Papezova H (2017) Dissatisfaction with own body makes patients with eating disorders more sensitive to pain. J Pain Res 10:1667–1675. https://doi.org/10.2147/JPR.S133425

Yellowlees R, Dingemans AE, Veldhuis J, de Vaate AB (2019) Face yourself (ie): investigating selfie-behavior in females with severe eating disorder symptoms. Comput Hum Behav 101:77–83

Zickgraf HF, Richard E, Zucker NL, Wallace GL (2022) Rigidity and sensory sensitivity: independent contributions to selective eating in children, adolescents, and young adults. J Clin Child Adolesc Psychol 51(5):675–687. https://doi.org/10.1080/15374416.2020.1738236

Zopf R, Contini E, Fowler C, Mondraty N, Williams MA (2016) Body distortions in anorexia nervosa: evidence for changed processing of multisensory bodily signals. Psychiatry Res 245: 473–481

Zucker NL, Bulik CM (2020) On bells, saliva, and abdominal pain or discomfort: early aversive visceral conditioning and vulnerability for anorexia nervosa. Int J Eat Disord 53(4):508–512. https://doi.org/10.1002/eat.23255

Zucker NL, Merwin RM, Bulik CM, Moskovich A, Wildes JE, Groh J (2013) Subjective experience of sensation in anorexia nervosa. Behav Res Ther 51(6):256–265

Zucker NL, LaVia MC, Craske MG et al (2019) Feeling and body investigators (FBI): ARFID division – an acceptance-based interoceptive exposure treatment for children with ARFID. Int J Eat Disord 52(4):466–472. https://doi.org/10.1002/eat.22996

Emerging Theories of Allostatic-Interoceptive Overload in Neurodegeneration

Daniel Franco-O'Byrne, Hernando Santamaría-García, Joaquín Migeot, and Agustín Ibáñez

Contents

D. Franco-O'Byrne and J. Migeot
Latin American Brain Health Institute (BrainLat), Universidad Adolfo Ibáñez, Santiago, Chile

Center for Social and Cognitive Neuroscience (CSCN), School of Psychology, Universidad Adolfo Ibáñez, Santiago, Chile

H. Santamaría-García
Global Brain Health Institute, University of California-San Francisco, San Francisco, CA, USA

Trinity College Dublin, Dublin, Ireland

Department of Psychiatry, Pontificia Universidad Javeriana, Bogotá, Colombia

Center of Memory and Cognition Intellectus, Hospital Universitario San Ignacio, Bogotá, Colombia

A. Ibáñez (✉)
Latin American Brain Health Institute (BrainLat), Universidad Adolfo Ibáñez, Santiago, Chile

Global Brain Health Institute, University of California-San Francisco, San Francisco, CA, USA

Trinity College Dublin, Dublin, Ireland

Cognitive Neuroscience Center (CNC), Universidad de San Andrés, Buenos Aires, Argentina

Trinity College Institute of Neuroscience (TCIN), Trinity College Dublin, Dublin, Ireland
e-mail: agustin.ibanez@gbhi.org

Curr Topics Behav Neurosci (2025) 74: 191–216
https://doi.org/10.1007/7854_2024_471
Published Online: 19 April 2024

Abstract Recent integrative multilevel models offer novel insights into the etiology and course of neurodegenerative conditions. The predictive coding of allostatic-interoception theory posits that the brain adapts to environmental demands by modulating internal bodily signals through the allostatic-interoceptive system. Specifically, a domain-general allostatic-interoceptive network exerts adaptive physiological control by fine-tuning initial top-down predictions and bottom-up peripheral signaling. In this context, adequate adaptation implies the minimization of prediction errors thereby optimizing energy expenditure. Abnormalities in top-down interoceptive predictions or peripheral signaling can trigger allostatic overload states, ultimately leading to dysregulated interoceptive and bodily systems (endocrine, immunological, circulatory, etc.). In this context, environmental stress, social determinants of health, and harmful exposomes (i.e., the cumulative life-course exposition to different environmental stressors) may interact with physiological and genetic factors, dysregulating allostatic interoception and precipitating neurodegenerative processes. We review the allostatic-interoceptive overload framework across different neurodegenerative diseases, particularly in the behavioral variant frontotemporal dementia (bvFTD). We describe how concepts of allostasis and interoception could be integrated with principles of predictive coding to explain how the brain optimizes adaptive responses, while maintaining physiological stability through feedback loops with multiple organismic systems. Then, we introduce the model of allostatic-interoceptive overload of bvFTD and discuss its implications for the understanding of pathophysiological and neurocognitive abnormalities in multiple neurodegenerative conditions.

Keywords Allostatic-interoceptive framework · Allostatic overload · bvFTD · Neurodegeneration · Predictive coding

1 Introduction

Classical clinical neuroscience approaches to dementias have contributed greatly to understanding the neurocognitive and behavioral characteristics across neurodegenerative spectra (Birba et al. 2022a; Chiu et al. 2006; Díaz-Rivera et al. 2023). However, various challenges remain on the complexity of the brain structure-function links, the difficulties of establishing one-to-one mappings between cognition and brain dynamics, the multiplicity of mechanisms yielding the same function and dysfunctions at multiple levels, and the heterogeneous brain phenotypes across neurodegenerative diseases (Anusha-Kiran et al. 2022; Kamaleddin 2022; Khachaturian et al. 2023; Li et al. 2022; Seenivasan and Narayanan 2022; Toledo et al. 2022). Another major limitation comes from the lack of integrative models to approach the study of brain activity, cognition, and disease in a multilevel system sustained by complex and interconnected mechanisms. Some solutions to these challenges have been proposed via synergistic frameworks offering insights focused on brain-behavior and brain-phenotype dynamics (Ibanez 2022). Here, we discuss one novel framework aiming to overcome those challenges and better understand dementia: the allostatic-interoceptive predictive coding model (Birba et al. 2022b; Kleckner et al. 2017; Migeot et al. 2022).

We first describe the allostatic and interoceptive processes and discuss how the integration of these domains with predictive coding theories has resulted in novel brain-body-world models (Seth et al. 2012; Seth and Friston 2016; Tsakiris 2017). Then we describe how these models integrate the dynamic interplay between neurophysiological, autonomic, and interoceptive domains and the environment. We explain how structural and functional hierarchies within the nervous system (particularly in allostatic-interoceptive brain circuits such as the cingulate cortex, amygdala, and insula) (Migeot et al. 2022, 2023) achieve equilibrium through the modulation of visceromotor and peripheral bodily states to face environmental demands, a process termed allostatic interoception. We also describe how environmental demands, multisystemic dysregulation, and changes in brain structure and function can explain multilevel pathological manifestations in neurodegenerative diseases (Birba et al. 2022b; De Felice et al. 2022; Martin et al. 2022; Ribeiro et al. 2019; Vineis et al. 2020). Finally, we envision how this framework can be applied to advance the understanding and treatment of neurodegenerative conditions.

2 Allostasis

Allostasis refers to the process by which the body's various systems adaptively regulate their activity in response to environmental demands (i.e., changes in blood pressure, muscle tension, and glucose level) in an energy-efficient way (Schulkin and Sterling 2019; Sterling 2012). For example, when facing a potentially dangerous situation, allostatic processes may involve a cascade of predictive and autonomic reflexes to prepare the body for freeze, fight, or flight responses, including the release of neurotransmitters, cortisol, and other neurohormones, activation of smooth and skeletomotor muscle fibers. High volumes of environmental demands encountered on a daily basis saturate the body's regulatory systems and hamper its ability to face those demands, also referred to as *allostatic overload* (McEwen 2012; McEwen and Akil 2020; Schulkin and Sterling 2019; Sterling 2012; Yuen et al. 2009). Allostatic overload is commonly measured through biomarkers of chronic multisystem dysregulation such as plasma concentrations of c-reactive protein, albumin, high density lipoprotein (HDL), and microglial waste (Beese et al. 2022; Shiels et al. 2017; Wright et al. 2021). Biological dysregulation in allostatic overload (i.e., albumin <3.8 mg/dl; total cholesterol >240 mg/dl, HDL < 40 mg/dl, glycated hemoglobin $>6.4\%$, etc.) is known to be related to pathological outcomes such as cardiovascular disease (Borrell et al. 2020; Logan and Barksdale 2008; Mazgelytė et al. 2019; Mocayar Marón et al. 2019), impaired insulin signaling leading to neurodegeneration (De Felice et al. 2022), physical decline, increased mortality (Beckie 2012; Juster et al. 2010), among others (see Guidi et al. 2021).

3 Interoception

Interoception is a fundamental factor for allostatic regulation, defined as the processing (conscious or unconscious) of internal bodily stimuli by the nervous system (Khalsa et al. 2018), instantiated by afferent signaling of visceromotor and autonomic information and integrated within the brain (i.e., insular and limbic cortices, frontal lobes) to form a model of bodily states (Kleckner et al. 2017). Interoceptive signaling is an important precursor to allostatic regulatory processes since it instantiates calibration of top-down brain interoceptive predictions, thereby optimizing adaptive responses and energy expenditure (Kleckner et al. 2017; Migeot et al. 2022).

Cognitive measurements of interoceptive awareness (i.e., conscious interoceptive processes) have typically focused on perception of cardiovascular sensations, via the utilization of a heartbeat perception task (Khalsa et al. 2018), assessing different dimensions, namely interoceptive accuracy (objective task performance); interoceptive learning (measuring performance after feedback); interoceptive sensibility (reflecting the participant's state or trait level beliefs about their performance), and interoceptive insight (metacognitive evaluation of task performance) (de la Fuente et al. 2019; García-Cordero et al. 2016; Garfinkel et al. 2015; Khalsa et al. 2018). Other systems also provide biomarkers of conscious interoceptive processing (Khalsa et al. 2018), such as nociceptive (Simons et al. 2014), pulmonary (von Leupoldt et al. 2013), thermoregulatory (Fealey 2013), and gastrointestinal systems (Drake et al. 2010; Mayer et al. 2006), nevertheless, the cardiovascular system is the standard approach to interoceptive research in neuroscience and the most characterized (Coll et al. 2021), being ideal to extend neurocognitive models based on existing evidence.

At the neurophysiological level, measurements have typically focused on the changes in neural activity evoked by peripheral signals. One of the most studied measures within the cardiovascular system is the heartbeat-evoked potential (HEP), putatively reflecting electrical brain activity in response to the conscious detection of one's heartbeat (Coll et al. 2021; Pollatos and Schandry 2004). Conversely, the HEP can also indicate unconscious processing of heartbeats and modulation of regulatory reflexes. Previous research has shown that the HEP is underlined by multiple physiological feedback loops starting from baroreceptors, cardiac neurons, and cutaneous receptors (Buot et al. 2021; Park and Blanke 2019). For instance, baroreceptors capture changes in blood pressure which are projected to viscerosensory cortices where programs for activation of sympathetic/parasympathetic nerve activity modulate (via brainstem activation) regulation of blood pressure (Park and Blanke 2019).

Brain processing of cardiac signaling (i.e., cardiac interoception) is key for the regulation of internal processes (e.g., cognition, biological regulatory processes) to achieve adaptation. For example, the processing of incoming arterial baroreceptor information (i.e., heartbeat rate, strength) is associated with the deployment of cognitive and peripheral responses to environmental threats (Azevedo et al. 2017;

Garfinkel and Critchley 2016). This evidence indicates that interoceptive processing is associated with contextual adaptation via brain–body integration. Recent theoretical models have integrated central and peripheral mechanisms with dynamic principles to offer insightful explanations for a range of embodied processes.

4 Allostasis, Interoception, and Predictive Coding

Integrative models linking predictive coding principles to allostatic and interoceptive processes have begun to show their potential to provide synergetic approaches to neuroscience. Such models are based on Bayesian models of perception (Chennu et al. 2013; Kok and de Lange 2014), borrowing upon the idea that the brain is an active inference generator rather than a passive receptor. Within this view, the brain makes *predictions* about what is more likely to be perceived. The accuracy of such prediction is then contrasted with actual sensory input and potential errors are corrected in the reformulated prediction (Chennu et al. 2013; Kok and de Lange 2014). This predictive coding tenet was extended to include the role of interoceptive inferences/predictions about the internal milieu in perceptual and adaptive processes (Gu et al. 2013; Seth and Friston 2016; Singer et al. 2009). The integration of interoceptive inference sparked complex theoretical models elucidating the multidimensional nature of interoception and its explanatory role in neuropsychiatric symptoms (Seth et al. 2012; Seth and Friston 2016), self-awareness (Tsakiris 2017), and more recently the allostatic-interoceptive predictive coding framework (Birba et al. 2022b; Migeot et al. 2022).

5 Predictive Coding of Allostatic Interoception

According to the allostatic-interoceptive predictive coding model (Fig. 1), successful allostatic regulation implies the smooth coupling of top-down interoceptive predictions and bottom-up visceromotor/peripheral feedback, leading to a minimization of prediction error (e.g., adequately adjusting blood pressure when changing body posture). The integration of interoceptive and visceromotor/peripheral processes is sustained by a domain-general allostatic-interoceptive network (AIN) featuring connections with key hubs for interoception (dorsal mid to posterior insula), cognitive control (fronto-temporo-parietal), visceromotor control (limbic nodes, motor, and premotor cortices), and allostatic processes (midbrain structures) (Birba et al. 2022b; Kleckner et al. 2017; Migeot et al. 2022). When predictive interoceptive and visceromotor signals match the upcoming peripheral feedback, the system does not require a major allocation of resources for the re-evaluation of initial predictions, and there is congruency within the system.

The uppermost level of the AIN hierarchy comprises various regions such as the prefrontal and orbitofrontal cortex, anterior mid-cingulate, pregenual cingulate,

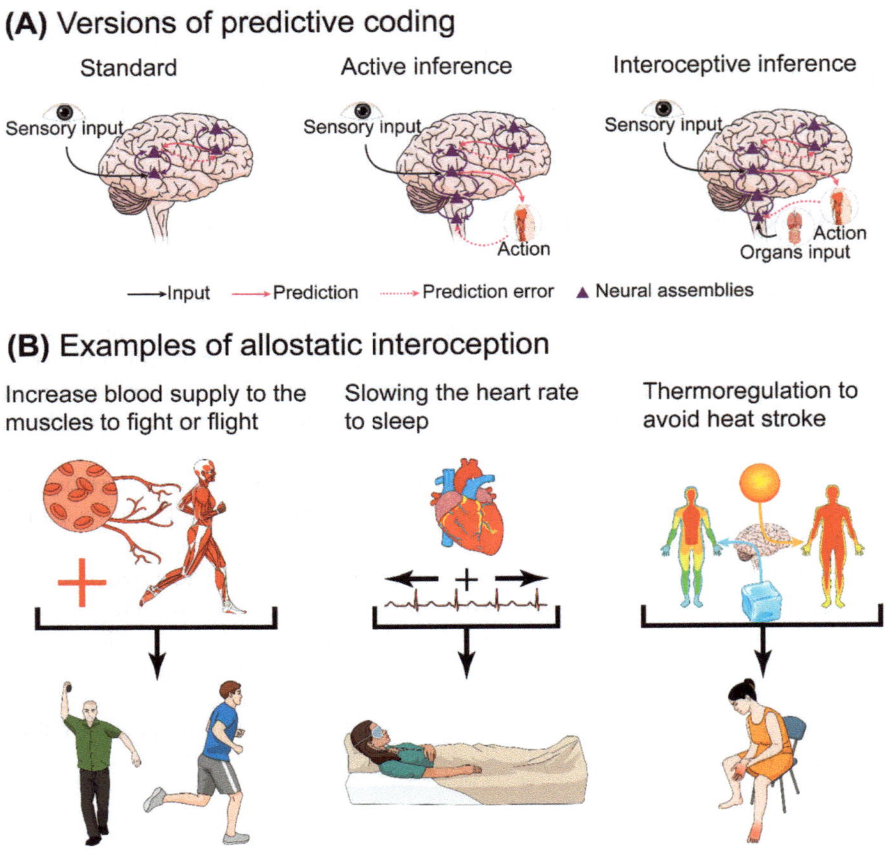

Fig. 1 Predictive coding and allostatic interoception. (**a**) Predictive coding is a general algorithm that can be instantiated among different biological substrates and their hierarchies. It has been widely applied to cortical brain activity (i.e., standard predictive coding), brain activity and proprioception (i.e., active inference), and brain activity and interoception (i.e., interoceptive inference). Three versions of predictive coding are schematically illustrated. In the first case, the minimization of prediction error is given by the updating of predictions to accommodate unexpected sensory signals. In the second case, the minimization is through performing actions that confirm predictions about sensory inputs. Finally, in the third case, the minimization is reached by performing actions to confirm predictions of interoceptive signals. (**b**) Allostasis refers to a general mechanism of bodily regulation by adaptation and changes. Allostasis can be instantiated at different levels of the body. Three examples of interoceptive allostasis are illustrated. Left: greater oxygen requirement in the muscles during a fight or flight situation, leading to the increased blood supply to the relevant muscles and the mobilization of resources needed to perform that redistribution (e.g., increasing cardiac input) to face the threat, or run to avoid it; Middle: reduction of heart rate when predisposing to sleep to align with the reduced metabolic needs. Right: vasoconstriction (i.e., narrowing of blood vessels) when the body is facing low temperatures to conserve core temperature by reducing the blood flow to the skin capillaries, thus avoiding heat stroke or hypothermia. Reproduced and modified with authorization from Migeot et al. (2022)

subgenual anterior cingulate, agranular insula, and dorsal amygdala. Activity within this level generates visceromotor and interoceptive predictions, which are then sent to relay regions (Barrett and Simmons 2015; Kleckner et al. 2017; Migeot et al. 2022). At the relay level, areas such as the dorsal mid and posterior insula, the thalamus, and the hypothalamus integrate and contrast predictions from top areas with corrective signals ascending from peripheral regions (via the vagus nerve). If top-down predictions fail to accommodate peripheral needs (i.e., due to interoceptive dysregulation), relay areas receive a prediction error, which is then projected to top regions for re-evaluation of initial inaccurate prediction. Re-evaluation comes at the cost of further allocation of resources, which is energetically costly for the organism, reduces energy efficiency, and ultimately leads to allostatic overload (Kleckner et al. 2017; Migeot et al. 2022). In sum, interactions between the components of the predictive coding allostatic-interoceptive system support anticipatory regulation of internal bodily systems leading to effective contextual adaptation while maintaining homeostasis. However, dysregulation at any of the levels increases the prediction error rate accompanied by allostatic overload states leading to pathophysiological consequences (Abrevaya et al. 2020; Birba et al. 2022b; De Felice et al. 2022; Migeot et al. 2022).

6 Measures of Allostatic-Interoceptive Overload

Cerebral, cardiocerebral, peripheral, and psychological domains can be targeted to evaluate allostatic-interoceptive overload manifestations in the organism. At the cerebral level, abnormal AIN structure and connectivity is associated with impaired interoceptive activity (measured at the individual level in a task (Kleckner et al. 2017)), autonomic imbalance, and behavioral maladjustment (Birba et al. 2022b; Migeot et al. 2022).

At the cardiocerebral level, interoceptive processing can be measured via HEP modulations reflecting the cortical processing of cardiac interoceptive activity (Birba et al. 2022b; Coll et al. 2021). For instance, abnormally high HEP modulations during resting state EEG are associated with hypervigilance to interoceptive signals and stress-induced allostatic overload in hypertensive and healthy controls (Legaz et al. 2022).

At the peripheral level, allostatic-interoceptive processes can be assessed through biomarkers of multisystem physiological imbalances including glucocorticoids (e.g., cortisol) (Williams et al. 2022; Woolley et al. 2014); cardiovascular: resting systolic and diastolic blood pressure, arterial tension (Seeman et al. 2008; Zeki Al Hazzouri et al. 2021); metabolic: cholesterol, body mass index, glucose, triglycerides (Seeman et al. 2008; Soysal et al. 2020); inflammatory: tumor necrosis factor-α, tumor growth factor-β, C-reactive protein (Bright et al. 2019; Cagnin et al. 2004; Dik et al. 2007; Duran-Aniotz et al. 2021; Rentzos et al. 2006; Seeman et al. 2008; Sjögren et al. 2004; Yaffe et al. 2004; Zhang 2015); and neural recycling parameters: neurofilament light chain, progranulin (Duran-Aniotz et al. 2021; Kaat et al. 2018;

Rohrer et al. 2016; Rojas et al. 2016; Steinacker et al. 2018; Takada et al. 2016; Yuan et al. 2017), among others (Shiels et al. 2017; Wright et al. 2021). The above measures are usually integrated into an "allostatic load battery" (Beese et al. 2022; Guidi et al. 2021; McCrory et al. 2019; Savransky et al. 2017).

At the psychological level, allostatic-interoceptive processes have been assessed via behavioral paradigms measuring the convergence between objective (skin conductance) and subjective (self-reported arousal) responses to relevant visual stimuli (i.e., evocative images conveying viscerosensory states) (Kleckner et al. 2017). Specifically, subjects with stronger intrinsic connectivity within the AIN exhibit stronger concordance between objective and self-reported arousal, thus putatively indicating normal allostatic states (Kleckner et al. 2017). Psychosocial measures of psychosomatic and dysregulated responses to environmental demands have also been used, including the diagnostic criteria for psychosomatic research (Fava et al. 2017) and the psychosocial index (Piolanti et al. 2016; Sonino and Fava 1998).

Previous research has shown that chronic allostatic overload symptoms start to emerge decades before diseases manifest and offer a greater predictive value beyond traditional detection methods (Juster et al. 2010; Karlamangla et al. 2014; Matos and De Souza-Talarico 2019). This is especially relevant for dementias due to their inconspicuous early symptoms, potentially providing a comprehensive understanding of the multisystemic physiological dysregulation and disparate symptomatology.

7 Social Determinants of Health, Allostatic Overload, and Neurodegeneration

Neurodegenerative physiopathology and symptomatology can be considered via the allostatic framework by accounting for the interaction between environmental demands, body responses, and brain function (Birba et al. 2022b; De Felice et al. 2022; García-Cordero et al. 2016; Migeot et al. 2022). In this sense, cumulative exposure to environmental stress factors associated with adverse social determinants of health (SDH) (e.g., adverse childhood experiences, social exclusion, pollution) can induce allostatic overload states, which interact with genetic predispositions and potentially lead to neurodegenerative diseases (see Figs. 2 and 3). Previous research has shown that negative SDH have a considerable multilevel impact (Alvarez et al. 2022; De Felice et al. 2022; Díaz-Rivera et al. 2023; Johnson et al. 2016; Tani et al. 2021). For example, subjects from low socioeconomic status show upregulated activity within the AIN in response to emotional stimuli indicating maladaptive allocation of resources, potentially leading to allostatic-interoceptive overload (Alvarez et al. 2022). The structure and function of the prefrontal cortex and amygdala (key hubs of the AIN) seem to be particularly vulnerable to socioeconomic-related stress (Johnson et al. 2016). In fact, amygdala hyperactivity is associated with increased inflammation and risk of adverse cardiac events in subjects living in deprived neighborhoods (Tawakol et al. 2017, 2019).

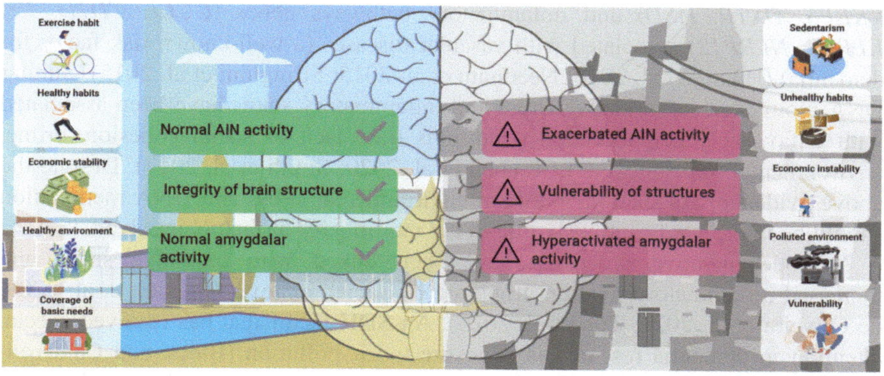

Fig. 2 Environmental stressors influencing allostatic overload in the brain. Lifestyle habits and environmental factors differentially affect the organism in individuals living in high-SES- vs. low-SES conditions (far left and right columns). High-SES individuals are exposed to a fewer number and intensity of environmental stressors than low-SES ones (i.e., better economic stability, more secure surroundings, better access to basic needs, etc.), maintaining stability in the brain as those stressors are manageable (left-middle column). In contrast, individuals living in low-SES conditions may commonly be chronically exposed to environmental stressors, damaging the brain structure and function as it tries to handle those stressors to maintain stability and provide adaptive responses. These characteristics are associated with dysregulated AIN functioning, increased structural and functional vulnerability of the prefrontal cortex and amygdala, and dysregulated amygdalar activity (right-middle column). AIN: allostatic-interoceptive network

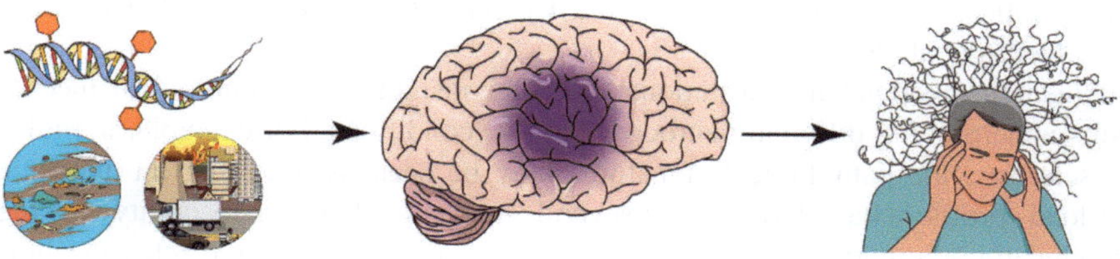

Fig. 3 Social epigenomics influences the brain and behavior. Different genomic compositions and environmental factors interact and impact the brain function and structure throughout the lifespan, influencing behavior

The stress-related effects of negative SDH are also noticeable at the peripheral level. Low socioeconomic status is associated with physiological dysregulations across the cardiovascular, inflammatory, metabolic, and neuroendocrine systems, indexed by a battery of allostatic load biomarkers (Ribeiro et al. 2018, 2019). Social epigenomic studies also illuminate the link between environmental stressors and epigenetic modifications leading to diseases (Martin et al. 2022). Specifically, individuals who faced early life socioeconomic disadvantage (i.e., low parental education and socioeconomic condition) showed decreased cardiometabolic health later in life as evidenced by abnormal DNA methylation of stress-related (*AVP,*

FKBP5, OXTR, HSD) and inflammatory pathways genes (*CCL1, CD1D, F8, KLRG1, NLRP12*) associated with vascular diseases (a well-known risk factor for dementia) (Huang et al. 2016; Needham et al. 2015; Stringhini et al. 2015). Similar DNA biomarkers of oxidative stress and inflammatory status have been associated with unfavorable neighborhood conditions (i.e., lack of social cohesion, crime, environmental toxicants, etc.) (Reuben et al. 2020; Smith et al. 2017). Despite the above evidence, associations between environmental stress and DNA methylation should be interpreted cautiously due to the multiplicity of factors involved.

Neurodegenerative processes have been linked with oxidative stress and increased inflammatory markers triggered by social contexts (Bennett et al. 2009; de Oliveira et al. 2021; Ton et al. 2020). An SDH factor that has received special attention in dementia research is exposure to air pollution (Block and Calderón-Garcidueñas 2009; Calderón-Garcidueñas et al. 2004; Levesque et al. 2011; Moulton and Yang 2012; Oudin et al. 2016; Peters et al. 2019). Several studies have shown that inhalation of air pollutants (i.e., ozone) causes immediate stress responses (mediated by HPA activation) leading to a cascade of glucocorticoid-related inflammatory and immunological dysregulation (Kodavanti 2016). Similarly, inhalation of $No2 + No_x$ polluted air is associated with elevated pro-inflammatory cytokines and reactive oxygen species in the brain, leading to microglial neurotoxic activity in neurodegeneration (Block and Calderón-Garcidueñas 2009; Jayaraj et al. 2017) and increased risk of cerebrovascular disease (Fiorito et al. 2018; Vineis et al. 2009, 2020). This is also consistent with post-mortem studies in pollution exposed individuals showing increased pro-inflammatory cytokines, microglia activation markers, accumulation of Aβ42, as well as brain lesions to the frontal cortex, hippocampus, and midbrain (Block and Calderón-Garcidueñas 2009; Calderón-Garcidueñas et al. 2004).

Allostatic overload manifestations can offer insights into pathological mechanisms underpinning different neurodegenerative conditions. For example, in multiple sclerosis, the pathological signatures may be seen as reflecting an allostatic overload profile characterized by several markers of hyper-reactivity of the hypothalamic-pituitary-thyroid axis, namely increased corticotropic releasing hormone (CRH)-expressing neurons, enlarged adrenal glands, elevated resting cortisol in plasma, elevated pro-inflammatory cytokines (Huitinga et al. 2003; Waliszewska-Prosół et al. 2022). Arguably, neurodegenerative-related lesions are believed to produce pro-inflammatory cytokines known to activate the HPA axis (Rivest 2001; Turnbull and Rivier 1999). Constant HPA activation would increase CRH, ACTH, and cortisol secretion. These excessive levels of glucocorticoids may worsen hippocampal damage (Gold et al. 2010) further exacerbating HPA activity. Moreover, prolonged HPA axis activity results in a decreased ratio of anti-to-pro inflammatory mediators (Straub and Besedovsky 2003). Interestingly, the increased number of CRH-expressing neurons is not significantly associated with age, sex, disease state, or age at onset, indicating a potential role of external environmental stressors in the onset of the disease (Huitinga et al. 2003). This is consistent with previous studies on the effects of air toxicant exposure (among other environmental factors) on inflammatory and immunological stress during preclinical stages of disease (Jayaraj et al. 2017; Olsson et al. 2017).

In Parkinson's disease (PD), allostatic overload interacts with individual vulnerability and catecholamine oxidation, increasing the death of dopaminergic neurons and autonomic failure (Goldstein and Kopin 2018). Residual catecholamine remaining after brain response to contextual stressors is reabsorbed into cytoplasmic pools, where it is oxidized to cytotoxic metabolites potentially leading to the death of dopaminergic neurons (Follmer et al. 2015; Li et al. 2001; Marchitti et al. 2007). In this scenario, various self-preservation mechanisms are triggered to neutralize the toxic effects of the oxidized metabolites (Fitzmaurice et al. 2013; Guillot and Miller 2009). However, repeated stress episodes may overwhelm such preservation mechanisms, ultimately leading to neuron loss (via autotoxicity). To maintain neurotransmitter delivery, allostatic processes upregulate the efflux of norepinephrine (Snyder et al. 1990) while paying the cost of allostatic load from chronically increased autotoxic metabolites cycle (Goldstein and Kopin 2018).

In the case of Alzheimer's disease (AD), the allostasis model offers an alternative explanation to the seemingly outdated Aβ theories (Luczkowski 2016), consistent with the tau hypothesis and the acceleration of AD pathogenesis linked with chronic stress exposure (Westfall et al. 2019; Ross et al. 2018; Machado et al. 2014). Previous work suggests that the primary factor for neurodegenerative alteration in AD is an overload-triggered breakdown of anatomical systems and pathways including inflammation, oxidative, mitochondrial, and glucose metabolism dysfunctions (Luczkowski 2016). In support of this view, recent work has posited that stress-mediated alterations in glucose metabolism are key for AD onset and progression (De Felice et al. 2022). Specifically, socioeconomic, lifestyle, and biological factors' interaction triggers chronic inflammation and cellular stress responses leading to dysregulated insulin signaling in the brain (De Felice et al. 2022). Decreased intraneuronal insulin is associated with decreased hippocampal plasticity and microglial dysfunctions underpinning cognitive decline and brain atrophy in AD (De Felice et al. 2022; Matos and De Souza-Talarico 2019).

8 Interoceptive Deficits in Dementias

According to the allostatic-interoceptive predictive coding model, allostatic overload is intrinsically associated with interoceptive dysregulations leading to a maladaptive cycle of imprecise predictions (Kleckner et al. 2017; Migeot et al. 2022, 2023). Accordingly, interoceptive dysfunctions are also observed across neurodegenerative diseases.

In multiple sclerosis (MS), symptomatology encompasses several alterations in interoceptive body signal processing, including temperature perception, taste, smell, the experience of pain, and highly disabling fatigue symptoms (Yoris et al. 2018). Particularly, fatigue symptoms can be triggered via bottom-up (i.e., peripheral inflammation triggering altered reactivity in interoceptive cortices) and top-down (i.e., impaired capacity to control bodily states) interoceptive processes (Harrison et al. 2009; Quadt et al. 2018; Stephan et al. 2016). Consequently, evidence indicates that decreased interoceptive accuracy associated with abnormal structure and functional connectivity in interoceptive regions (i.e., anterior insula, ACC) could reliably discriminate MS patients with fatigue from those without fatigue symptoms (Gonzalez Campo et al. 2019). Moreover, patients with multiple sclerosis show abnormal monitoring of autonomic bodily signals as evidenced by electrophysiological and fMRI measures (Salamone et al. 2018). Specifically, it was found that MS patients showed no difference in HEP modulations between interoceptive and non-interoceptive states, which was associated with atrophy in the insula and ACC. Similarly, MS patients showed no change in functional connectivity between interoceptive hubs when switching to an interoceptive task (Salamone et al. 2018).

In PD, neurodegeneration affects the structure and connectivity of limbic areas, particularly the insula and anterior cingulate cortex (ACC), key areas for interoception (Christopher et al. 2014). In fact, interoceptive performance might provide a tool for discriminating PD subtypes with particular neurodegenerative patterns (i.e., tremor dominant vs postural instability) (Santangelo et al. 2018). Overall, compared to controls, PD patients show decreased interoceptive sensitivity as measured by the heartbeat detection task (Ricciardi et al. 2016). This is consistent with a large body of research showing a range of interoceptive-related impairments in PD including sensory/viscerosensory perception, emotion regulation, and social cognition (Christopher et al. 2014). More recently, a study showed that compared to controls, PD patients have poor facial emotion recognition after interoceptive priming, which was also associated with decreased HEP modulation (during task) and aberrant structure and connectivity in insular and ACC (Salamone et al. 2021).

Regarding AD, widespread atrophy and aberrant connectivity in fronto-temporo-insular network hubs have been consistently associated with cardiac interoceptive impairments, impaired emotion recognition, and importantly, poor self-awareness (Abrevaya et al. 2020; Bateman et al. 2021; García-Cordero et al. 2016, 2021; Moon et al. 2014; Shany-Ur et al. 2014). Previous evidence shows that compared to HCs, AD (and bvFTD) patients evidence impairments in interoceptive accuracy and awareness, mainly associated with attenuated HEP modulations, as well as the abnormal structure and connectivity within the fronto-temporo-insular network (García-Cordero et al. 2016). Interestingly, a recent study showed that despite marked deficits in interoceptive accuracy and learning in AD, interoceptive awareness holds the strongest predictive value on AD (Abrevaya et al. 2020). Indeed, a lack of awareness of cognitive impairments and poor interoceptive monitoring seem to be prevalent in AD patients. For example, AD patients overestimate their performance on the HBD (García-Cordero et al. 2016) and emotion recognition (Garcia-Cordero et al. 2021) tasks, as well as their performance on daily life cognitive (i.e., remembering names of familiar people) and emotional (i.e., accepting criticisms from others) functioning (Shany-Ur et al. 2014). Interestingly, such insight deficits in AD have been consistently associated with high levels of agitation, hostility, apathy, and tension (Harwood et al. 2000; Horning et al. 2014).

9 Allostatic Interceptive Overload in Frontotemporal Lobar Degeneration Syndromes

Behavioral variant frontotemporal dementia (bvFTD) is a neurodegenerative condition encompassing particular characteristics at the behavioral (i.e. personality changes, disinhibition, impulsivity), cognitive (i.e. pronounced language and executive impairment as well as relative memory preservation), and neural (i.e., frontolobar degeneration) levels (Dodich et al. 2021; García-Cordero et al. 2016; Geraudie et al. 2021; Sedeño et al. 2016). Beyond other neurodegenerative conditions, bvFTD is characterized by systematic impairments in both interoceptive and allostatic domains (Birba et al. 2022b; García-Cordero et al. 2016; Migeot et al. 2022). Such pathological manifestations converge selectively with the allostatic-interoceptive predictive coding framework via cerebral, cardiocerebral, peripheral, and psychological manifestations as illustrated in Fig. 4 (Abrevaya et al. 2020; Beydoun et al. 2022; Bright et al. 2019; Duran-Aniotz et al. 2021; García-Cordero et al. 2016; Kaat et al. 2018).

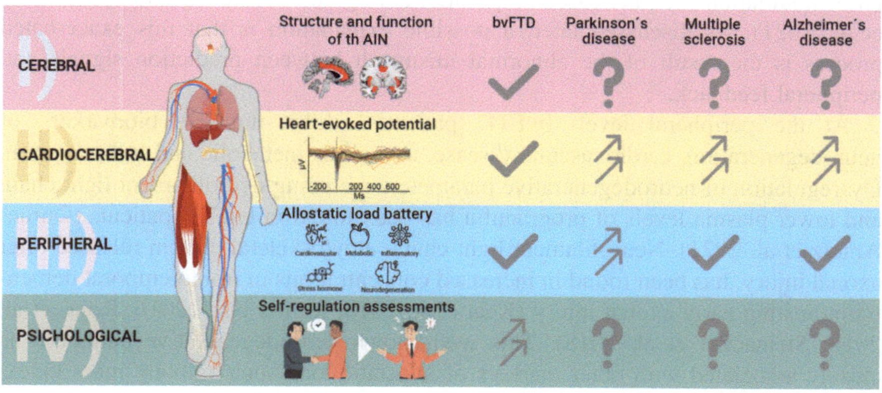

Fig. 4 Multimodal characterization of the allostatic-interoceptive system in neurodegenerative diseases. Different measures of allostatic load can be employed at the cerebral, cardiocerebral, peripheral, and psychological levels. At the cerebral level, the structure of the AIN hubs can be evaluated by voxel-based morphometry and functional connectivity techniques. The cardiocerebral level can be evaluated via electrophysiological measures targeting the heartbeat-evoked potential. Also, the allostatic load battery can assess the peripheral level, including cardiovascular, metabolic, inflammatory, stress hormone, and neurodegenerative biomarkers. Lastly, self-regulation and social adjustment measures can evaluate the psychological level. Notably, emerging evidence has shown the specific affectation of these measures in bvFTD, which seems to characterize the symptomatology and physiopathology associated with these conditions beyond other neurodegenerative diseases, such as Parkinson's disease, multiple sclerosis, and Alzheimer's disease. Tickets reflect the presence of direct evidence regarding the alteration of each measure for each neurodegenerative condition, parallel arrows reflect the presence of indirect evidence, and question marks suggest no evidence. AIN: allostatic-interoceptive network; bvFTD: behavioral variant frontotemporal dementia

At the cerebral level, subjects with bvFTD present structural and functional abnormalities in key hubs of the AIN supporting both interoceptive and allostatic processes, including the anterior insula, amygdala, and ACC (Baez et al. 2017; Kamalian et al. 2022; Van den Stock and Kumfor 2019). Further, resting state studies have shown that patients with bvFTD show aberrant connectivity within the default mode (Whitwell et al. 2011) and salience networks (Filippi et al. 2013; Seeley et al. 2008, 2012), two crucial networks interacting also with AIN (Kleckner et al. 2017). Activity in key hubs of default mode networks is strongly associated with allostatic-interoceptive processes (Kleckner et al. 2017; Ruiz-Rizzo et al. 2019), while salience network activity supports continuous tracking of bodily states (Seeley et al. 2012). The AIN's selective structural and functional abnormalities in bvFTD (Birba et al. 2022b) may be associated with prediction and prediction error signals within the system (Kleckner et al. 2017).

At the cardiocerebral level, bvFTD patients exhibit abnormal HEP modulations during resting state (Birba et al. 2022b) and active tasks (Abrevaya et al. 2020; García-Cordero et al. 2016; Salamone et al. 2021). Further, HEP sources such as the insula and amygdala (Azzalini et al. 2019; Legaz et al. 2022; Pollatos et al. 2005; Zhang 2015) seem to be structurally and functionally impaired in bvFTD (García-Cordero et al. 2016). Thus, it is not surprising that bvFTD patients show abnormal HEP modulation (i.e., increased amplitude) during emotion recognition (Salamone et al. 2021). As discussed later, a possible explanation is that this exacerbated process is the result of the abnormal mismatch between prediction signals and peripheral feedback.

At the peripheral level, bvFTD patients exhibit multiple biomarkers of neurodegeneration, cardiovascular disease, as well as metabolic and inflammation. Dysregulation of neurodegenerative parameters including neurofilament light chain and lower plasma levels of progranulin has been found in bvFTD patients (Duran-Aniotz et al. 2021). Neurofilament light chain, a cytoskeletal protein released after axonal injury, has been found in increased concentrations in frontotemporal dementia indicating early axonal injury (Kaat et al. 2018; Rohrer et al. 2016; Rojas et al. 2016; Steinacker et al. 2018). This well-known neurodegenerative parameter is closely associated with other markers of allostatic load such as body mass index, cardiovascular, metabolic, and inflammatory biomarkers (Beydoun et al. 2022).

Regarding psychological manifestations, the bvFTD-related cognitive and socioemotional deficits result in behavioral impairment and discordant responses to environmental stimuli (Healey et al. 2015; Hsieh et al. 2012; Ibáñez et al. 2018; McMillan et al. 2012). Prototypical maladaptive behavior includes underreacting to emotional stimuli (Chen et al. 2022; Marshall et al. 2019; Van den Stock et al. 2020) and aberrant responses to social situations (Van den Stock et al. 2021). Such cognitive and behavioral manifestations may be modulated by allostatic-interoceptive mechanisms (Cartwright et al. 2019; Kleckner et al. 2017) and regarded as manifestations of allostatic overload (D'Amico et al. 2020; Evans et al. 2021; Gray et al. 2017; Ottino-González et al. 2019; Ruisoto and Contador 2019; Seth and Friston 2016). From a predictive coding perspective of allostatic interoception, prototypical behavioral abnormalities in bvFTD are explained by malfunctions in the allostatic-interoceptive system, characterized by discordant predictions and prediction errors. Such systematic malfunctions conveniently account for emotion recognition deficits (Birba et al. 2021; Salamone et al. 2021) as well as cognitive and behavioral disturbances in bvFTD (Birba et al. 2021; Tawakol et al. 2017). Moreover, the characterization of allostatic-interoceptive deregulation profiles may potentially inform genetic-phenotype-endophenotype relationships and alterations across bvFTD subtypes, but further research is needed to fully understand the complexity and clinical implications of these profiles.

In sum, multilevel allostatic-interoceptive dysregulation presents a theoretical explanatory structure underlying the bvFTD's disparate manifestations. As we will further discuss, damage to such components of the allostatic-interoceptive system may explain core neurocognitive and psychological manifestations of bvFTD.

9.1 Toward a Model of Allostatic-Interoceptive Predictive Coding of bvFTD

The model states that neurodegenerative triggered distortion in interoceptive and peripheral signals leads to imprecise predictions worsened by inaccurate prediction errors, consequently drifting away from error minimization. The increasingly misguided predictions and feedback eventually lead to discordant top-down and bottom-up signaling that could explain some of the dysregulated behaviors of bvFTD patients, such as over/underreacting to environmental stimuli, as well as aberrant modulation of HEP amplitudes (Birba et al. 2022b; García-Cordero et al. 2016; Salamone et al. 2021; Salvato et al. 2018). Such dysfunctions, accompanied by progressive neurodegeneration would strengthen selective affectations of structural and functional characteristics of the AIN in bvFTD (Birba et al. 2022b) and account for the neurocognitive and psychological manifestations of the disease. In this line, the above biomarkers related to peripheral and immunological stress as well as psychosocial measures may track dysfunctions at these levels.

Another implication of the model accounts for the progression of the disease, which could be based on the circular nature of the neurodegeneration and allostatic-interoceptive interaction (Cedernaes et al. 2017; De Felice et al. 2022; Wang and Holtzman 2020). Specifically, it is arguable that early impairments associated with degeneration interact with environmental demands and behavioral dysregulation to perpetuate allostatic overload. As overload ensues, the loop of prediction and prediction error is altered, generating predictions with low values of precision and overcharged prediction errors, which in turn demands the relocation of resources to reevaluate the predictions (Migeot et al. 2022). Thus, as predictions are unprecise, the expected action instantiated by neurocognitive processes (e.g., inhibiting an inadequate action, inferring emotion on another's face) is erratic, generating increased prediction error in a vicious cycle. In terms of the etiology of bvFTD, it is possible to establish a causal link between the initial overload of the allostatic-interoceptive system and subsequent pathophysiological mechanisms.

In summary, the allostatic-interoceptive overload model of bvFTD implies that the observed symptomatology may be related to different cascades of predictions and prediction errors generated in the interactive loops (i.e., top-down and/or bottom-up interactions). Importantly, the model could contribute to prevention and intervention practices through the employment of a battery of biomarkers and psychosocial measures, yielding a composite risk index for vulnerability to bvFTD. Further studies should evaluate the predictive value of the composite index or individual markers on the subsequent incidence of bvFTD.

10 Conclusions and Future Research

The allostatic-interoceptive predictive coding model proposes a complex system sustained by multiple interactions at various levels. This model offers comprehensive explanations for etiological and progression factors associated with neurodegeneration, especially frontotemporal dementia. Future research should proceed cautiously and limit the biological, contextual, and exposome levels that need to be analyzed. Relatedly, measurements and manipulation of interoceptive signals are currently limited due to the anatomy of interoceptive afferents and the multiplicity of sites of sensory integration (Quigley et al. 2021). In this sense, dynamic system models (Irwin and Wang 2017) may be useful to assess how multiple interacting components of the model unfold simultaneously over time.

The validation of neurocognitive, cardiovascular, and peripheral markers as significant predictors of symptomatology is imperative. Successful validation would yield significant implications for characterizing and further understanding allostatic-interoceptive functioning and pathophysiology, for instance providing the evidence to reinforce the role of practices (e.g., mediation (Laneri et al. 2017; Tang et al. 2015)), early interventions (e.g., non-invasive stimulation of interoceptive cortices (Ibrahim et al. 2019)) as relevant applications promoting brain health.

References

Abrevaya S, Fittipaldi S, García AM, Dottori M, Santamaria-Garcia H, Birba A, Yoris A, Hildebrandt MK, Salamone P, De la Fuente A (2020) At the heart of neurological dimensionality: cross-nosological and multimodal cardiac interoceptive deficits. Psychosom Med

Alvarez GM, Rudolph MD, Cohen JR, Muscatell KA (2022) Lower socioeconomic position is associated with greater activity in and integration within an allostatic-interoceptive brain network in response to affective stimuli. J Cogn Neurosci 34(10):1906–1927

Anusha-Kiran Y, Mol P, Dey G, Bhat FA, Chatterjee O, Deolankar SC, Philip M, Prasad TSK, Bharath MMS, Mahadevan A (2022) Regional heterogeneity in mitochondrial function

underlies region specific vulnerability in human brain ageing: implications for neurodegeneration. Free Radic Biol Med 193:34–57
Azevedo RT, Garfinkel SN, Critchley HD, Tsakiris M (2017) Cardiac afferent activity modulates the expression of racial stereotypes. Nat Commun 8(1):13854
Azzalini D, Rebollo I, Tallon-Baudry C (2019) Visceral signals shape brain dynamics and cognition. Trends Cogn Sci 23(6):488–509. https://doi.org/10.1016/j.tics.2019.03.007
Baez S, García AM, Ibanez A (2017) The social context network model in psychiatric and neurological diseases. Social behavior from rodents to humans: neural foundations and clinical implications, pp 379–396
Barrett LF, Simmons WK (2015) Interoceptive predictions in the brain. Nat Rev Neurosci 16(7): 419–429. https://doi.org/10.1038/nrn3950
Bateman JR, Kawas MI, Craft S, Lockhart SN, Kim J, Sai KKS, Lindquist KA, Whitlow CT (2021) Difference in allostatic interoceptive network connectivity by cognitive status. Alzheimers Dement 17:e056634
Beckie TM (2012) A systematic review of allostatic load, health, and health disparities. Biol Res Nurs 14(4):311–346. https://doi.org/10.1177/1099800412455688
Beese S, Postma J, Graves JM (2022) Allostatic load measurement: a systematic review of reviews, database inventory, and considerations for neighborhood research. Int J Environ Res Public Health 19(24):17006
Bennett S, Grant MM, Aldred S (2009) Oxidative stress in vascular dementia and Alzheimer's disease: a common pathology. J Alzheimers Dis 17(2):245–257
Beydoun MA, Noren Hooten N, Maldonado AI, Beydoun HA, Weiss J, Evans MK, Zonderman AB (2022) BMI and allostatic load are directly associated with longitudinal increase in plasma neurofilament light among urban middle-aged adults. J Nutr 152(2):535–549
Birba A, Fittipaldi S, Cediel Escobar JC, Gonzalez Campo C, Legaz A, Galiani A, Díaz Rivera MN, Martorell Caro M, Alifano F, Piña-Escudero SD, Cardona JF, Neely A, Forno G, Carpinella M, Slachevsky A, Serrano C, Sedeño L, Ibáñez A, García AM (2021) Multimodal neurocognitive markers of naturalistic discourse typify diverse neurodegenerative diseases. Cereb Cortex. https://doi.org/10.1093/CERCOR/BHAB421
Birba A, Fittipaldi S, Cediel Escobar JC, Gonzalez Campo C, Legaz A, Galiani A, Díaz Rivera MN, Martorell Caro M, Alifano F, Piña-Escudero SD (2022a) Multimodal neurocognitive markers of naturalistic discourse typify diverse neurodegenerative diseases. Cereb Cortex 32(16): 3377–3391
Birba A, Santamaría-García H, Prado P, Cruzat J, Ballesteros AS, Legaz A, Fittipaldi S, Duran-Aniotz C, Slachevsky A, Santibañez R (2022b) Allostatic-interoceptive overload in frontotemporal dementia. Biol Psychiatry 92(1):54–67
Block ML, Calderón-Garcidueñas L (2009) Air pollution: mechanisms of neuroinflammation and CNS disease. Trends Neurosci 32(9):506–516. https://doi.org/10.1016/j.tins.2009.05.009
Borrell LN, Rodríguez-Álvarez E, Dallo FJ (2020) Racial/ethnic inequities in the associations of allostatic load with all-cause and cardiovascular-specific mortality risk in US adults. PLoS One 15(2):e0228336
Bright F, Werry EL, Dobson-Stone C, Piguet O, Ittner LM, Halliday GM, Hodges JR, Kiernan MC, Loy CT, Kassiou M, Kril JJ (2019) Neuroinflammation in frontotemporal dementia. Nat Rev Neurol 15(9):540–555. https://doi.org/10.1038/s41582-019-0231-z
Buot A, Azzalini D, Chaumon M, Tallon-Baudry C (2021) Does stroke volume influence heartbeat evoked responses? Biol Psychol 165:108165
Cagnin A, Rossor M, Sampson EL, MacKinnon T, Banati RB (2004) In vivo detection of microglial activation in frontotemporal dementia. Ann Neurol 56(6):894–897
Calderón-Garcidueñas L, Reed W, Maronpot RR, Henriquez-Roldán C, Delgado-Chavez R, calderon-Garciduenas ANA, Dragustinovis I, Franco-Lira M, Aragón-Flores M, Solt AC (2004) Brain inflammation and Alzheimer's-like pathology in individuals exposed to severe air pollution. Toxicol Pathol 32(6):650–658

Cartwright KB, Marshall TR, Huemer CM, Payne JB (2019) Executive function in the classroom: cognitive flexibility supports reading fluency for typical readers and teacher-identified low-achieving readers. Res Dev Disabil. https://doi.org/10.1016/j.ridd.2019.01.011

Cedernaes J, Osorio RS, Varga AW, Kam K, Schiöth HB, Benedict C (2017) Candidate mechanisms underlying the association between sleep-wake disruptions and Alzheimer's disease. Sleep Med Rev 31:102–111

Chen K-H, Hua AY, Toller G, Lwi SJ, Otero MC, Haase CM, Rankin KP, Rosen HJ, Miller BL, Levenson RW (2022) Diminished preparatory physiological responses in frontotemporal lobar degeneration syndromes. Brain Commun 4(2):fcac075

Chennu S, Noreika V, Gueorguiev D, Blenkmann A, Kochen S, Ibánez A, Owen AM, Bekinschtein TA (2013) Expectation and attention in hierarchical auditory prediction. J Neurosci 33(27): 11194–11205

Chiu M-J, Chen T-F, Yip P-K, Hua M-S, Tang L-Y (2006) Behavioral and psychologic symptoms in different types of dementia. J Formos Med Assoc 105(7):556–562

Christopher L, Koshimori Y, Lang AE, Criaud M, Strafella AP (2014) Uncovering the role of the insula in non-motor symptoms of Parkinson's disease. Brain 137(8):2143–2154. https://doi.org/10.1093/brain/awu084

Coll M-P, Hobson H, Bird G, Murphy J (2021) Systematic review and meta-analysis of the relationship between the heartbeat-evoked potential and interoception. Neurosci Biobehav Rev 122:190–200. https://doi.org/10.1016/j.neubiorev.2020.12.012

D'Amico D, Amestoy ME, Fiocco AJ (2020) The association between allostatic load and cognitive function: a systematic and meta-analytic review. Psychoneuroendocrinology 121:104849. https://doi.org/10.1016/j.psyneuen.2020.104849

De Felice FG, Gonçalves RA, Ferreira ST (2022) Impaired insulin signalling and allostatic load in Alzheimer disease. Nat Rev Neurosci 23(4):215–230

de la Fuente A, Sedeño L, Vignaga SS, Ellmann C, Sonzogni S, Belluscio L, García-Cordero I, Castagnaro E, Boano M, Cetkovich M, Torralva T, Cánepa ET, Tagliazucchi E, Garcia AM, Ibañez A (2019) Multimodal neurocognitive markers of interoceptive tuning in smoked cocaine. Neuropsychopharmacology. https://doi.org/10.1038/s41386-019-0370-3

de Oliveira J, Kucharska E, Garcez ML, Rodrigues MS, Quevedo J, Moreno-Gonzalez I, Budni J (2021) Inflammatory cascade in Alzheimer's disease pathogenesis: a review of experimental findings. Cells 10(10):2581

Díaz-Rivera MN, Birba A, Fittipaldi S, Mola D, Morera Y, de Vega M, Moguilner S, Lillo P, Slachevsky A, González Campo C (2023) Multidimensional inhibitory signatures of sentential negation in behavioral variant frontotemporal dementia. Cereb Cortex 33(2):403–420

Dik MG, Jonker C, Comijs HC, Deeg DJH, Kok A, Yaffe K, Penninx BW (2007) Contribution of metabolic syndrome components to cognition in older individuals. Diabetes Care 30(10): 2655–2660

Dodich A, Crespi C, Santi GC, Cappa SF, Cerami C (2021) Evaluation of discriminative detection abilities of social cognition measures for the diagnosis of the behavioral variant of frontotemporal dementia: a systematic review. Neuropsychol Rev 31:251–266

Drake MJ, Fowler CJ, Griffiths D, Mayer E, Paton JFR, Birder L (2010) Neural control of the lower urinary and gastrointestinal tracts: supraspinal CNS mechanisms. NeurourolUrodyn 29(1): 119–127

Duran-Aniotz C, Orellana P, Leon Rodriguez T, Henriquez F, Cabello V, Aguirre-Pinto MF, Escobedo T, Takada LT, Pina-Escudero SD, Lopez O (2021) Systematic review: genetic, neuroimaging, and fluids biomarkers for frontotemporal dementia across Latin America countries. Front Neurol 12:1034

Evans GW, Farah MJ, Hackman DA (2021) Early childhood poverty and adult executive functioning: distinct, mediating pathways for different domains of executive functioning. Dev Sci 24(5): e13084

Fava GA, Cosci F, Sonino N (2017) Current psychosomatic practice. Psychother Psychosom 86(1): 13–30

Fealey RD (2013) Interoception and autonomic nervous system reflexes thermoregulation. Handb Clin Neurol 117:79–88

Filippi M, Agosta F, Scola E, Canu E, Magnani G, Marcone A, Valsasina P, Caso F, Copetti M, Comi G (2013) Functional network connectivity in the behavioral variant of frontotemporal dementia. Cortex 49(9):2389–2401

Fiorito G, Vlaanderen J, Polidoro S, Gulliver J, Galassi C, Ranzi A, Krogh V, Grioni S, Agnoli C, Sacerdote C (2018) Oxidative stress and inflammation mediate the effect of air pollution on cardio-and cerebrovascular disease: a prospective study in nonsmokers. Environ Mol Mutagen 59(3):234–246

Fitzmaurice AG, Rhodes SL, Lulla A, Murphy NP, Lam HA, O'Donnell KC, Barnhill L, Casida JE, Cockburn M, Sagasti A (2013) Aldehyde dehydrogenase inhibition as a pathogenic mechanism in Parkinson disease. Proc Natl Acad Sci 110(2):636–641

Follmer C, Coelho-Cerqueira E, Yatabe-Franco DY, Araujo GDT, Pinheiro AS, Domont GB, Eliezer D (2015) Oligomerization and membrane-binding properties of covalent adducts formed by the interaction of α-synuclein with the toxic dopamine metabolite 3, 4-dihydroxyphenylacetaldehyde (DOPAL). J Biol Chem 290(46):27660–27679

García-Cordero I, Sedeño L, de la Fuente L, Slachevsky A, Forno G, Klein F, Lillo P, Ferrari J, Rodriguez C, Bustin J, Torralva T, Baez S, Yoris A, Esteves S, Melloni M, Salamone P, Huepe D, Manes F, García AM, Ibañez A (2016) Feeling, learning from and being aware of inner states: interoceptive dimensions in neurodegeneration and stroke. Philos Trans R Soc B Biol Sci. https://doi.org/10.1098/rstb.2016.0006

Garcia-Cordero I, Migeot J, Fittipaldi S, Aquino A, Campo CG, García A, Ibáñez A (2021) Metacognition of emotion recognition across neurodegenerative diseases. Cortex 137:93–107. https://doi.org/10.1016/J.CORTEX.2020.12.023

Garfinkel SN, Critchley HD (2016) Threat and the body: how the heart supports fear processing. Trends Cogn Sci 20(1):34–46

Garfinkel SN, Seth AK, Barrett AB, Suzuki K, Critchley HD (2015) Knowing your own heart: distinguishing interoceptive accuracy from interoceptive awareness. Biol Psychol 104:65–74

Geraudie A, Battista P, García AM, Allen IE, Miller ZA, Gorno-Tempini ML, Montembeault M (2021) Speech and language impairments in behavioral variant frontotemporal dementia: a systematic review. Neurosci Biobehav Rev 131:1076–1095

Gold SM, Kern KC, O'Connor M-F, Montag MJ, Kim A, Yoo YS, Giesser BS, Sicotte NL (2010) Smaller cornu ammonis 2–3/dentate gyrus volumes and elevated cortisol in multiple sclerosis patients with depressive symptoms. Biol Psychiatry 68(6):553–559

Goldstein DS, Kopin IJ (2018) Linking stress, catecholamine autotoxicity, and allostatic load with neurodegenerative diseases: a focused review in memory of Richard Kvetnansky. Cell Mol Neurobiol 38:13–24

Gonzalez Campo C, Salamone PC, Rodríguez-Arriagada N, Richter F, Herrera E, Bruno D, Pagani Cassara F, Sinay V, García AM, Ibáñez A, Sedeño L (2019) Fatigue in multiple sclerosis is associated with multimodal interoceptive abnormalities. Mult Scler J 26(14):1845–1853. https://doi.org/10.1177/1352458519888881

Gray JD, Kogan JF, Marrocco J, McEwen BS (2017) Genomic and epigenomic mechanisms of glucocorticoids in the brain. Nat Rev Endocrinol 13(11):661–673. https://doi.org/10.1038/nrendo.2017.97

Gu X, Hof PR, Friston KJ, Fan J (2013) Anterior insular cortex and emotional awareness. J Comp Neurol 521(15):3371–3388

Guidi J, Lucente M, Sonino N, Fava GA (2021) Allostatic load and its impact on health: a systematic review. Psychother Psychosom 90(1):11–27

Guillot TS, Miller GW (2009) Protective actions of the vesicular monoamine transporter 2 (VMAT2) in monoaminergic neurons. Mol Neurobiol 39:149–170

Harrison NA, Brydon L, Walker C, Gray MA, Steptoe A, Critchley HD (2009) Inflammation causes mood changes through alterations in subgenual cingulate activity and mesolimbic connectivity. Biol Psychiatry 66(5):407–414

Harwood DG, Sultzer DL, Wheatley MV (2000) Impaired insight in Alzheimer disease: association with cognitive deficits, psychiatric symptoms, and behavioral disturbances. Neuropsychiatry Neuropsychol Behav Neurol 13(2):83–88

Healey ML, McMillan CT, Golob S, Spotorno N, Rascovsky K, Irwin DJ, Clark R, Grossman M (2015) Getting on the same page: the neural basis for social coordination deficits in behavioral variant frontotemporal degeneration. Neuropsychologia 69:56–66

Horning SM, Melrose R, Sultzer D (2014) Insight in Alzheimer's disease and its relation to psychiatric and behavioral disturbances. Int J Geriatr Psychiatry 29(1):77–84

Hsieh S, Foxe D, Leslie F, Savage S, Piguet O, Hodges JR (2012) Grief and joy: emotion word comprehension in the dementias. Neuropsychology 26(5):624

Huang JY, Gavin AR, Richardson TS, Rowhani-Rahbar A, Siscovick DS, Hochner H, Friedlander Y, Enquobahrie DA (2016) Accounting for life-course exposures in epigenetic biomarker association studies: early life socioeconomic position, candidate gene DNA methylation, and adult cardiometabolic risk. Am J Epidemiol 184(7):520–531. https://doi.org/10.1093/aje/kww014

Huitinga I, Erkut ZA, van Beurden D, Swaab DF (2003) The hypothalamo-pituitary-adrenal axis in multiple sclerosis. Ann N Y Acad Sci 992(1):118–128

Ibanez A (2022) The mind's golden cage and cognition in the wild. Trends Cogn Sci

Ibáñez A, García AM, Esteves S, Yoris A, Muñoz E, Reynaldo L, Pietto ML, Adolfi F, Manes F (2018) Social neuroscience: undoing the schism between neurology and psychiatry. Soc Neurosci 13(1):1–39

Ibrahim C, Rubin-Kahana DS, Pushparaj A, Musiol M, Blumberger DM, Daskalakis ZJ, Zangen A, Le Foll B (2019) The insula: a brain stimulation target for the treatment of addiction. Front Pharmacol 10:720. https://doi.org/10.3389/fphar.2019.00720

Irwin M, Wang Z (2017) Dynamic systems modeling. The international encyclopedia of communication research methods, pp 1–12

Jayaraj RL, Rodriguez EA, Wang Y, Block ML (2017) Outdoor ambient air pollution and neurodegenerative diseases: the neuroinflammation hypothesis. Curr Environ Health Rep 4(2):166–179. https://doi.org/10.1007/s40572-017-0142-3

Johnson SB, Riis JL, Noble KG (2016) State of the art review: poverty and the developing brain. Pediatrics 137(4)

Juster R-P, McEwen BS, Lupien SJ (2010) Allostatic load biomarkers of chronic stress and impact on health and cognition. Neurosci Biobehav Rev 35(1):2–16. https://doi.org/10.1016/j.neubiorev.2009.10.002

Kaat LD, Meeter LH, Chiu WZ, Melhem S, Boon AJW, Blennow K, Zetterberg H, van Swieten JC (2018) Serum neurofilament light chain in progressive supranuclear palsy. Parkinsonism Relat Disord 56:98–101

Kamaleddin MA (2022) Degeneracy in the nervous system: from neuronal excitability to neural coding. Bioessays 44(1):2100148

Kamalian A, Khodadadifar T, Saberi A, Masoudi M, Camilleri JA, Eickhoff CR, Zarei M, Pasquini L, Laird AR, Fox PT (2022) Convergent regional brain abnormalities in behavioral variant frontotemporal dementia: a neuroimaging meta-analysis of 73 studies. Alzheimer's Dementia (Amst) 14(1):e12318

Karlamangla AS, Miller-Martinez D, Lachman ME, Tun PA, Koretz BK, Seeman TE (2014) Biological correlates of adult cognition: midlife in the United States (MIDUS). Neurobiol Aging 35(2):387–394. https://doi.org/10.1016/j.neurobiolaging.2013.07.028

Khachaturian AS, Dengel A, Dočkal V, Hroboň P, Tolar M (2023) Accelerating innovations for enhanced brain health. Can artificial intelligence advance new pathways for drug discovery for Alzheimer's and other neurodegenerative disorders? J Prev Alzheimers Dis:1–4

Khalsa SS, Adolphs R, Cameron OG, Critchley HD, Davenport PW, Feinstein JS, Feusner JD, Garfinkel SN, Lane RD, Mehling WE, Meuret AE, Nemeroff CB, Oppenheimer S, Petzschner FH, Pollatos O, Rhudy JL, Schramm LP, Simmons WK, Stein MB et al (2018) Interoception and mental health: a roadmap. Biol Psychiatry Cogn Neurosci Neuroimaging 3(6):501–513. https://doi.org/10.1016/j.bpsc.2017.12.004

Kleckner IR, Zhang J, Touroutoglou A, Chanes L, Xia C, Simmons WK, Quigley KS, Dickerson BC, Feldman Barrett L (2017) Evidence for a large-scale brain system supporting allostasis and interoception in humans. Nat Hum Behav 1(5):69

Kodavanti UP (2016) Stretching the stress boundary: linking air pollution health effects to a neurohormonal stress response. Biochim Biophys Acta Gen Subj 1860(12):2880–2890. https://doi.org/10.1016/j.bbagen.2016.05.010

Kok P, de Lange FP (2014) Shape perception simultaneously up-and downregulates neural activity in the primary visual cortex. Curr Biol 24(13):1531–1535

Laneri D, Krach S, Paulus FM, Kanske P, Schuster V, Sommer J, Müller-Pinzler L (2017) Mindfulness meditation regulates anterior insula activity during empathy for social pain. Hum Brain Mapp 38(8):4034–4046. https://doi.org/10.1002/hbm.23646

Legaz A, Yoris A, Sedeño L, Abrevaya S, Martorell M, Alifano F, Garcia AM, Ibanez A (2022) Heart–brain interactions during social and cognitive stress in hypertensive disease: a multidimensional approach. Eur J Neurosci 55(9–10):2836–2850

Levesque S, Surace MJ, McDonald J, Block ML (2011) Air pollution & the brain: subchronic diesel exhaust exposure causes neuroinflammation and elevates early markers of neurodegenerative disease. J Neuroinflammation 8(1):105. https://doi.org/10.1186/1742-2094-8-105

Li Ren J, Zhang M, Wang H, Yi F, Wu J, Tang Y (2022) The heterogeneity of microglial activation and its epigenetic and non-coding RNA regulations in the immunopathogenesis of neurodegenerative diseases. Cell Mol Life Sci 79(10):511

Li SW, Lin T-S, Minteer S, Burke WJ (2001) 3,4-Dihydroxyphenylacetaldehyde and hydrogen peroxide generate a hydroxyl radical: possible role in Parkinson's disease pathogenesis. Mol Brain Res 93(1):1–7

Logan JG, Barksdale DJ (2008) Allostasis and allostatic load: expanding the discourse on stress and cardiovascular disease. J Clin Nurs 17(7b):201–208

Luczkowski M (2016) "No screams and cries will convince us that white is white and black is black", an ode to the defenders of amyloid cascade hypothesis of Alzheimer's disease. Coord Chem Rev 327–328:35–42. https://doi.org/10.1016/j.ccr.2016.03.001

Machado A, Herrera A, de Pablos R, Espinosa-Oliva A, Sarmiento M, Ayala A, Venero J, Santiago M, Villarán R, Delgado-Cortés M, Argüelles S, Cano J (2014) Chronic stress as a risk factor for Alzheimer's disease. Rev Neurosci 25(6):785–804. https://doi.org/10.1515/revneuro-2014-0035

Marchitti SA, Deitrich RA, Vasiliou V (2007) Neurotoxicity and metabolism of the catecholamine-derived 3,4-dihydroxyphenylacetaldehyde and 3,4-dihydroxyphenylglycolaldehyde: the role of aldehyde dehydrogenase. Pharmacol Rev 59(2):125–150

Marshall CR, Hardy CJD, Russell LL, Bond RL, Sivasathiaseelan H, Greaves C, Moore KM, Agustus JL, van Leeuwen JEP, Wastling SJ (2019) The functional neuroanatomy of emotion processing in frontotemporal dementias. Brain 142(9):2873–2887

Martin CL, Ghastine L, Lodge EK, Dhingra R, Ward-Caviness CK (2022) Understanding health inequalities through the lens of social epigenetics. Annu Rev Public Health 43:235–254

Matos TM, De Souza-Talarico JN (2019) How stress mediators can cumulatively contribute to Alzheimer's disease an allostatic load approach. Dement Neuropsychol 13:11–21

Mayer EA, Naliboff BD, Craig ADB (2006) Neuroimaging of the brain-gut axis: from basic understanding to treatment of functional GI disorders. Gastroenterology 131(6):1925–1942

Mazgelytė E, Karčiauskaitė D, Linkevičiūtė A, Mažeikienė A, Burokienė N, Matuzevičienė R, Radzevičius M, Janiulionienė A, Jakaitienė A, Dindienė L (2019) Association of hair cortisol concentration with prevalence of major cardiovascular risk factors and allostatic load. Med Sci Monit 25:3573

McCrory C, Fiorito G, Cheallaigh CN, Polidoro S, Karisola P, Alenius H, Layte R, Seeman T, Vineis P, Kenny RA (2019) How does socio-economic position (SEP) get biologically

embedded? A comparison of allostatic load and the epigenetic clock(s). Psychoneuroendocrinology 104:64–73
McEwen BS (2012) Brain on stress: how the social environment gets under the skin. Proc Natl Acad Sci U S A 109(Suppl 2):17180–17185. https://doi.org/10.1073/pnas.1121254109
McEwen BS, Akil H (2020) Revisiting the stress concept: implications for affective disorders. J Neurosci 40(1):12–21
McMillan CT, Rascovsky K, Khella MC, Clark R, Grossman M (2012) The neural basis for establishing a focal point in pure coordination games. Soc Cogn Affect Neurosci 7(8):881–887
Migeot J, Duran-Aniotz CA, Signorelli CM, Piguet O, Ibáñez A (2022) A predictive coding framework of allostatic–interoceptive overload in frontotemporal dementia. Trends Neurosci
Migeot J, Hesse E, Fittipaldi S, Mejía J, Fraile M, García AM, García M d C, Ortega R, Lawlor B, Lopez V, Ibáñez A (2023) Allostatic-interoceptive anticipation of social rejection. Neuroimage 276:120200. https://doi.org/10.1016/j.neuroimage.2023.120200
Mocayar Marón FJ, Ferder L, Saraví FD, Manucha W (2019) Hypertension linked to allostatic load: from psychosocial stress to inflammation and mitochondrial dysfunction. Stress 22(2):169–181. https://doi.org/10.1080/10253890.2018.1542683
Moon Y, Moon W-J, Kim H, Han S-H (2014) Regional atrophy of the insular cortex is associated with neuropsychiatric symptoms in Alzheimer's disease patients. Eur Neurol 71(5–6):223–229
Moulton PV, Yang W (2012) Air pollution, oxidative stress, and Alzheimer's disease. J Environ Public Health 2012:472751. https://doi.org/10.1155/2012/472751
Needham BL, Smith JA, Zhao W, Wang X, Mukherjee B, Kardia SLR, Shively CA, Seeman TE, Liu Y, Diez Roux AV (2015) Life course socioeconomic status and DNA methylation in genes related to stress reactivity and inflammation: the multi-ethnic study of atherosclerosis. Epigenetics 10(10):958–969. https://doi.org/10.1080/15592294.2015.1085139
Olsson T, Barcellos LF, Alfredsson L (2017) Interactions between genetic, lifestyle and environmental risk factors for multiple sclerosis. Nat Rev Neurol 13(1):25–36
Ottino-González J, Jurado MA, García-García I, Caldú X, Prats-Soteras X, Tor E, Sender-Palacios MJ, Garolera M (2019) Allostatic load and executive functions in overweight adults. Psychoneuroendocrinology 106:165–170. https://doi.org/10.1016/j.psyneuen.2019.04.009
Oudin A, Forsberg B, Adolfsson AN, Lind N, Modig L, Nordin M, Nordin S, Adolfsson R, Nilsson L-G (2016) Traffic-related air pollution and dementia incidence in northern Sweden: a longitudinal study. Environ Health Perspect 124(3):306–312
Park H-D, Blanke O (2019) Heartbeat-evoked cortical responses: underlying mechanisms, functional roles, and methodological considerations. Neuroimage 197:502–511
Peters R, Ee N, Peters J, Booth A, Mudway I, Anstey KJ (2019) Air pollution and dementia: a systematic review. J Alzheimers Dis 70:S145–S163. https://doi.org/10.3233/JAD-180631
Piolanti A, Offidani E, Guidi J, Gostoli S, Fava GA, Sonino N (2016) Use of the psychosocial index: a sensitive tool in research and practice. Psychother Psychosom 85(6):337–345
Pollatos O, Schandry R (2004) Accuracy of heartbeat perception is reflected in the amplitude of the heartbeat-evoked brain potential. Psychophysiology 41(3):476–482
Pollatos O, Kirsch W, Schandry R (2005) Brain structures involved in interoceptive awareness and cardioafferent signal processing: a dipole source localization study. Hum Brain Mapp 26(1): 54–64. https://doi.org/10.1002/hbm.20121
Quadt L, Critchley HD, Garfinkel SN (2018) The neurobiology of interoception in health and disease. Ann N Y Acad Sci 1428(1):112–128
Quigley KS, Kanoski S, Grill WM, Barrett LF, Tsakiris M (2021) Functions of interoception: from energy regulation to experience of the self. Trends Neurosci 44(1):29–38
Rentzos M, Paraskevas GP, Kapaki E, Nikolaou C, Zoga M, Rombos A, Tsoutsou A, Vassilopoulos D (2006) Interleukin-12 is reduced in cerebrospinal fluid of patients with Alzheimer's disease and frontotemporal dementia. J Neurol Sci 249(2):110–114
Reuben A, Sugden K, Arseneault L, Corcoran DL, Danese A, Fisher HL, Moffitt TE, Newbury JB, Odgers C, Prinz J (2020) Association of neighborhood disadvantage in childhood with DNA methylation in young adulthood. JAMA Netw Open 3(6):e206095

Ribeiro AI, Amaro J, Lisi C, Fraga S (2018) Neighborhood socioeconomic deprivation and allostatic load: a scoping review. Int J Environ Res Public Health 15(6):1092

Ribeiro AI, Fraga S, Kelly-Irving M, Delpierre C, Stringhini S, Kivimaki M, Joost S, Guessous I, Gandini M, Vineis P (2019) Neighbourhood socioeconomic deprivation and allostatic load: a multi-cohort study. Sci Rep 9(1):1–11

Ricciardi L, Ferrazzano G, Demartini B, Morgante F, Erro R, Ganos C, Bhatia KP, Berardelli A, Edwards M (2016) Know thyself: exploring interoceptive sensitivity in Parkinson's disease. J Neurol Sci 364:110–115. https://doi.org/10.1016/j.jns.2016.03.019

Rivest S (2001) How circulating cytokines trigger the neural circuits that control the hypothalamic–pituitary–adrenal axis. Psychoneuroendocrinology 26(8):761–788

Rohrer JD, Woollacott IOC, Dick KM, Brotherhood E, Gordon E, Fellows A, Toombs J, Druyeh R, Cardoso MJ, Ourselin S (2016) Serum neurofilament light chain protein is a measure of disease intensity in frontotemporal dementia. Neurology 87(13):1329–1336

Rojas JC, Karydas A, Bang J, Tsai RM, Blennow K, Liman V, Kramer JH, Rosen H, Miller BL, Zetterberg H (2016) Plasma neurofilament light chain predicts progression in progressive supranuclear palsy. Ann Clin Transl Neurol 3(3):216–225

Ross JA, Gliebus G, Van Bockstaele EJ (2018) Stress induced neural reorganization: a conceptual framework linking depression and Alzheimer's disease. Prog Neuropsychopharmacol Biol Psychiatry 85:136–151. https://doi.org/10.1016/j.pnpbp.2017.08.004

Ruisoto P, Contador I (2019) The role of stress in drug addiction. An integrative review. Physiol Behav 202:62–68. https://doi.org/10.1016/j.physbeh.2019.01.022

Ruiz-Rizzo AL, Sorg C, Napiórkowski N, Neitzel J, Menegaux A, Müller HJ, Vangkilde S, Finke K (2019) Decreased cingulo-opercular network functional connectivity mediates the impact of aging on visual processing speed. Neurobiol Aging 73:50–60. https://doi.org/10.1016/j.neurobiolaging.2018.09.014

Salamone PC, Esteves S, Sinay VJ, García-Cordero I, Abrevaya S, Couto B, Adolfi F, Martorell M, Petroni A, Yoris A, Torquati K, Alifano F, Legaz A, Cassará FP, Bruno D, Kemp AH, Herrera E, García AM, Ibáñez A, Sedeño L (2018) Altered neural signatures of interoception in multiple sclerosis. Hum Brain Mapp. https://doi.org/10.1002/hbm.24319

Salamone PC, Legaz A, Sedeño L, Moguilner S, Fraile-Vazquez M, Campo CG, Fittipaldi S, Yoris A, Miranda M, Birba A (2021) Interoception primes emotional processing: multimodal evidence from neurodegeneration. J Neurosci 41(19):4276–4292

Salvato G, Mercurio M, Sberna M, Paulesu E, Bottini G (2018) A very light lunch: interoceptive deficits and food aversion at onset in a case of behavioral variant frontotemporal dementia. Alzheimer's & Dementia: Diagnosis, Assessment & Disease Monitoring 10:750–754. https://doi.org/10.1016/j.dadm.2018.10.001

Santangelo G, Vitale C, Baiano C, D'Iorio A, Longo K, Barone P, Amboni M, Conson M (2018) Interoceptive processing deficit: A behavioral marker for subtyping Parkinson's disease. Parkinsonism Relat Disord 53:64–69

Savransky A, Chiappelli J, Rowland LM, Wisner K, Shukla DK, Kochunov P, Hong LE (2017) Fornix structural connectivity and allostatic load: empirical evidence from schizophrenia patients and healthy controls. Psychosom Med 79(7):770

Schulkin J, Sterling P (2019) Allostasis: a brain-Centered, predictive mode of physiological regulation. Trends Neurosci 42(10):740–752. https://doi.org/10.1016/j.tins.2019.07.010

Sedeño L, Couto B, García-Cordero I, Melloni M, Baez S, Sepúlveda JPM, Fraiman D, Huepe D, Hurtado E, Matallana D, Kuljis R, Torralva T, Chialvo D, Sigman M, Piguet O, Manes F, Ibanez A (2016) Brain network organization and social executive performance in frontotemporal dementia. J Int Neuropsychol Soc. https://doi.org/10.1017/S1355617715000703

Seeley WW, Crawford R, Rascovsky K, Kramer JH, Weiner M, Miller BL, Gorno-Tempini ML (2008) Frontal paralimbic network atrophy in very mild behavioral variant frontotemporal dementia. Arch Neurol 65(2):249–255

Seeley WW, Zhou J, Kim EJ (2012) Frontotemporal dementia: what can the behavioral variant teach us about human brain organization? Neuroscientist. https://doi.org/10.1177/1073858411410354

Seeman T, Merkin SS, Crimmins E, Koretz B, Charette S, Karlamangla A (2008) Education, income and ethnic differences in cumulative biological risk profiles in a national sample of US adults: NHANES III (1988–1994). Soc Sci Med 66(1):72–87. https://doi.org/10.1016/j.socscimed.2007.08.027

Seenivasan P, Narayanan R (2022) Efficient information coding and degeneracy in the nervous system. Curr Opin Neurobiol 76:102620

Seth AK, Friston KJ (2016) Active interoceptive inference and the emotional brain. Philos Trans R Soc B Biol Sci 371(1708):20160007

Seth AK, Suzuki K, Critchley HD (2012) An interoceptive predictive coding model of conscious presence. Front Psychol 2:395

Shany-Ur T, Lin N, Rosen HJ, Sollberger M, Miller BL, Rankin KP (2014) Self-awareness in neurodegenerative disease relies on neural structures mediating reward-driven attention. Brain 137(8):2368–2381. https://doi.org/10.1093/brain/awu161

Shiels PG, Stenvinkel P, Kooman JP, McGuinness D (2017) Circulating markers of ageing and allostatic load: a slow train coming. Pract Lab Med 7:49–54. https://doi.org/10.1016/j.plabm.2016.04.002

Simons LE, Elman I, Borsook D (2014) Psychological processing in chronic pain: a neural systems approach. Neurosci Biobehav Rev 39:61–78

Singer T, Critchley HD, Preuschoff K (2009) A common role of insula in feelings, empathy and uncertainty. Trends Cogn Sci 13(8):334–340

Sjögren M, Folkesson S, Blennow K, Tarkowski E (2004) Increased intrathecal inflammatory activity in frontotemporal dementia: pathophysiological implications. J Neurol Neurosurg Psychiatry 75(8):1107–1111

Smith JA, Zhao W, Wang X, Ratliff SM, Mukherjee B, Kardia SLR, Liu Y, Roux AVD, Needham BL (2017) Neighborhood characteristics influence DNA methylation of genes involved in stress response and inflammation: the multi-ethnic study of atherosclerosis. Epigenetics 12(8):662–673. https://doi.org/10.1080/15592294.2017.1341026

Snyder GL, Keller RW, Zigmond MJ (1990) Dopamine efflux from striatal slices after intracerebral 6-hydroxydopamine: evidence for compensatory hyperactivity of residual terminals. J Pharmacol Exp Ther 253(2):867–876

Sonino N, Fava GA (1998) A simple instrument for assessing stress in clinical practice. Postgrad Med J 74(873):408–410. https://doi.org/10.1136/pgmj.74.873.408

Soysal P, Dokuzlar O, Erken N, Günay FSD, Isik AT (2020) The relationship between dementia subtypes and nutritional parameters in older adults. J Am Med Dir Assoc 21(10):1430–1435

Steinacker P, Anderl-Straub S, Diehl-Schmid J, Semler E, Uttner I, von Arnim CAF, Barthel H, Danek A, Fassbender K, Fliessbach K (2018) Serum neurofilament light chain in behavioral variant frontotemporal dementia. Neurology 91(15):e1390–e1401

Stephan KE, Manjaly ZM, Mathys CD, Weber LAE, Paliwal S, Gard T, Tittgemeyer M, Fleming SM, Haker H, Seth AK (2016) Allostatic self-efficacy: A metacognitive theory of dyshomeostasis-induced fatigue and depression. Front Hum Neurosci 10:550

Sterling P (2012) Allostasis: a model of predictive regulation. Physiol Behav 106(1):5–15. https://doi.org/10.1016/j.physbeh.2011.06.004

Straub RH, Besedovsky HO (2003) Integrated evolutionary, immunological, and neuroendocrine framework for the pathogenesis of chronic disabling inflammatory diseases. FASEB J 17(15):2176–2183

Stringhini S, Polidoro S, Sacerdote C, Kelly RS, van Veldhoven K, Agnoli C, Grioni S, Tumino R, Giurdanella MC, Panico S, Mattiello A, Palli D, Masala G, Gallo V, Castagné R, Paccaud F, Campanella G, Chadeau-Hyam M, Vineis P (2015) Life-course socioeconomic status and DNA methylation of genes regulating inflammation. Int J Epidemiol 44(4):1320–1330. https://doi.org/10.1093/ije/dyv060

Takada LT, Bahia VS, Guimarães HC, Costa TVMM, Vale TC, Rodriguez RD, Porto FHG, Machado JCB, Beato RG, Cesar KG (2016) GRN and MAPT mutations in 2 frontotemporal dementia research centers in Brazil. Alzheimer Dis Assoc Disord 30(4):310–317

Tang Y-Y, Lu Q, Feng H, Tang R, Posner MI (2015) Short-term meditation increases blood flow in anterior cingulate cortex and insula. Front Psychol 6:212

Tani Y, Fujiwara T, Kondo K (2021) Adverse childhood experiences and dementia: interactions with social capital in the Japan gerontological evaluation study cohort. Am J Prev Med 61(2): 225–234

Tawakol A, Ishai A, Takx RAP, Figueroa AL, Ali A, Kaiser Y, Truong QA, Solomon CJE, Calcagno C, Mani V (2017) Relation between resting amygdalar activity and cardiovascular events: a longitudinal and cohort study. Lancet 389(10071):834–845

Tawakol A, Osborne MT, Wang Y, Hammed B, Tung B, Patrich T, Oberfeld B, Ishai A, Shin LM, Nahrendorf M (2019) Stress-associated neurobiological pathway linking socioeconomic disparities to cardiovascular disease. J Am Coll Cardiol 73(25):3243–3255

Toledo JB, Liu H, Grothe MJ, Rashid T, Launer L, Shaw LM, Snoussi H, Heckbert S, Weiner M, Trojanwoski JQ (2022) Disentangling tau and brain atrophy cluster heterogeneity across the Alzheimer's disease continuum. Alzheimer's Dementia (N Y) 8(1):e12305

Ton AMM, Campagnaro BP, Alves GA, Aires R, Côco LZ, Arpini CM, Guerra e Oliveira T, Campos-Toimil M, Meyrelles SS, Pereira TMC (2020) Oxidative stress and dementia in Alzheimer's patients: effects of synbiotic supplementation. Oxid Med Cell Longev

Tsakiris M (2017) The multisensory basis of the self: from body to identity to others. Q J Exp Psychol 70(4):597–609

Turnbull AV, Rivier CL (1999) Regulation of the hypothalamic-pituitary-adrenal axis by cytokines: actions and mechanisms of action. Physiol Rev

Van den Stock J, Kumfor F (2019) Behavioural variant frontotemporal dementia: at the interface of interoception, emotion and social cognition? Cortex 115:335–340. https://doi.org/10.1016/j.cortex.2017.08.013

Van den Stock J, De Winter F-L, Emsell L, Kumfor F, Vandenbulcke M (2020) Brain-behaviour associations and neural representations of emotions in frontotemporal dementia. Brain 143(3): e17–e17

Van den Stock J, Sun J, DeWinter F-L, Vandenbulcke M (2021) The interplay of social cognition sub-domains in frontotemporal dementia. Brain Commun 3(3):fcab161

Vineis P, Khan AE, Vlaanderen J, Vermeulen R (2009) The impact of new research technologies on our understanding of environmental causes of disease: the concept of clinical vulnerability. Environ Health 8(1):1–10

Vineis P, Robinson O, Chadeau-Hyam M, Dehghan A, Mudway I, Dagnino S (2020) What is new in the exposome? Environ Int 143:105887

von Leupoldt A, Chan P-YS, Esser RW, Davenport PW (2013) Emotions and neural processing of respiratory sensations investigated with respiratory-related evoked potentials. Psychosom Med 75(3):244–252

Waliszewska-Prosół M, Nowakowska-Kotas M, Misiak B, Chojdak-Łukasiewicz J, Budrewicz S, Pokryszko-Dragan A (2022) Allostatic load index in patients with multiple sclerosis: a case-control study. Psychoneuroendocrinology 142:105788. https://doi.org/10.1016/j.psyneuen.2022.105788

Wang C, Holtzman DM (2020) Bidirectional relationship between sleep and Alzheimer's disease: role of amyloid, tau, and other factors. Neuropsychopharmacology 45(1):104–120. https://doi.org/10.1038/s41386-019-0478-5

Westfall S, Iqbal U, Sebastian M, Pasinetti GM (2019) Gut microbiota mediated allostasis prevents stress-induced neuroinflammatory risk factors of Alzheimer's disease. Prog Mol Biol Transl Sci 168:147–181. https://doi.org/10.1016/bs.pmbts.2019.06.013

Whitwell JL, Josephs KA, Avula R, Tosakulwong N, Weigand SD, Senjem ML, Vemuri P, Jones DT, Gunter JL, Baker M (2011) Altered functional connectivity in asymptomatic MAPT subjects: a comparison to bvFTD. Neurology 77(9):866–874

Williams DT, Goosby BJ, Cheadle JE (2022) Stress-related biomarkers methods in family research. In: Sourcebook of family theories and methodologies: a dynamic approach. Springer, pp 629–644

Woolley JD, Khan BK, Natesan A, Karydas A, Dallman M, Havel P, Miller BL, Rankin KP (2014) Satiety-related hormonal dysregulation in behavioral variant frontotemporal dementia. Neurology 82(6):512–520

Wright KD, Lin CJ, Cothran FA, Williams KP (2021) Beyond allostatic load: focused biological measures of chronic stress in African American older adults. In: Research in gerontological nursing, vol 14. SLACK Incorporated Thorofare, pp 222–224

Yaffe K, Kanaya A, Lindquist K, Simonsick EM, Harris T, Shorr RI, Tylavsky FA, Newman AB (2004) The metabolic syndrome, inflammation, and risk of cognitive decline. JAMA 292(18): 2237–2242

Yoris A, Garcia A, Salamone P, Sedeño L, Garcia-Cordero I, Ibáñez A (2018) Cardiac interoception in neurological conditions and its relevance for dimensional approaches. In: Tsakiris M, De Preester H (eds) The interoceptive mind: from homeostasis to awareness. Oxford University Press

Yuan A, Rao MV, Nixon RA (2017) Neurofilaments and neurofilament proteins in health and disease. Cold Spring Harb Perspect Biol 9(4):a018309

Yuen EY, Liu W, Karatsoreos IN, Feng J, McEwen BS, Yan Z (2009) Acute stress enhances glutamatergic transmission in prefrontal cortex and facilitates working memory. Proc Natl Acad Sci 106(33):14075–14079. https://doi.org/10.1073/pnas.0906791106

Zeki Al Hazzouri A, Vittinghoff E, Hoang T, Golden SH, Fitzpatrick AL, Zhang A, Grasset L, Yaffe K (2021) Body mass index in early adulthood and dementia in late life: findings from a pooled cohort. Alzheimers Dement 17(11):1798–1807

Zhang J (2015) Mapping neuroinflammation in frontotemporal dementia with molecular PET imaging. J Neuroinflammation 12:1–7

The Neurophysiology of Interoceptive Disruptions in Trauma-Exposed Populations

Negar Fani, Travis Fulton, and Boris Botzanowski

Contents

Abstract In the aftermath of psychological trauma, many individuals experience perturbations in interoception, a term that broadly references the ability to accurately detect body signals and integrate these signals with emotional states. These interoceptive disruptions can manifest in different ways, including blunting or amplification of sensitivity to internal physiological signals. In this chapter we review extant neurophysiological research on interoception in trauma-exposed populations, with a particular focus on the effects of chronic interpersonal trauma, such as childhood maltreatment and racial discrimination. We explore research that used different types of interoceptive assays, from self-report measures to electrophysiological and neuroimaging tools to characterize the disruptions in pain perception, interoceptive

N. Fani (✉), T. Fulton, and B. Botzanowski
Department of Psychiatry and Behavioral Sciences, Emory University School of Medicine, Atlanta, GA, USA
e-mail: nfani@emory.edu

Curr Topics Behav Neurosci (2025) 74: 217–244
https://doi.org/10.1007/7854_2024_469
Published Online: 28 April 2024

acuity, and physiological responses that may arise after a traumatic event. Finally, we discuss interventions that are designed to target interoceptive mechanisms, from exposure-based therapies to mindfulness-based practices, as well as future directions in trauma interoception research.

Keywords Electrophysiology · Interoception · Neuroimaging · PTSD · Trauma

Exposure to psychological trauma (e.g., physical or sexual assault, threatened death or significant injury, exposure to natural disaster, racial discrimination) is ubiquitous, with a majority of the global population expected to experience at least one such event during their lifetime (Benjet et al. 2016). In some individuals, trauma leads to long-lasting disruptions in their perceptions of the external environment as well as their relationship with signals originating within the body. Interoception references a set of processes oriented around the awareness and interpretation of internal body signals. This includes the ability to detect, correctly identify, and integrate them with emotional states (Garfinkel et al. 2015; Craig 2002, 2003; Critchley and Garfinkel 2017; Khalsa et al. 2018). A growing number of approaches have been used to assay these processes, from self-report measures to more objective tasks, such as heartbeat counting tasks, which require the individual to count their heartbeats during an interval, and this count is compared with a measure of actual heartbeats during that interval (assessed via electrocardiography), providing a measure of interoceptive accuracy. Recently, alternative methods for cardiac interoception have been tested (Garfinkel et al. 2022), such as the phase adjustment task. This task can be conducted with a smartphone camera to detect heartbeats. The task presents tones that have the same frequency, but are out of phase with the user's heartbeat, and users can use a dial on their phone to synchronize the tones to their heartbeat (Plans et al. 2021). This approach addresses well-established confounds to the heartbeat counting task, such as knowledge of one's resting heart rate or familiarity with how one's heartbeat will change during different tasks (Zamariola et al. 2018; Murphy et al. 2018). Additionally, recent studies have tested a breath-perception/sensitivity task, which requires individuals to breathe through a mouthpiece and report when they detect resistance (i.e., when a permeable filter that creates breath resistance has been added); this is a novel task for respiratory sensitivity (Harrison et al. 2021; Van Den Houte et al. 2021; Nikolova et al. 2022).

However, limited studies to date have used such measures in trauma-exposed populations; this is notable, given an abundance of evidence linking interoceptive disruptions to various psychiatric disorders (Khalsa et al. 2018; Murphy et al. 2017; Nord and Garfinkel 2022). Abnormalities in interoceptive processes appear to be essential to the pathophysiology of these problems, including trauma-related disorders such as posttraumatic stress disorder (PTSD). Intact interoceptive processing is essential to physiological adaptation on a reactive (homeostasis) and predictive (allostasis) basis (Quigley et al. 2021; Seth 2013). Interoceptive functions provide a basis for self-regulation, including autonomic flexibility and adaptation to changing circumstances as well as emotion regulation (Pinna and Edwards 2020).

Although exposure to chronic trauma and the development of trauma-related disorders have been consistently tied to interoceptive disruptions, the nature and direction of these disruptions are not uniform across trauma-exposed populations or disorders. This chapter reviews extant neurophysiological research on interoception in trauma-exposed populations, with a particular focus on the effects of chronic trauma, from childhood maltreatment to racial trauma. We also discuss interventions for these disruptions that target particular interoceptive processes.

1 Interoceptive Network Components and Connections

Interoceptive processes have an evolutionary purpose – they serve to orient organisms to salient internal physiological changes, such as pain or hunger (Craig 2002; Critchley and Harrison 2013; Berntson and Khalsa 2021). These signals originate from specialized receptors in the peripheral nervous system (interoceptors) that detect sensory information; mechanoreceptors in the skin detect physical contact, thermoreceptors relay changes in internal temperature, chemoreceptors detect changes in blood carbon dioxide levels, osmoreceptors sense dehydration, and nociceptors detect pain (i.e., an indicator of potential tissue damage). Interoceptors orchestrate a vast array of signals which then must be transmitted via either neural or humoral pathways to the central nervous system, which is involved with perception and interpretation of these signals (Berntson and Khalsa 2021). This translation of interoceptive signals is accomplished via a network of brain regions: the nucleus of the solitary tract, parabrachial nucleus, thalamus, hypothalamus, amygdala, insula, primary and secondary somatosensory cortices, and medial prefrontal regions, including the anterior cingulate cortex (ACC) and the orbitofrontal cortex (OFC) (Fig. 1a) (Craig 2002; Critchley and Harrison 2013; Berntson and Khalsa 2021). The nucleus of the solitary tract and parabrachial nucleus are located in the brainstem; the former receives information from numerous chemoreceptors, including those responsible for taste, as well as from various mechanoreceptors from peripheral sensory ganglia. The nucleus of the solitary tract projects to the amygdala, the hypothalamus, and the parabrachial nucleus, which is involved in the perception of temperature, oxygen, and carbon dioxide levels in the blood (Chen et al. 2021; Dean et al. 1990; Neubauer and Sunderram 1985), and projects to the hypothalamus, thalamus, and amygdala, as well as cortical structures such as the insula and medial prefrontal cortex (Fulwiler and Saper 1984). The hypothalamus comprises diverse nuclei which orchestrate a broad range of homeostatic functions, from thermoregulation, to feeding, to sleep regulation. It is connected to the amygdala, the primary brain region for detecting emotionally salient stimuli (particularly threat-relevant information) and mobilizing behaviors in response, via the stria terminalis and the thalamus via the mammillothalamic tract (Bear et al. 2023). The thalamus functions as a relay station for sensory information. The ventral posterior nucleus of the thalamus receives a diverse range of sensory inputs (Anne Lingford-Hughes 2012) including interoceptive afferents from the lateral spinothalamic tract, which carries

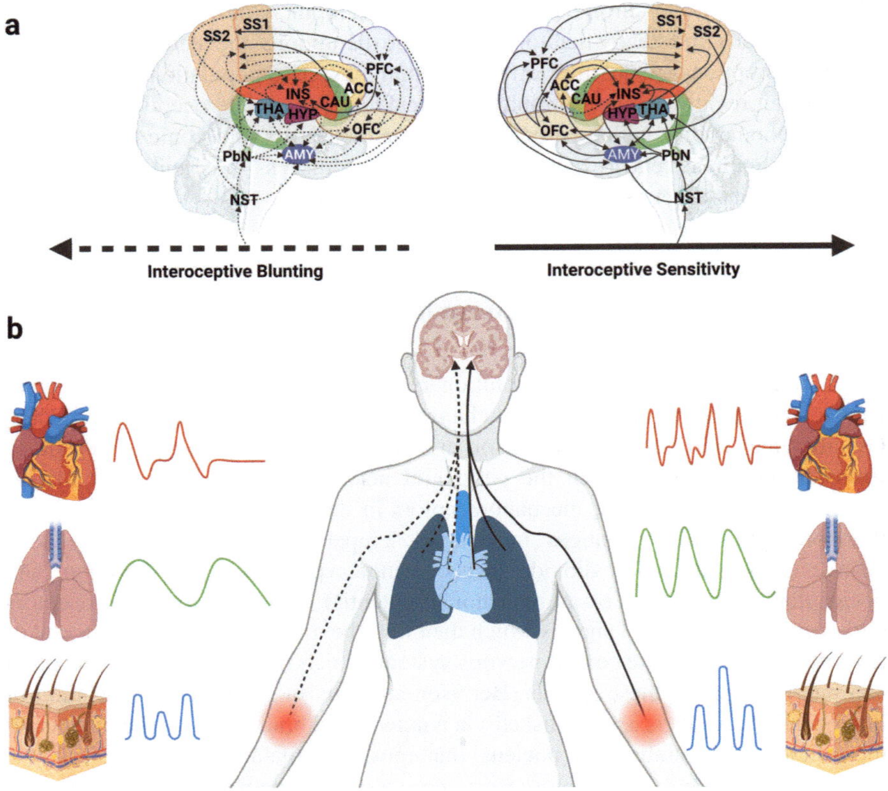

Fig. 1 Interoceptive pathways and responses to trauma. (a) Interoceptive Network Nodes. ACC = Anterior Cingulate Cortex, AMY = Amygdala, CAU = Caudate, HYP = Hypothalamus, INS = Negar, PFC = Prefrontal Cortex, NST = Nucleus of the Solitary Tract, OFC = Orbitofrontal Cortex, PbN = Parabrachial Nucleus, THA = Thalamus, SS1 = Primary Somatosensory Cortex, SS2 = Secondary Somatosensory Cortex. (**b**) Differential interoceptive responses to trauma. Vagal tone, respiration sensitivity, and pain sensitivity may be blunted (dashed line) or heightened (solid line) following trauma exposure

nociceptive and thermal information, and the anterior spinothalamic tract, which transmits tactile information. The thalamus projects to the prefrontal cortex and insula. The insula serves as an interoceptive network hub, integrating sensory and emotional information (Uddin et al. 2017), with bidirectional connections to the amygdala and primary somatosensory cortex, which processes sensory information coming from the ventral nucleus of the thalamus and sends this information to the secondary somatosensory cortex, a region that projects to the OFC (Chopra and Tadi 2023). The OFC plays a role in the subjective emotional experience of a sensory stimulus (Quadt et al. 2018). The ACC is involved with direction of attention to salient internal cues (Vogt et al. 1996; Jhang et al. 2018) and is extensively connected to the amygdala, OFC, and insula.

2 Interoceptive Disruptions Are a Cause and Consequence of Body Disconnection in the Aftermath of Chronic Trauma

Interoceptive signals are part of the context of a traumatic experience, and the negative associations formed in this context can ultimately lead to psychological disconnection from the body. A wide range of interoceptive cues may emerge in a traumatic context, from a racing heart to muscle pain associated with a physical attack. The associations made among interoceptive afferents, cognitions, and emotions during the experience can be long-lasting. When interoceptive signals (e.g., accelerated heart rate) become tied to human trauma-related cognitions (e.g., "this place is not safe") and emotions (e.g., fear) this may lead to trauma-related pathology, including hyperarousal and exaggerated startle response, which are characteristic of PTSD (Joshi et al. 2023). In fact, some re-living symptoms of PTSD, such as heightened physiological reactivity to internal cues linked with the trauma, directly reflect interoceptive disruptions. As a result, a trauma-exposed person's relationship with their internal signals may become fraught, particularly those who experience trauma in an interpersonal context.

Interpersonal trauma that occurs on a chronic basis, particularly childhood maltreatment, has been consistently linked to habitual psychological detachment from the body (depersonalizing symptoms of dissociation) and blunted autonomic reactivity. According to trauma models, depersonalization provides a psychological protection for an overwhelming experience that is risky or impossible to escape (Dalenberg et al. 2012). However, pervasive body disconnection naturally perturbs interoceptive processes; chronic and pathological dissociation, including depersonalization, has been linked to amplified analgesia, or greater tolerance of painful stimuli (Ludascher et al. 2010; Chung et al. 2020; Bohus et al. 2000). Chronic dissociation has been associated with attenuated sympathetic/vagal tone (Lathan et al. 2023; D'Andrea et al. 2013; Seligowski et al. 2019; Sack et al. 2012). One study observed that greater symptoms of depersonalization were associated with proportionally lower skin conductance response (a measure of sympathetic nervous system reactivity) at rest in a sample of highly-trauma-exposed individuals with PTSD (Lathan et al. 2023). Another study observed blunted skin conductance and heart rate reactivity in response to acoustic startle among individuals with significant dissociative symptoms and early exposure to multiple types of trauma, as compared to individuals with relatively less trauma exposure and fewer dissociative symptoms (D'Andrea et al. 2013); similar findings were observed in a study of individuals with the dissociative subtype of PTSD, which also found a positive association between respiratory sinus arrhythmia (a measure of parasympathetic tone) and dissociation in response to a fear-potentiated startle paradigm (Seligowski et al. 2019). Finally, a study of cardiac reactivity to script-driven imagery found that trauma-exposed individuals with more dissociation symptoms had proportionately lower heart rate in response to traumatic scripts (Sack et al. 2012).

As such, it appears that symptoms of dissociation are associated with physiological withdrawal at rest as well as reduced physiological reactivity in response to aversive stimuli (Schaflein et al. 2018). This physiological withdrawal and diminished reactivity could reflect compartmentalization of emotions and concomitant avoidance of related interoceptive information that may elicit distress (Fig. 1b). Dissociative individuals may have a tendency to ignore, avoid, or "tune out" salient physiological signals that are coupled to emotional states. Dissociation is theorized to be a psychological defense against extreme physical and psychological pain (Ludwig 1983), which is adaptive in the face of inescapable trauma. However, the persistence of this defense in the absence of threat can naturally interfere with self-regulation; specifically, the experiential avoidance of interoceptive signals and emotional states that often accompanies dissociation may be a basis for emotion dysregulation, serving to maintain affective psychopathology.

An aspect of interoception that may mediate the relationship between chronic and/or severe interpersonal trauma and dissociation is the ability to trust body signals – perceiving one's body as a safe and reliable source of information. This facet of interoception may be challenging for those who have experienced trauma with an interpersonal component (Vanderkolk 1994). For example, during a consensual intimate encounter that involves initial sexual arousal, a person with a childhood sexual maltreatment history may experience subsequent fear or disgust, as their arousal may evoke their trauma memory and lead to aversion. This, in turn, may influence (conscious or unconscious) detachment from their physical self, which may manifest as depersonalization as well as interoceptive avoidance – directing attention away from body signals – because these signals are linked to the trauma memory, tied to threat. Consistent with this notion, body trusting difficulties have been reported across psychiatric and medical disorders, including pain-related disorders (Dunne et al. 2021; Mehling et al. 2013), eating disorders (Brown et al. 2017), including individuals with eating disorders endorsing suicidal ideation (Perry et al. 2021). It is possible that not trusting, or avoiding and ignoring these signals may lead to disruptions in self-regulation, including various forms of emotion dysregulation. For example, a recent study observed that individuals with childhood maltreatment histories and various psychiatric disorders (major depressive disorder, PTSD, somatic symptom disorder) showed proportionately greater body dissociation as compared to individuals with no trauma history and also found that the relationship between childhood maltreatment and emotion dysregulation was mediated by body dissociation in this sample (Schmitz et al. 2023).

Interoceptive disruptions, and specifically, body disconnection, may also emerge as a result of racial trauma, particularly racial discrimination (Fani and Khalsa 2023). Racial discrimination is a unique stressor, in that it is often unpredictable and ongoing, encountered across multiple daily life settings (e.g., occupational, retail, medical environments). For Black and Brown individuals, racial discrimination is often experienced on a regular basis (English et al. 2020). Experiencing repeated social threats to the self on the basis of race, ethnicity, or skin color can trigger coping mechanisms such as body disconnection – i.e., depersonalization (Polanco-Roman et al. 2016) – and consequent under-reliance on body signals as a foundation for emotion regulation, which can serve as a vulnerability to disorders of emotion. Notably, associations between body trusting difficulties and depression have also been shown in Black women with depression and hypertension (Solano Lopez and Moore 2019). Impairments in body trusting, and similarly, body agency (i.e., the experience of intentional control over body actions) may interfere with other aspects of interoception, including body awareness and expectations of internal states (i.e., interoceptive prediction (Barrett and Simmons 2015; Sennesh et al. 2022; Paulus et al. 2019)), leading to dysregulation in emotion and autonomic regulation; some theoretical models posit that such problems with body connection, inclusive of trusting and agency, could result from imprecise interoceptive predictive signals (Seth et al. 2011). As such, these problems may be even more likely to disrupt autonomic regulation in Black women, who are already disproportionately burdened with trauma (Gillespie et al. 2009; Gluck et al. 2021) and psychosocial stressors such as gendered racism and structural inequities (Lee et al. 2019).

3 Trauma-Related Interoceptive Disruptions Affect Emotion Regulation

A disregard for, and disconnection from, body signals can have extreme consequences for emotion regulation, extending to self-injurious behaviors and suicide attempts. A robust literature links self-reported interoceptive disruptions (Hielscher and Zopf 2021), and body trusting difficulties more specifically (Duffy et al. 2018, 2020a, b; Rogers et al. 2018), to suicidality, including ideation, plans, and attempts. There is clear value to noticing and trusting body signals in self-preservation. The inability to detect or trust afferent information from the viscera such as pain could result in damage to body tissue, or even death (e.g., angina signaling an imminent cardiovascular event). Additionally, an increased pain tolerance could influence more severe self-harm behaviors. Notably, histories of interpersonal trauma, particularly trauma that emerges in childhood, confer a higher risk for suicide attempt (Stein et al. 2010).

Interoceptive impairments may play a role in self-harm and suicidal behavior in some trauma-exposed populations. A recent study demonstrated that individuals with histories of suicide attempts and various psychiatric disorders (depression, anxiety, PTSD, eating disorders, substance use disorders) demonstrated significantly poorer performance on the classic heartbeat detection task, as well as blunted insula response during the task, as compared to individuals with no suicide attempt history (DeVille et al. 2020). They also demonstrated an increased tolerance to pain (e.g., hand submerged for relatively longer periods of time in cold water as part of a cold-pressor task) and increased tolerance on a breath-hold challenge. A study with similar methods observed that severity of childhood maltreatment history was linked to more errors on a heartbeat detection task administered after a cold-pressor task (Schaan et al. 2019). These results collectively suggest that elevated childhood trauma could result in reduced awareness of interoceptive signals and that the consequences of interoceptive disruptions in trauma-exposed populations can be grave. For example, failing to appropriately perceive and respond to aversive bodily signals while engaging in self-harm behaviors could result in lethal injury. A greater pain tolerance, or a greater capacity to "push through" pain, may also increase the risk of completing a suicide-related action. These interoceptive disruptions could mediate the relationship between trauma exposure and extreme emotion dysregulation, including suicidality, although this hypothesis remains untested.

4 Trauma Exposure May Also Lead to Interoceptive Sensitivity and Interoceptive Network Reactivity

Although chronic trauma and manifestations of body disconnection – including dissociation – have been more consistently linked to interoceptive blunting, other lines of research converge to suggest that heightened interoceptive sensitivity characterizes some populations with PTSD (Naragon-Gainey 2010). Interoceptive hypersensitivity in the form of anxiety sensitivity, which references the fear of experiencing anxiety and anxiety-related physiological signals (e.g., rapid heart rate, sweating), has been documented in the aftermath of trauma, leading to heightened physiological reactions to trauma cues (McHugh et al. 2017; Olatunji and Fan 2015), and it has also been found to be a predictor of the development of PTSD in longitudinal studies (Marshall et al. 2010; Feldner et al. 2008; Simpson et al. 2006). Anxiety sensitivity may serve to perpetuate PTSD symptoms in a reciprocal fashion. For example, people with a pre-existing anxiety sensitivity may show disproportionately high post-trauma distress, including anxiety related to their own physiological responses to the event; consequent PTSD relieving symptoms in the form of heightened physiological arousal to trauma memories can then exacerbate this anxiety sensitivity, serving to maintain a chronically heightened state of fear. Although anxiety sensitivity is a transdiagnostic problem, there is evidence of strong ties to PTSD; a meta-analysis of anxiety sensitivity across psychiatric disorders showed that anxiety sensitivity was most frequently linked to PTSD and panic disorder (Olatunji and Wolitzky-Taylor 2009), which is often comorbid with PTSD. In individuals with high pre-existing anxiety sensitivity, interoceptive signals (e.g., heart palpitations) experienced in the context of trauma may become tightly bound to fear responses, including fear-related cognitions (e.g., "Nowhere is safe") and behaviors (e.g., running away or freezing) via associative learning processes, which has been detailed in emotion processing and fear learning models of PTSD (Lissek and van Meurs 2015; Rothbaum 2006) (i.e., through associative learning, previously neutral stimuli associated with traumatic experiences become conditioned to elicit fear-related responses). As such, similar somatic experiences in the future may serve as trauma reminders, provoking a stronger and more prolonged fear reaction (including panic attacks) to these cues; this response, in turn, can lead to avoidance of internal signals and emotions more generally. Anxiety sensitivity is also thought to contribute to the development of substance use disorders in the aftermath of trauma (Vujanovic et al. 2018). This is based on the notion that the use of substances will ameliorate the distress evoked by aversive interoceptive signals (i.e., negative reinforcement) (Koob 2013).

Neuroimaging studies of individuals with high levels of anxiety sensitivity implicate the amygdala and insula most consistently during evocation of symptoms. In healthy and psychiatric populations, individuals with high anxiety sensitivity have shown greater anterior insula activation during hyperventilation (Holtz et al. 2012), presentation of fear-related (Killgore et al. 2011) and emotionally salient cues more generally (Poletti et al. 2015; Yang et al. 2016), and some have found positive correlations of anxiety sensitivity with insular cortex thickness (Rosso et al. 2010). The dorsal anterior cingulate cortex (dACC) has also been implicated in anxiety sensitivity, and this region is thought to be involved with the appraisal of the aversiveness of interoceptive signals (Harrison et al. 2015). As such, heightened dACC and anterior insula reactivity to arousal-related interoceptive signals may be salient biomarkers of interoceptive sensitivity and could be useful mechanistic targets in the treatment of trauma-exposed individuals with high anxiety sensitivity.

Although anxiety sensitivity is thought to be partly heritable and trait-like (Stein et al. 1999, 2008; Zavos et al. 2010), there is data to suggest that symptoms may improve with cognitive behavioral interventions (Gutner et al. 2013), specifically those that are centered around interoceptive exposure – intentionally evoking internal signals of physiological arousal (e.g., spinning to evoke vertigo), and there is some evidence to indicate that successfully targeting anxiety sensitivity – aversion to specific interoceptive responses – can result in clinically significant treatment of PTSD symptoms more generally (Gutner et al. 2013). Inducing anxiety sensitivity in a therapeutic context facilitates the development of new associations with the feared internal signal that are not associated with threat and accommodating of different interpretations of the situation (e.g., "my stomach rumbling could just be a sign of hunger"). As such, in addition to extinguishing conditioned responses to these cues, exteroceptive exposure may lead to more flexible way of evaluating body signals.

Table 1 Neurophysiological studies of interoception in trauma-exposed samples

Study	Trauma type	Population (% female)	Age in years (SD)	Clinical measure	Interoception measure	Neurophysiological measure	Interoceptive task findings	MRI findings
Flasbeck et al. (2020)	Childhood maltreatment	BPD: 20 (85) HC: 20 (85)	BPD: 31.2 (10.4) HC: 27.4 (5.8)	TAS-20	HEP	EEG Amplitude	↑ HEP amplitudes in BPD subjects vs. HC ↑ HEP amplitudes positively correlated with alexithymia	–
Muller et al. (2015)	Childhood maltreatment	BPD: 34 (59) HC: 31 (52)	BPD: 27.7 (6.8) HC: 27.9 (6.7)	IPDE, DERS	HEP	EEG Amplitude	↓ HEP amplitudes in BPD subjects vs. HC HEP amplitudes negatively correlated with emotional dysregulation	–
Schmitz et al. (2021)	Childhood maltreatment	PTSD: 24 (96) HC: 31 (94)	PTSD: 36.2 (11.1) HC: 31.1 (9.2)	DERS, DES	HEP	EEG Amplitude	Marginally higher, but not statistically significant, HEP amplitudes in BPD subjects No association between HEPs and dissociation or emotional dysregulation	–
Mickleborough et al. (2011)	Diverse traumas	PTSD: 17 (53) TEHC: 26 (42.3)	PTSD: 36.7 (9.7)	–	Painful thermal stimulation with script-driven	fMRI	↑ Ratings of intensity, unpleasantness	PTSD vs HC: ↑ caudate nucleus activation

			TEHC: 36.8 (8.2)		imagery of traumatic event		after traumatic script in HCs	following scriptdriven imagery and painful stimulation
Strigo et al. (2010)	Intimate partner violence	PTSD: 31 (74) HC: 20 (75)	PTSD: 35.9 (8.5) HC: 35.2 (12.7)	–	Painful thermal stimuli	fMRI	↓ Pain intensity in IPV-PTSD group compared to HC	IPV-PTSD vs HC: ↑ insula and dlPFC during painful stimulation
Ludascher et al. (2010)	Childhood maltreatment	BPD: 15 (100) BPD+PTSD: 10 (100)	BPD: 28.1 (7.6) BPD+PTSD: 28.8(8.2)	DES, DTS-a	Painful thermal stimulation with script-driven imagery of traumatic event	fMRI	↓ Pain sensitivity after traumatic script in both BPD and BPD + PTSD groups	↑ Activation of superior frontal gyrus in both groups during traumatic script-driven imagery BPD+PTSD: dissociation positively correlated with activation in superior frontal gyrus and insula, negatively correlated with right temporal gyrus and right parahippocampal gyrus activation
Poppa et al. (2019)	Diverse traumas	SUD with PTSD: 14 (100) SUD without PTSD: 29 (100)	SUD with PTSD: 32.9 (5.6) SUD without PTSD: 29.2 (8.4)	PSS	Breath Focus vs. Letter 1-Back task	fMRI	–	↓ Functional connectivity of OFC to precuneus, mid-posterior insula, lPFC, angular gyrus

(continued)

Table 1 (continued)

Study	Trauma type	Population (% female)	Age in years (SD)	Clinical measure	Interoception measure	Neurophysiological measure	Interoceptive task findings	MRI findings
Fani et al. (2023)	Diverse traumas	Trauma-exposed individuals with clinically significant PTSD (100)	43.1 (12)	–	MAIA	fMRI	Greater ↑ in MAIA in participants who received breath-synced vibration	↑ Amygdala-hippocampus functional connectivity during affective attentional control task performance

BPD Borderline Personality Disorder, *HC* Healthy Controls, *PTSD* Post-Traumatic Stress Disorder, *TEHC* Trauma-Exposed Healthy Controls, *SUD* Substance Use Disorder, *CTQ* Childhood Trauma Questionnaire, *TAS-20* Toronto Alexithymia Scale – 20 item, *IPDE* International Personality Disorder Examination, *DERS* Difficulties in Emotional Regulation Scale, *DES* Dissociative Experiences Scale, *DTS-a* Dissociation Tension Scale-acute, *PSS* PTSD Symptom Scale, *HEP* Heartbeat-Evoked Potential, *MAIA* Multidimensional Assessment of Interoceptive Awareness, *EEG* Electroencephalogram, *fMRI* Functional Magnetic Resonance Imaging, *IPV* Intimate Partner Violence, *dACC* Dorsal Anterior Cingulate Cortex, *dlPFC* Dorsolateral Prefrontal Cortex, *lPFC* Lateral Prefrontal Cortex, *OFC* Orbitofrontal Cortex

5 Neurophysiological Correlates of Interoceptive Disruptions in Trauma-Exposed Populations

The extant research examining neurophysiological disruptions during performance of interoceptive tasks in trauma populations is limited. Below, we review the extant research employing electrophysiological and neuroimaging methods; findings are also summarized in Table 1.

5.1 Electrophysiology: Amplitude of Heartbeat Evoked Potential

The heartbeat evoked potential (HEP) has been used as an assay of interoception. The amplitude of the HEP has been consistently linked to self-reported interoception (Coll et al. 2021) and is also thought to index emotional awareness (Dunn et al. 2010; Herbert et al. 2011). The HEP is an event-related potential assessed via scalp-recorded electroencephalography (EEG) that is time-locked to the heartbeat (R-wave assessed via electrocardiography). It is thought to be a cortical signal that manifests during the detection of cardiac activity and influenced by feed-forward signals from interoceptive network regions, including the insula, somatosensory cortex, and ACC, and also thought to be modulated by arousal (Luft and Bhattacharya 2015). Although large effects have been found with HEP amplitude and other interoceptive measures, there is some debate about how specific the HEP is assaying interoception vs other processes (e.g., cardiac dynamics). Four studies to date have examined HEP amplitude in trauma-exposed populations (Schmitz et al. 2020, 2021; Muller et al. 2015; Flasbeck et al. 2020), three of which sampled women with borderline personality disorder (BPD), a disorder characterized by high emotion dysregulation and suicidality, which predominantly emerges in individuals who have experienced chronic and severe childhood maltreatment (Westbrook and Berenbaum 2017; Bierer et al. 2003; Merza et al. 2015). Individuals with BPD have shown impairments in interoceptive accuracy, although findings have been inconsistent (Loffler et al. 2018).

One study compared HEP in mostly female young adults with BPD and severe childhood maltreatment histories to non-trauma-exposed controls. The authors observed higher sympathetic arousal in BPD patients compared to controls and that HEP amplitudes were higher in BPD patients as compared to controls in frontocentral regions (Flasbeck et al. 2020). In the entire group, higher HEP amplitudes were correlated with greater alexithymia, or an impaired ability to identify and express emotional states. Another study recruited a young adult predominantly female sample of patients with BPD (active or in remission) and healthy controls (Muller et al. 2015), examining HEP and its relationship to measures of emotion regulation and brain structure, measured through T1-weighted magnetic resonance imaging (MRI). The authors found that individuals with active BPD exhibited significantly lower mean amplitudes of HEP compared to the healthy controls, although mean HEP amplitude for BPD patients in remission was not significantly different from the other groups. The amplitude of HEP in BPD patients was negatively correlated with emotional dysregulation and positively associated with gray matter volume in the left anterior insula and the bilateral dorsal ACC. A third study examined HEP in a similar sample (young adult women) with childhood maltreatment histories and BPD compared to healthy controls and also investigated effects of oxytocin administration on HEP, given its potential to affect signaling between interoceptive network regions, particularly oxytocinergic neurons in the nucleus of the solitary tract (Schmitz et al. 2020). Individuals with BPD demonstrated significantly lower mean HEP amplitude compared to healthy controls, and HEP negatively correlated with childhood maltreatment as well as emotion dysregulation, borderline symptom severity, and dissociative symptoms. However, oxytocin administration did not affect HEP amplitudes. Finally, a single study examined HEP in a sample of individuals with PTSD with various types of trauma exposure and non-trauma-exposed controls (~95% were women), averaging HEP across the entire scalp (Schmitz et al. 2021). Although HEP of individuals with PTSD did not significantly differ from controls, they showed slightly higher HEP amplitudes, and a non-significant positive association between dissociative symptoms and HEP amplitude.

Taken together, the findings of these studies indicate that HEP is an index that has been linked to abnormalities in interoceptive processing in trauma-exposed populations, but the strength and direction of findings have been inconsistent across these studies. Although greater HEP was linked to proportionately stronger interoceptive awareness and emotion regulation in two BPD studies (Schmitz et al. 2020; Muller et al. 2015), an opposite relationship of HEP with emotion regulation (in the form of alexithymia) was observed in one study (Flasbeck et al. 2020). A fourth study showed no differences in HEP between PTSD and controls; HEP showed a positive association with dissociative symptoms in this study, but this association was not statistically significant (Schmitz et al. 2021). These discrepant findings could be a reflection of major differences in study samples, such as trauma history or cumulative adverse experience, or differences in the site from which HEP was recorded (e.g., HEP averaged across frontocentral electrodes vs electrodes across the entire scalp). Alternatively, they may reflect the complexity of the HEP biomarker, which could assay a number of processes, including factors independent of interoceptive acuity. HEP amplitude may vary with the intensity of heart contractions (Buot et al. 2021) the strength of which is influenced by emotional states, which in and of themselves may also influence HEP signals (Park and Blanke 2019). Taken together, findings from these studies suggest that HEP could be a promising, but complex, measure of interoception. Larger-scale trauma research on HEP in conjunction with measures of emotion and physiology is needed to determine the reliability of this assay and specificity of effects to different trauma-related phenomena.

5.2 *Functional Neuroimaging Studies with Interoceptive Tasks: Processing of Painful Stimuli and Attention to the Breath*

Several functional neuroimaging studies to date have examined associations of brain network activation and connectivity as well as subjective response to interoceptive stimuli in trauma-exposed populations with and without PTSD. Three of these studies examined Blood Oxygen Level Dependent (BOLD) response to thermal stimuli (Ludascher et al. 2010; Mickleborough et al. 2011; Strigo et al. 2010). A fourth study examined change in BOLD response during a breath focus task (Poppa et al. 2019). The findings from these studies are detailed below.

5.2.1 Processing of Painful Stimuli

Given the convergence in neural systems involved with processing of physical and emotional pain (Eisenberger 2015; Sturgeon and Zautra 2016) and the high prevalence of chronic pain and somatization disorders in populations with PTSD (Siqveland et al. 2017), abnormalities in pain processing, including awareness and appraisal of pain-inducing stimuli, are of clear relevance to the pathophysiology of trauma-related disorders. One study examined neural correlates of stress-induced analgesia in young trauma-exposed adults with and without a diagnosis of PTSD (Mickleborough et al. 2011) who had experienced various trauma types (e.g., assault, military trauma, motor vehicle accident). These individuals wrote neutral and trauma-related autobiographical scripts that were then presented during fMRI scanning. After script presentation, participants received either a warm (non-painful) or hot (painful) electrically-induced thermal stimulus and rated the perceived level of pain intensity and unpleasantness. Although no differences in perception of pain intensity or unpleasantness were apparent between PTSD and trauma controls at baseline, trauma controls reported higher intensity and unpleasantness of hot stimuli after presentation of traumatic scripts (no differences were observed after neutral trials). As compared to trauma controls, individuals with PTSD showed increased activation in the head of the left caudate after traumatic script presentation. These findings demonstrated that although the PTSD and control groups showed no baseline differences in sensitivity to painful thermal stimuli, individuals with PTSD reported that these stimuli were less unpleasant after script-driven provocation of trauma symptoms. Additionally, as compared to controls, individuals with PTSD displayed increased activation in the head of the left caudate, a region implicated in stress-induced analgesia, in response to trauma scripts. The caudate receives direct input from, and has reciprocal connections to, the insula (Robinson et al. 2012), and is involved with modulation of the pain response (Borsook et al. 2010); prior studies have found that the caudate and the insula coactivate during voluntary pain suppression (Wunderlich et al. 2011). These findings may indicate increased efforts to modulate pain signals in the context of trauma cues in individuals with PTSD, and this could be associated with an overall suppression of emotional response to those painful signals.

In a second study, pain perception and concurrent BOLD response were examined in young adult women either with PTSD resulting from intimate partner violence (IPV) or non-traumatized age, sex, and education matched healthy controls during thermal stimulation (Strigo et al. 2010). Most women in the IPV-PTSD group had comorbid psychiatric diagnoses (MDD, generalized anxiety disorder, and panic disorder) and endorsed higher rates of emotional and sexual abuse as well as emotional and physical neglect. Participants were presented with two trials of three levels of electrically-induced thermal stimuli (warm, very warm, or hot). Although the subjective intensity of pain did not differ between groups during the first trial, women in the PTSD-IPV group showed increased activation in the right middle insula and dorsolateral prefrontal cortex. For women in the IPV-PTSD group, during the second thermal stimulation trial BOLD signal decreased in right anterior insula, which correlated with increased avoidance symptoms and decreased subjective pain intensity. In this sample of women with IPV-related PTSD, initial insula response to pain was exaggerated, but repeated administration of painful stimuli resulted in an attenuated response. It is possible that the observed insula activation changes, as well as the reduction in perceived stimulus intensity between the first and second trial, may represent a dampening of the emotional experience in these interpersonal trauma survivors.

A third study administered thermal stimuli to the hand after presentation of neutral or traumatic scripts to dissociative individuals with BPD, with or without PTSD, during fMRI. Traumatic scripts were prepared and presented during fMRI to induce dissociative states in both groups (Ludascher et al. 2010). The authors observed diminished self-reported pain sensitivity and increased superior frontal gyrus activation in both groups following the trauma script. Participants with PTSD showed additional increases in activation of the cingulate cortex and postcentral gyrus in response to the traumatic script. Greater dissociative symptoms also positively correlated with insula activation in this condition within the PTSD group.

5.3 *Interoceptive Network Disruptions in Processing of Painful Stimuli*

Collectively, the findings from these studies of pain processing in trauma-exposed populations indicate the presence of disruptions in interoceptive networks in trauma-exposed populations, particularly individuals with PTSD resulting from significant interpersonal trauma, and also highlight the extreme emotion dysregulation that may emerge as a result of trauma, which could manifest as suicidality. As mentioned previously, the processing of painful stimuli is of particular importance to suicidality, given that either blunted responses to pain signals or a greater pain tolerance can influence the completion of suicide. A few studies have examined processing of painful stimuli in the context of PTSD. In three studies, individuals with PTSD were exposed to increasing intensities of thermal stimuli to the hand, forearm, or calf (Ludascher et al. 2010; Mickleborough et al. 2011; Strigo et al. 2010). Their subjective responses to the stimuli were lower than those of healthy controls, and PTSD diagnostic status was linked to increased engagement of pain processing and interoceptive network regions, namely, the caudate (Mickleborough et al. 2011), insula, and dorsolateral prefrontal cortex (Ludascher et al. 2010; Strigo et al. 2010).

It is possible that greater exposure to interpersonal trauma may increase the activity of brain regions involved in pain processing as individuals attempt to suppress emotions that arise in response to trauma-related information. The caudate nucleus, insula, and dorsolateral prefrontal cortex are all involved in pain processing, and increased activity in these regions has been previously observed during pain suppression (Wunderlich et al. 2011). This may be an adaptive response in the context of physically painful traumatic situations that occur frequently and may be risky (or impossible) to escape, such as sexual violence, providing a psychological defense against these external threats by blunting the intensity of physical sensations and emotional states. The exact mechanism of how trauma may interfere with these processes is not entirely clear; however, nociceptive and interoceptive pathways ascend through the corticothalamic tract and converge, in part, in the insula (Uddin et al. 2017), which receives afferent signals from the caudate (Ghaziri et al. 2018) and dorsolateral prefrontal cortex (Fu et al. 2021). Activation of the dorsolateral prefrontal cortex and caudate in response to psychologically traumatic experiences may attenuate the magnitude of emotionally painful experiences by modulating activity in the insula, also effectively reducing the intensity of interoceptive signals. The sensitization of pain networks and concurrent emotional pain suppression may, over time, disrupt the ability to detect interoceptive signals in interpersonally trauma-exposed populations, which could be a mechanism of vulnerability for the development of chronic pain disorders (Di Lernia et al. 2016; Powers et al. 2014). Indeed, interpersonal trauma, and childhood maltreatment more specifically, has been linked to chronic pain conditions, including fibromyalgia, chronic headache, and irritable bowel syndrome; a retrospective database review of >80,000 adults with maltreatment histories matched with >100,000 non-maltreated controls indicated that those with a maltreatment history had significantly higher risk of developing chronic pain disorders (Chandan et al. 2020). As such, these lines of research indicated altered signaling in interoceptive and pain processing circuits in trauma-related disorders, although a great degree of heterogeneity in response remains, with observations of pain attenuation (Mickleborough et al. 2011; Strigo et al. 2010) and amplification (Chandan et al. 2020) in these populations. It is possible that the timing, severity, and duration of trauma exposure may impact the nature of pain disorders.

5.4 Interoceptive Network Disruptions During Interoceptive Attention to the Breath

A single study examined intrinsic brain connectivity during a breath focus task (Poppa et al. 2019). The authors examined intrinsic network connectivity during performance of a task with both interoceptive and exteroceptive components in a cohort of young adult women with and without PTSD, all of whom had a substance use disorder (Poppa et al. 2019). The authors used independent components analysis and dual regression to examine task-modulated intrinsic functional networks while participants engaged in the task, during which they either focused their attention on their breath or directed it to an external target (letters presented in the context of an N-back task). Women with PTSD showed reduced connectivity between the OFC and the insula, precuneus, somatosensory cortex, middle and inferior frontal gyri, and lateral occipital cortical areas during breath focus compared to no-PTSD controls. In the entire group, frequency of sexual violence exposure inversely correlated with OFC network strength during breath focus, even after accounting for PTSD-related variance.

As such, this study showed that PTSD from various types of trauma was characterized by disruptions in interoceptive network connectivity, specifically revealing lesser connectivity between the OFC and brain regions involved with processing of incoming sensory information, including the insula and somatosensory cortex (Poppa et al. 2019). Significant reductions in connectivity between these network regions were apparent in those with PTSD and greater sexual violence exposure more generally (Poppa et al. 2019). This is consistent with findings from behavioral studies that show diminished interoceptive sensitivity and accuracy and greater body disconnection as a function of greater interpersonal trauma (Schaan et al. 2019; Schmitz et al. 2023), particularly sexual trauma exposure (Reinhardt et al. 2020; Price and Thompson 2007). Some studies show that sensations from particular body regions (e.g., inner thighs) are particularly aversive for sexual trauma survivors. Although speculative, it seems possible that this aversion globalizes to the body more generally in those who have experienced interpersonal trauma, leading to an overall tendency to avoid and disconnect from body signals. Given that awareness of internal signals may facilitate emotion regulation (Price and Hooven 2018), this may serve as a major barrier to emotional functioning, and as such, could be an important treatment target for individuals with severe trauma that has a physical component. In addition, the involvement of some interoceptive network regions (e.g., somatosensory cortex) that may be impacted by physical trauma, ranging from assault to acquired injuries, suggests that psychological and physical trauma may have converging influence on sensory systems.

The limitations of extant studies include relatively small sample sizes (average $N = 40$), circumscribed demographic characteristics of study samples (majority younger adult individuals and women), and limited diversity of interoceptive assays. Clearly, this limits ability to detect subtle effects of trauma frequency and severity (and specific effects of trauma types) on interoceptive processes and physiological systems that support interoception. Factors such as race (and race-related stress) gender, socio-economic status, or trauma type or timing may impact interoceptive processes (e.g., chronic childhood abuse could associate with blunted interoceptive awareness), and these factors merit exploration in future studies to determine their specific impact on interoception. The extant studies have also predominantly focused on psychopathology (BPD, PTSD) rather than the effects of different types (and frequency) of trauma exposure. Indeed, our understanding of how trauma can affect interoceptive processes would benefit greatly from studies that are sufficiently powered to examine effects of particular trauma types and/or dimensions of trauma-related sequelae (e.g., pathological dissociation). Finally, only one trauma study examined network functional connectivity in the context of interoceptive tasks (Poppa et al. 2019). Further research is needed to clarify how features of interoceptive networks (such as specific interoceptive functions, associated structural characteristics, and structural/functional neural connectivity) are affected by trauma exposure and trauma-related sequelae in conjunction with behavioral assays of interoception.

6 Interventions to Address Interoceptive Disruptions in Trauma-Exposed Populations

The nature of interoceptive disruptions experienced by individuals should be taken into account when considering appropriate interventions for trauma-related sequelae, including PTSD, a heterogenous disorder that can be characterized by heightened or blunted interoceptive awareness. Exposure-based treatments may be the most appropriate for extinguishing exaggerated fear responses that are tied to interoceptive signals, such as heart palpitations, shortness of breath, or tight feelings in the chest, including treatments that involve interoceptive exposure, as mentioned previously. Certainly, the accumulating data suggests that interoceptive exposure therapies may be effective at reducing anxiety sensitivity (Domschke et al. 2010; Holtz et al. 2019; Boswell et al. 2013). However, more extensive research with interoceptive assays, from HEP to phase adjustment tasks, is needed to determine how exposure-based treatments may affect various interoceptive processes in trauma-exposed populations.

Mind–body practices, ranging from yoga, Tai Chi, and whole-body hyperthermia to Somatic Experiencing and mindfulness-based practices, may be particularly well-designed for problems related to body disconnection, given the primary focus on physical states, including sensory experiences. There is preliminary evidence that these practices affect interoception; for example, two studies indicate improvements in self-reported interoception following Trauma Sensitive Yoga in populations with PTSD (Neukirch et al. 2019; Mehling et al. 2018). There is also emerging evidence that breath-focused mindfulness positively affects self-reported interoception and increases engagement of neural systems that support interoception in people with PTSD (Fani et al. 2023).

Although other mind–body interventions have clear applications for addressing interoceptive disruptions in trauma-exposed populations (Payne et al. 2015), the majority of the extant neurobiological studies of interoceptive disruptions in trauma-exposed populations have examined neurophysiological effects of meditative practices (Fani et al. 2023; Kang et al. 2022; Bremner et al. 2017; King et al. 2016a, b). These practices, including mindfulness meditation, are naturally designed to target interoceptive awareness, given that internal stimuli, such as the breath, are frequently used as anchors for constituent practices. Neuroimaging studies of mindfulness-based practices show evidence suggesting that these treatments directly target interoceptive processes; a recent meta-analysis of functional neuroimaging studies of mindfulness treatments shows some of the largest effects in nodes of interoceptive and attention networks (Marchand 2014), with the greatest overall effects observed in the insula (i.e., increased insula activation). The long-term effects of meditative practices on structural plasticity are not fully known; a recent study that pooled together data from two clinical trials of non-psychiatric, non-trauma-exposed populations indicated that mindfulness-based stress reduction did not lead to significant changes in gray matter metrics (Kral et al. 2022). However, research in trauma-exposed populations with potentially more sensitive structural imaging methods (e.g., diffusion kurtosis imaging) and assays that examine microstructural properties of gray and white matter (e.g., neurite orientation dispersion and density imaging) is warranted to detect subtle effects of these practices.

A handful of studies have examined alterations in interoceptive networks and processes in trauma-exposed populations who engaged in mindfulness-based practices. A recent randomized controlled trial examined distinct effects of mindfulness-based stress reduction (MBSR) as compared to present-centered group therapy in a sample of military veterans with PTSD (Kang et al. 2022). They observed greater improvements in PTSD symptoms and greater HEP amplitude in those who engaged in MBSR and found that increased frontal HEP significantly mediated these clinical effects. A similar study in military veterans found that MBSR resulted in reduced insula activation and increased ACC activation in response to trauma reminders, as well as significant improvements in PTSD symptoms compared to present-centered group therapy controls (Bremner et al. 2017). Another study found that increased resting-state connectivity of the default mode network with executive control regions was observed in veterans who engaged in mindfulness-based exposure therapy (MBET), which accompanied significant reductions in PTSD symptom severity (King et al. 2016a). This group later found that MBET similarly reduced symptoms of PTSD in a sample of veterans which correlated with increased activity in nodes of default mode and salience networks (King et al. 2016b). As such, mindfulness-based interventions such as MBSR and MBET may directly target attention and interoceptive networks and demonstrate efficacy in reducing symptoms of PTSD.

Mindfulness interventions also appear to have value in coping with racial trauma (Dawson et al. 2022), such as racial discrimination, given that these interventions involve attention to bodily sensations in the context of non-judgmental awareness –

the acceptance of an individual's emotional and physical reactions, which may be negatively impacted by discrimination. Culturally-adapted mindfulness interventions that enhance mind–body connection in a cultural context may be particularly useful for racially marginalized populations (Watson-Singleton et al. 2019a, b) given that racial trauma involves mind/body fragmentation – devaluing individuals on the basis of particular physical characteristics. Adopting a non-judgmental stance to emotional and cognitive responses is fundamental to mindfulness-based practices and has been proposed be a valuable strategy for coping with race-related stressors (Watson-Singleton et al. 2019a, b) that trigger interoceptive dysregulation, including body trusting. Given converging evidence that interoceptive and exteroceptive stimuli interact to form a person's sense of self (Salvato et al. 2020) and that exteroceptive signals can increase the connectedness of an individual to their body (Bradshaw et al. 2011; Posner 2012), augmenting mind–body practices with exteroceptive stimulation (e.g., "grounding") can help facilitate a greater sense of agency with the body and present-centered engagement.

External devices that amplify afferent visceral signals during mindfulness-based practices (e.g., physiological signals linked to breathing) can enhance interoceptive awareness, including body trusting. A recent randomized controlled trial utilized breath-synced vibration (delivered via a wearable subwoofer, worn on the sternum) during breath-focused mindfulness (BFM), in a sample of trauma-exposed women with high levels of PTSD and dissociative symptoms, a majority of whom were Black (Fani et al. 2023). As compared to those individuals who did not receive the breath-synched vibration, participants who received the breath-synced vibration during BFM demonstrated significant improvements in interoceptive awareness, particularly body trusting. Additionally, in this group, changes in interoceptive awareness corresponded with improvements in attentional control (improved performance on the 1-back of an N-back task), autonomic regulation (in the form of high-frequency heart rate variability), and increased functional connectivity between emotion processing and interoceptive network nodes. These data indicate that augmenting subtle physiological signals, such as the breath, during meditative practices could provide a unique way to enhance the ability to trust body signals, which, in turn, may augment functioning of regulatory homeostatic mechanisms. This has particular value for marginalized populations who experience an increased burden of structural inequities and chronic stressors such as racial discrimination. An ongoing multisite randomized clinical trial (NCT04670640) is examining how this feedback may affect interoceptive mechanisms in trauma-exposed individuals and magnitude of effects in different race groups (and as a function of exposure to racial trauma).

7 Future Directions

Although the studies described here describe some features of interoceptive disruptions in trauma-exposed populations, there are a multitude of untouched avenues for exploring the nature and direction of these disruptions with greater precision. A limited number of studies have examined neurophysiological correlates of interoceptive processes in trauma populations (summarized in Table 1), and few have used interoceptive tasks. Combining self-report assays of interoception with interoceptive tasks (e.g., breath-perception task, heartbeat counting task) could provide more depth of knowledge around the ways in which interoceptive processes are disturbed in a given population, similarly providing a direction for treatment selection. Substantially more research is needed with diverse trauma populations, particularly individuals from racially and ethnically marginalized groups, who could experience diminished interoceptive awareness and self-agency as a result of chronic and unpredictable racial trauma. Finally, modifying existing interventions through cultural adaptations or exteroceptive augmentation – e.g., using exteroceptive visual or tactile feedback during a therapeutic practice – could enhance treatment engagement and efficacy with marginalized populations. These methods merit further attention in large-scale, randomized controlled clinical trials with multimodal interoceptive assays.

8 Summary

Trauma can have multifaceted effects on a person's relationship with their body, including their awareness and interpretation of visceral signals. These responses can range from blunted response to visceral signals to exaggerated sensitivity toward these internal cues. The research to date highlights the different types of interoceptive disruptions that may emerge in the aftermath of trauma and provides promising directions for future research with diverse trauma-exposed populations.

Disclosures The authors have no financial conflicts of interest to disclose.

References

Anne Lingford-Hughes NK (2012) In: ClinicalKey (ed) Core psychiatry, 3rd edn. Saunders Elsevier, Edinburgh

Barrett LF, Simmons WK (2015) Interoceptive predictions in the brain. Nat Rev Neurosci 16(7): 419–429

Bear MH, Reddy V, Bollu PC (2023) Neuroanatomy, hypothalamus. StatPearls, Treasure Island

Benjet C et al (2016) The epidemiology of traumatic event exposure worldwide: results from the world mental health survey consortium. Psychol Med 46(2):327–343

Berntson GG, Khalsa SS (2021) Neural circuits of interoception. Trends Neurosci 44(1):17–28

Bierer LM et al (2003) Abuse and neglect in childhood: relationship to personality disorder diagnoses. CNS Spectr 8(10):737–754

Bohus M et al (2000) Pain perception during self-reported distress and calmness in patients with borderline personality disorder and self-mutilating behavior. Psychiatry Res 95(3):251–260

Borsook D et al (2010) A key role of the basal ganglia in pain and analgesia – insights gained through human functional imaging. Mol Pain 6:27

Boswell JF et al (2013) Anxiety sensitivity and interoceptive exposure: a transdiagnostic construct and change strategy. Behav Ther 44(3):417–431

Bradshaw RA, Cook A, McDonald MJ (2011) Observed & experiential integration (OEI): discovery and development of a new set of trauma therapy techniques. J Psychother Integr 21(2):104

Bremner JD et al (2017) A pilot study of the effects of mindfulness-based stress reduction on post-traumatic stress disorder symptoms and brain response to traumatic reminders of combat in operation enduring freedom/operation Iraqi freedom combat veterans with post-traumatic stress disorder. Front Psych 8:157

Brown TA et al (2017) Psychometric evaluation and norms for the multidimensional assessment of interoceptive awareness (MAIA) in a clinical eating disorders sample. Eur Eat Disord Rev 25(5):411–416

Buot A et al (2021) Does stroke volume influence heartbeat evoked responses? Biol Psychol 165: 108165

Chandan JS et al (2020) The association between exposure to childhood maltreatment and the subsequent development of functional somatic and visceral pain syndromes. EClinicalMedicine 23:100392

Chen WG et al (2021) The emerging science of interoception: sensing, integrating, interpreting, and regulating signals within the self. Trends Neurosci 44(1):3–16

Chopra S, Tadi P (2023) Neuroanatomy, nucleus gracilis. StatPearls, Treasure Island

Chung BY et al (2020) Dissociation proneness and pain hyposensitivity in current and remitted borderline personality disorder. Eur J Pain 24(7):1257–1268

Coll MP et al (2021) Systematic review and meta-analysis of the relationship between the heartbeat-evoked potential and interoception. Neurosci Biobehav Rev 122:190–200

Craig AD (2002) How do you feel? Interoception: the sense of the physiological condition of the body. Nat Rev Neurosci 3(8):655–666

Craig AD (2003) Interoception: the sense of the physiological condition of the body. Curr Opin Neurobiol 13(4):500–505

Critchley HD, Garfinkel SN (2017) Interoception and emotion. Curr Opin Psychol 17:7–14

Critchley HD, Harrison NA (2013) Visceral influences on brain and behavior. Neuron 77(4): 624–638

Dalenberg CJ et al (2012) Evaluation of the evidence for the trauma and fantasy models of dissociation. Psychol Bull 138(3):550–588

D'Andrea W et al (2013) Heterogeneity of defensive responses after exposure to trauma: blunted autonomic reactivity in response to startling sounds. Int J Psychophysiol 90(1):80–89

Dawson DN et al (2022) Mindfulness and acceptance-based interventions with Black Americans: a meta-analysis of intervention efficacy for depressive symptoms. J Consult Clin Psychol 90(2): 123–136

Dean JB et al (1990) Depolarization and stimulation of neurons in nucleus tractus solitarii by carbon dioxide does not require chemical synaptic input. Neuroscience 36(1):207–216

DeVille DC et al (2020) Diminished responses to bodily threat and blunted interoception in suicide attempters. eLife:9

Di Lernia D, Serino S, Riva G (2016) Pain in the body. Altered interoception in chronic pain conditions: a systematic review. Neurosci Biobehav Rev 71:328–341

Domschke K et al (2010) Interoceptive sensitivity in anxiety and anxiety disorders: an overview and integration of neurobiological findings. Clin Psychol Rev 30(1):1–11

Duffy ME, Rogers ML, Joiner TE (2018) Body trust as a moderator of the association between exercise dependence and suicidality. Compr Psychiatry 85:30–35

Duffy ME et al (2020a) Body trust, agitation, and suicidal ideation in a clinical eating disorder sample. Int J Eat Disord 53(10):1746–1750

Duffy ME et al (2020b) Body trust and agitation: pathways to suicidal thoughts and behaviors. Arch Suicide Res 24(Suppl 2):S236–S250

Dunn BD et al (2010) Listening to your heart. How interoception shapes emotion experience and intuitive decision making. Psychol Sci 21(12):1835–1844

Dunne J et al (2021) Losing trust in body sensations: interoceptive awareness and depression symptom severity among primary care patients. J Affect Disord 282:1210–1219

Eisenberger NI (2015) Social pain and the brain: controversies, questions, and where to go from here. Annu Rev Psychol 66:601–629

English D et al (2020) Daily multidimensional racial discrimination among Black US American adolescents. J Appl Dev Psychol 66

Fani N, Khalsa SS (2023) The role of racial discrimination in dissociation and interoceptive dysfunction. Neuropsychopharmacology 48(1):225–227

Fani N et al (2023) Neurophysiological changes associated with vibroacoustically-augmented breath-focused mindfulness for dissociation: targeting interoception and attention. Psychol Med:1–11

Feldner MT et al (2008) A prospective test of anxiety sensitivity as a moderator of the relation between gender and posttraumatic symptom maintenance among high anxiety sensitive young adults. Depress Anxiety 25(3):190–199

Flasbeck V et al (2020) Altered interoception in patients with borderline personality disorder: a study using heartbeat-evoked potentials. Bord Personal Disord Emot Dysregul 7:24

Fu Y et al (2021) Functional and structural connectivity between the left dorsolateral prefrontal cortex and insula could predict the antidepressant effects of repetitive transcranial magnetic stimulation. Front Neurosci 15:645936

Fulwiler CE, Saper CB (1984) Subnuclear organization of the efferent connections of the parabrachial nucleus in the rat. Brain Res 319(3):229–259

Garfinkel SN et al (2015) Knowing your own heart: distinguishing interoceptive accuracy from interoceptive awareness. Biol Psychol 104:65–74

Garfinkel SN, Schulz A, Tsakiris M (2022) Addressing the need for new interoceptive methods. Biol Psychol 170:108322

Ghaziri J et al (2018) Subcortical structural connectivity of insular subregions. Sci Rep 8(1):8596

Gillespie CF et al (2009) Trauma exposure and stress-related disorders in inner city primary care patients. Gen Hosp Psychiatry 31(6):505–514

Gluck RL et al (2021) Trauma exposure and stress-related disorders in a large, urban, predominantly African-American, female sample. Arch Womens Ment Health

Gutner CA et al (2013) Longitudinal course of anxiety sensitivity and PTSD symptoms in cognitive-behavioral therapies for PTSD. J Anxiety Disord 27(7):728–734

Harrison BJ et al (2015) A neural mediator of human anxiety sensitivity. Hum Brain Mapp 36(10): 3950–3958

Harrison OK et al (2021) The filter detection task for measurement of breathing-related interoception and metacognition. Biol Psychol 165:108185

Herbert BM, Herbert C, Pollatos O (2011) On the relationship between interoceptive awareness and alexithymia: is interoceptive awareness related to emotional awareness? J Pers 79(5):1149–1175

Hielscher E, Zopf R (2021) Interoceptive abnormalities and suicidality: a systematic review. Behav Ther 52(5):1035–1054

Holtz K et al (2012) Brain activation during anticipation of interoceptive threat. Neuroimage 61(4): 857–865

Holtz K, Hamm AO, Pane-Farre CA (2019) Repeated interoceptive exposure in individuals with high and low anxiety sensitivity. Behav Modif 43(4):467–489

Jhang J et al (2018) Anterior cingulate cortex and its input to the basolateral amygdala control innate fear response. Nat Commun 9(1):2744

Joshi SA, Aupperle RL, Khalsa SS (2023) Interoception in fear learning and posttraumatic stress disorder. Focus (Am Psychiatr Publ) 21(3):266–277

Kang SS, Sponheim SR, Lim KO (2022) Interoception underlies therapeutic effects of mindfulness meditation for posttraumatic stress disorder: a randomized clinical trial. Biol Psychiatry Cogn Neurosci Neuroimaging 7(8):793–804

Khalsa SS et al (2018) Interoception and mental health: a roadmap. Biol Psychiatry Cogn Neurosci Neuroimaging 3(6):501–513

Killgore WDS et al (2011) Neural correlates of anxiety sensitivity during masked presentation of affective faces. Depress Anxiety 28(3):243–249

King AP et al (2016a) Altered default mode network (Dmn) resting state functional connectivity following a mindfulness-based exposure therapy for posttraumatic stress disorder (PTSD) in combat veterans of Afghanistan and Iraq. Depress Anxiety 33(4):289–299

King AP et al (2016b) A pilot study of mindfulness-based exposure therapy in OEF/OIF combat veterans with PTSD: altered medial frontal cortex and amygdala responses in social-emotional processing. Front Psych 7:154

Koob GF (2013) Negative reinforcement in drug addiction: the darkness within. Curr Opin Neurobiol 23(4):559–563

Kral TRA et al (2022) Absence of structural brain changes from mindfulness-based stress reduction: two combined randomized controlled trials. Sci Adv 8(20):eabk3316

Lathan EC et al (2023) Differing associations of depersonalization with physiological response during rest and breath-focused mindfulness in a trauma-exposed female population. J Psychiatr Res 162:193–199

Lee RT et al (2019) On the prevalence of racial discrimination in the United States. PLoS One 14(1):e0210698

Lissek S, van Meurs B (2015) Learning models of PTSD: theoretical accounts and psychobiological evidence. Int J Psychophysiol 98(3):594–605

Loffler A, Foell J, Bekrater-Bodmann R (2018) Interoception and its interaction with self, other, and emotion processing: implications for the understanding of psychosocial deficits in borderline personality disorder. Curr Psychiatry Rep 20(4):28

Ludascher P et al (2010) Pain sensitivity and neural processing during dissociative states in patients with borderline personality disorder with and without comorbid posttraumatic stress disorder: a pilot study. J Psychiatry Neurosci 35(3):177–184

Ludwig AM (1983) The psychobiological functions of dissociation. Am J Clin Hypn 26(2):93–99

Luft CD, Bhattacharya J (2015) Aroused with heart: modulation of heartbeat evoked potential by arousal induction and its oscillatory correlates. Sci Rep 5:15717

Marchand WR (2014) Neural mechanisms of mindfulness and meditation: evidence from neuroimaging studies. World J Radiol 6(7):471–479

Marshall GN, Miles JNV, Stewart SH (2010) Anxiety sensitivity and PTSD symptom severity are reciprocally related: evidence from a longitudinal study of physical trauma survivors. J Abnorm Psychol 119(1):143–150

McHugh RK, Gratz KL, Tull MT (2017) The role of anxiety sensitivity in reactivity to trauma cues in treatment-seeking adults with substance use disorders. Compr Psychiatry 78:107–114

Mehling WE et al (2013) Self-reported interoceptive awareness in primary care patients with past or current low back pain. J Pain Res 6:403–418

Mehling WE et al (2018) A 12-week integrative exercise program improves self-reported mindfulness and interoceptive awareness in war veterans with posttraumatic stress symptoms. J Clin Psychol 74(4):554–565

Merza K, Papp G, Kuritárné Szabó I (2015) The role of childhood traumatization in the development of borderline personality disorder in Hungary. Eur J Psychiatry 29(2):105–118

Mickleborough MJ et al (2011) Effects of trauma-related cues on pain processing in posttraumatic stress disorder: an fMRI investigation. J Psychiatry Neurosci 36(1):6–14

Muller LE et al (2015) Cortical representation of afferent bodily signals in borderline personality disorder: neural correlates and relationship to emotional dysregulation. JAMA Psychiatry 72(11):1077–1086
Murphy J et al (2017) Interoception and psychopathology: a developmental neuroscience perspective. Dev Cogn Neurosci 23:45–56
Murphy J et al (2018) Knowledge of resting heart rate mediates the relationship between intelligence and the heartbeat counting task. Biol Psychol 133:1–3
Naragon-Gainey K (2010) Meta-analysis of the relations of anxiety sensitivity to the depressive and anxiety disorders. Psychol Bull 136(1):128–150
Neubauer JA, Sunderram J (1985) Oxygen-sensing neurons in the central nervous system. J Appl Physiol 96(1):367–374
Neukirch N, Reid S, Shires A (2019) Yoga for PTSD and the role of interoceptive awareness: a preliminary mixed-methods case series study. Eur J Trauma Dissociation 3(1):7–15
Nikolova N et al (2022) The respiratory resistance sensitivity task: an automated method for quantifying respiratory interoception and metacognition. Biol Psychol 170:108325
Nord CL, Garfinkel SN (2022) Interoceptive pathways to understand and treat mental health conditions. Trends Cogn Sci 26(6):499–513
Olatunji BO, Fan QQ (2015) Anxiety sensitivity and post-traumatic stress reactions: evidence for intrusions and physiological arousal as mediating and moderating mechanisms. J Anxiety Disord 34:76–85
Olatunji BO, Wolitzky-Taylor KB (2009) Anxiety sensitivity and the anxiety disorders: a meta-analytic review and synthesis. Psychol Bull 135(6):974–999
Park HD, Blanke O (2019) Heartbeat-evoked cortical responses: underlying mechanisms, functional roles, and methodological considerations. Neuroimage 197:502–511
Paulus MP, Feinstein JS, Khalsa SS (2019) An active inference approach to interoceptive psychopathology. Annu Rev Clin Psychol 15:97–122
Payne P, Levine PA, Crane-Godreau MA (2015) Somatic experiencing: using interoception and proprioception as core elements of trauma therapy. Front Psychol 6:93
Perry TR et al (2021) Interoceptive awareness and suicidal ideation in a clinical eating disorder sample: the role of body trust. Behav Ther 52(5):1105–1113
Pinna T, Edwards DJ (2020) A systematic review of associations between interoception, vagal tone, and emotional regulation: potential applications for mental health, wellbeing, psychological flexibility, and chronic conditions. Front Psychol 11
Plans D et al (2021) Measuring interoception: the phase adjustment task. Biol Psychol 165:108171
Polanco-Roman L, Danies A, Anglin DM (2016) Racial discrimination as race-based trauma, coping strategies, and dissociative symptoms among emerging adults. Psychol Trauma 8(5): 609–617
Poletti S et al (2015) Neural correlates of anxiety sensitivity in panic disorder: a functional magnetic resonance imaging study. Psychiatry Research-Neuroimaging 233(2):95–101
Poppa T et al (2019) Sexual trauma history is associated with reduced orbitofrontal network strength in substance-dependent women. Neuroimage Clin 24:101973
Posner MI (2012) Attentional networks and consciousness. Front Psychol 3:64
Powers A et al (2014) Childhood abuse and the experience of pain in adulthood: the mediating effects of PTSD and emotion dysregulation on pain levels and pain-related functional impairment. Psychosomatics 55(5):491–499
Price CJ, Hooven C (2018) Interoceptive awareness skills for emotion regulation: theory and approach of mindful awareness in body-oriented therapy (MABT). Front Psychol 9:798
Price CJ, Thompson EA (2007) Measuring dimensions of body connection: body awareness and bodily dissociation. J Altern Complement Med 13(9):945–953
Quadt L, Critchley HD, Garfinkel SN (2018) The neurobiology of interoception in health and disease. Ann N Y Acad Sci 1428(1):112–128
Quigley KS et al (2021) Functions of interoception: from energy regulation to experience of the self. Trends Neurosci 44(1):29–38

Reinhardt KM et al (2020) A multi-method assessment of interoception among sexual trauma survivors. Physiol Behav 226

Robinson JL et al (2012) The functional connectivity of the human caudate: an application of meta-analytic connectivity modeling with behavioral filtering. Neuroimage 60(1):117–129

Rogers ML, Hagan CR, Joiner TE (2018) Examination of interoception along the suicidality continuum. J Clin Psychol 74(6):1004–1016

Rosso IM et al (2010) Anxiety sensitivity correlates with two indices of right anterior insula structure in specific animal phobia. Depress Anxiety 27(12):1104–1110

Rothbaum BOE (2006) Pathological anxiety: emotional processing in etiology and treatment. The Guilford Press

Sack M, Cillien M, Hopper JW (2012) Acute dissociation and cardiac reactivity to script-driven imagery in trauma-related disorders. Eur J Psychotraumatol:3

Salvato G, Richter F, Sedeño L, Bottini G, Paulesu E (2020) Building the bodily self-awareness: evidence for the convergence between interoceptive and exteroceptive information in a multilevel kernel density analysis study. Hum Brain Mapp 41(2):401–418

Schaan VK et al (2019) Childhood trauma affects stress-related interoceptive accuracy. Front Psych 10:750

Schaflein E et al (2018) Disconnected – impaired interoceptive accuracy and its association with self-perception and cardiac vagal tone in patients with dissociative disorder. Front Psychol 9: 897

Schmitz M, Müller LE, Schulz A, Kleindienst N, Herpertz SC, Bertsch K (2020) Heart and brain: cortical representation of cardiac signals is disturbed in borderline personality disorder, but unaffected by oxytocin administration. J Affect Disord 264:24–28

Schmitz M et al (2021) Heartbeat evoked potentials in patients with post-traumatic stress disorder: an unaltered neurobiological regulation system? Eur J Psychotraumatol 12(1):1987686

Schmitz M et al (2023) The impact of traumatic childhood experiences on interoception: disregarding one's own body. Bord Personal Disord Emot Dysregul 10(1):5

Seligowski AV et al (2019) Autonomic responses to fear conditioning among women with PTSD and dissociation. Depress Anxiety 36(7):625–634

Sennesh E et al (2022) Interoception as modeling, allostasis as control. Biol Psychol 167

Seth AK (2013) Interoceptive inference, emotion, and the embodied self. Trends Cogn Sci 17(11): 565–573

Seth AK, Suzuki K, Critchley HD (2011) An interoceptive predictive coding model of conscious presence. Front Psychol 2:395

Simpson T, Jakupcak M, Luterek JA (2006) Fear and avoidance of internal experiences among patients with substance use disorders and PTSD: the centrality of anxiety sensitivity. J Trauma Stress 19(4):481–491

Siqveland J et al (2017) Prevalence of posttraumatic stress disorder in persons with chronic pain: a meta-analysis. Front Psych 8:164

Solano Lopez AL, Moore S (2019) Dimensions of body-awareness and depressed mood and anxiety. West J Nurs Res 41(6):834–853

Stein MB, Jang KL, Livesley WJ (1999) Heritability of anxiety sensitivity: a twin study. Am J Psychiatry 156(2):246–251

Stein MB, Schork NJ, Gelernter J (2008) Gene-by-environment (serotonin transporter and childhood maltreatment) interaction for anxiety sensitivity, an intermediate phenotype for anxiety disorders. Neuropsychopharmacology 33(2):312–319

Stein DJ et al (2010) Cross-national analysis of the associations between traumatic events and suicidal behavior: findings from the WHO world mental health surveys. PLoS One 5(5)

Strigo IA et al (2010) Neural correlates of altered pain response in women with posttraumatic stress disorder from intimate partner violence. Biol Psychiatry 68(5):442–450

Sturgeon JA, Zautra AJ (2016) Social pain and physical pain: shared paths to resilience. Pain Manag 6(1):63–74

Uddin LQ et al (2017) Structure and function of the human insula. J Clin Neurophysiol 34(4): 300–306
Van Den Houte M et al (2021) The respiratory occlusion discrimination task: a new paradigm to measure respiratory interoceptive accuracy. Psychophysiology 58(4):e13760
Vanderkolk BA (1994) The body keeps the score – memory and the evolving psychobiology of posttraumatic stress. Harv Rev Psychiatry 1(5):253–265
Vogt BA, Derbyshire S, Jones AK (1996) Pain processing in four regions of human cingulate cortex localized with co-registered PET and MR imaging. Eur J Neurosci 8(7):1461–1473
Vujanovic AA et al (2018) Anxiety sensitivity in the association between posttraumatic stress and substance use disorders: a systematic review. Clin Psychol Rev 62:37–55
Watson-Singleton NN, Black AR, Spivey BN (2019a) Recommendations for a culturally-responsive mindfulness-based intervention for African Americans. Complement Ther Clin Pract 34:132–138
Watson-Singleton NN, Hill LK, Case AD (2019b) Past discrimination, race-related vigilance, and depressive symptoms: the moderating role of mindfulness. Mindfulness (N Y) 10(9):1768–1778
Westbrook J, Berenbaum H (2017) Emotional awareness moderates the relationship between childhood abuse and borderline personality disorder symptom factors. J Clin Psychol 73(7): 910–921
Wunderlich AP et al (2011) Caudate nucleus and insular activation during a pain suppression paradigm comparing thermal and electrical stimulation. Open Neuroimaging J 5:1–8
Yang Y et al (2016) Neural correlates of individual differences in anxiety sensitivity: an fMRI study using semantic priming. Soc Cogn Affect Neurosci 11(8):1245–1254
Zamariola G et al (2018) Interoceptive accuracy scores from the heartbeat counting task are problematic: evidence from simple bivariate correlations. Biol Psychol 137:12–17
Zavos HMS et al (2010) Genetic influences on the cognitive biases associated with anxiety and depression symptoms in adolescents. J Affect Disord 124(1–2):45–53

Dissociative Symptoms and Interoceptive Integration

Sascha P. Woelk and Sarah N. Garfinkel

Contents

Abstract Dissociative symptoms and disorders of dissociation are characterised by disturbances in the experience of the self and the surrounding world, manifesting as a breakdown in the normal integration of consciousness, memory, identity, emotion, and perception. This paper aims to provide insights into dissociative symptoms from the perspective of interoception, the sense of the body's internal physiological state, adopting a transdiagnostic framework.

Dissociative symptoms are associated with a blunting of autonomic reactivity and a reduction in interoceptive precision. In addition to the central function of interoception in homeostasis, afferent visceral signals and their neural and mental representation have been shown to shape emotional feeling states, support memory encoding, and contribute to self-representation. Changes in interoceptive processing and disrupted integration of interoceptive signals into wider cognition may

S. P. Woelk (✉)
Institute of Cognitive Neuroscience, University College London, London, UK
e-mail: s.wolk.16@ucl.ac.uk

Curr Topics Behav Neurosci (2025) 74: 245–277
https://doi.org/10.1007/7854_2024_480
Published Online: 17 May 2024

contribute to detachment from the body and the world, blunted emotional experience, and altered subjective recall, as experienced by individuals who suffer from dissociation.

A better understanding of the role of altered interoceptive integration across the symptom areas of dissociation could thus provide insights into the neurophysiological mechanisms underlying dissociative disorders. As new therapeutic approaches targeting interoceptive processing emerge, recognising the significance of interoceptive mechanisms in dissociation holds potential implications for future treatment targets.

Keywords Dissociation · Dissociative Disorders · Interoception · Mental Health · Minimal Self · Body Ownership · Sense of Agency · Self-Other Distinction · Emotional Numbing · Multisensory Integration

1 Introduction

Dissociation is a complex psychological and clinical phenomenon that has garnered considerable interest across a range of disciplines including psychology, psychiatry, neuroscience, and philosophy. Typically defined as a disruption in the normal integration of consciousness, memory, identity, and perception (American Psychiatric Association 2022), dissociation can manifest in different ways, ranging from mild detachment to profound alterations in one's sense of self and reality.

Characteristic to dissociative phenomena is a perceptual distancing, detachment, or estrangement from one's body, world, or identity. In the literature, dissociative symptoms are sometimes treated on a continuum – from mild, transient dissociative states (which are suggested to be 'normal' experiences, shared by most people) to pathological, chronic dissociative conditions. Common examples of mild, transient dissociative states include disorientation or detached states of consciousness due to fatigue (Tibubos et al. 2018) or sleep deprivation (van Heugten-van der Kloet et al. 2015). Heightened levels of anxiety and stress can potentially also induce temporary states of dissociation, such as those documented in first-time skydivers (Sterlini and Bryant 2002). More extreme transient states of dissociation are known effects of recreational psychoactive drugs, such as ketamine, psilocybin, and lysergic acid diethylamide, which can lead to feelings of disconnection from one's body and identity, and can distort one's sense of time and space (Luckenbaugh et al. 2014; Schmid et al. 2015; Duerler et al. 2022). However, it remains unclear whether these transient dissociative states share the same underlying mechanisms or possess the same subjective qualities as pathological dissociation.

Interoception is an internal sense of the condition of the body, implicated in a range of cognitive, perceptual, and affective processes (Khalsa et al. 2018). Delineated across different hierarchical levels, interoception can include the nature of afferent signals, their neural processing, and the precision with which they can be detected. Impaired interoception and reduced integration of interoceptive signals with cognitive and perceptual processing have been shown to play important roles across a broad range of mental health conditions (Critchley and Garfinkel 2018; Khalsa et al. 2018). In this review, we adopt a transdiagnostic discussion of dissociative symptoms, addressing their association with changes in interoceptive processing and interoceptive integration, and consider the potential implications for the main dimensions of dissociative symptomatology.

1.1 The Current Clinical Perspective on Dissociative Disorders

Pathological, chronic states of dissociation are associated with poor mental health and social functioning. Dissociative disorders constitute an independent category of mental health conditions, recognised in both the American Psychiatric Association's Diagnostic and Statistical Manual of Mental Disorders (DSM-5-TR, American Psychiatric Association 2022) and the World Health Organization's International Statistical Classification of Diseases and Related Health Problems (ICD-11, World Health Organization 2019).

The unifying theme underlying all dissociative disorders in both DSM and ICD is that individuals experience partial or full breakdown of integration and/or continuity of cognitive functioning, which can occur in multiple areas of experience, including identity, memory, perception, emotion, and control over one's body. In their respective definitions, both manuals take a syndrome-centric perspective and do not focus on underlying mechanisms. This definition allows for a broad spectrum of dissociative conditions and implicated cognitive functions to be grouped under the same umbrella term. Dissociative disorders listed in the DSM-5-TR and the ICD-11 include dissociative identity disorder (DID), dissociative amnesia (DA), depersonalisation/derealisation (DPD), other specified dissociative disorder, and unspecified dissociative disorder. The DSM-5-TR, moreover, specifically lists a dissociative subtype of post-traumatic stress disorder (PTSD-DS), as research has shown that 14-30% of individuals with PTSD differ in their defensive response and underlying neural mechanisms from the main hyperarousal PTSD type, displaying a passive response characterised by emotional blunting, tonic immobility, and concomitant depersonalisation/derealisation (a dissociative condition characterised by a sense of unreality and detachment from self or world) (Lanius et al. 2012; Stein et al. 2013; Terpou et al. 2019; Harricharan et al. 2021). Neither temporary dissociative states due to religious or spiritual practices nor temporarily altered states of consciousness due to medication or psychoactive substances are sufficient to qualify as dissociative disorders. Any condition that can be better explained by another diagnosis is also excluded.

Reliable epidemiological data on dissociative disorders remain scarce and estimates of prevalence rates vary widely. Among the general population, the prevalence of dissociative disorders is likely around 1-3% (Hunter et al. 2004; Lynn et al. 2022). Critically, prevalence among clinical populations appears to be much higher – as high as 50% in cases of depression and abuse, and up to 20% in anxiety (Yang et al. 2022). After patients with dissociative disorders, the highest dissociation scores are observed in PTSD, borderline personality disorder (BPD), and conversion disorder, followed by somatic disorders, addiction, schizophrenia, anxiety disorders, and obsessive compulsive disorder (Lyssenko et al. 2018). Dissociative symptoms are usually a predictor of poorer outcomes for standard treatments, recurrent hospitalisation, and higher suicidality and may potentially contribute to hallucinations, paranoia, cognitive disorganisation, and insomnia (Langeland et al. 2020; Černis et al. 2021). Socio-developmental antecedents include disrupted attachment, family conflict, and heightened anxiety in childhood (Lee et al. 2012; American Psychiatric Association 2022). Moreover, a history of trauma has often been linked to dissociation, and it has been suggested that dissociation may be a natural defence mechanism in the face of extreme, inescapable danger (Kozlowska et al. 2015). Yet, findings on the strength of correlation between trauma and dissociation vary (Dalenberg et al. 2012; Patihis and Lynn 2017), while the relationship with psychological distress and negative affect (such as in anxiety and depression) may be more consistent (Buchnik-Daniely et al. 2021; Černis et al. 2021).

Overall, dissociative disorders remain poorly understood in terms of diagnosis, treatment, and underlying aetiology. Misdiagnosis is common and no effective pharmacological treatment exists (Indelli et al. 2018; Reinders et al. 2019; Ganslev et al. 2020; Lynn et al. 2022). There is a lack of reliable biomarkers and insights into neurobiological mechanisms of dissociative conditions and symptoms (Reinders et al. 2019; Roydeva and Reinders 2021). As mentioned above, it remains unclear whether transient states of dissociation and clinical dissociation have similar or distinct underlying mechanisms.

1.2 Shortcomings of a Syndrome-Focused Classification of Dissociative Disorders

A challenge with syndrome-centric disorder classifications as applied in the DSM and the ICD is that they do not facilitate the identification of common underlying mechanisms. This perpetuates the defined disease categories not only in clinical practice but also in research, for example by contributing to homogeneous participant recruitment, despite substantial co-morbidity across mental health conditions (Morris et al. 2022). Future research into dissociation may thus benefit from taking a holistic, transdiagnostic perspective.

One such initiative is the Research Domain Criteria (RDoC) framework, developed with support of the National Institute of Mental Health (Insel et al. 2010). The RDoC provides a framework for investigating mental health across the major domains of neurobehavioural functioning, while at the same time encouraging measurement across multiple units of analysis (ranging from genes and molecules to behaviour and self-report). A more nuanced understanding of the distinct genetic and pathophysiological mechanisms paves the way for 'precision medicine', i.e. highly-targeted, personalised treatments (Cuthbert and Insel 2013).

Adopting a transdiagnostic framework may be particularly advantageous in the context of dissociation, as it has been proposed that dissociative disorders may not represent a separate class of mental health conditions, but that dissociative symptoms may instead represent a transdiagnostic feature of psychopathology (Ellickson-Larew et al. 2020). Notably, two network analyses highlighted extensive connections between dissociative symptoms and a broad range of mental health conditions, as well as several foundational processes underlying the regulation of affect (Černis et al. 2021, 2022). To better understand dissociative symptomatology, various exploratory and confirmatory factor analyses have been conducted to determine the underlying constructs measured in the main clinical assessment tools of dissociation, including the Dissociative Experiences Scale (Bernstein and Putnam 1986), the Cambridge Depersonalisation Scale (Sierra and Berrios 2000), and the Multidimensional Inventory of Dissociation (Dell 2006). These factor analyses identified a set of four to five interrelated factors, pertaining to (1) reduced self-representation, (2) impaired self-other/self-world differentiation, (3) reduced emotional experience, and (4) altered subjective recall (Briere et al. 2005; Sierra et al. 2005; Simeon et al. 2008). These four factors relate to four out of the six major domains of basic human behavioural functioning identified in the RDoC, namely: sensorimotor systems (agency and ownership), social processes (understanding of self and others), valence systems (emotion), and cognitive systems (memory) (Indelli et al. 2018). This further underlines the relevance of a transdiagnostic perspective for a better understanding of dissociation.

At present, interoception is not directly specified within the RDoC. However, physiological metrics such as heart rate variability, heart rate changes, vagal tone, skin conductance, and respiration are listed as relevant for a number of the fundamental psychological and behavioural constructs across RDoC domains. Moreover, impaired interoception has been implicated across a broad and diverse range of mental health and neurodevelopmental conditions, including anxiety, depression, PTSD, eating disorders, autism, addiction, and DPD (Khalsa et al. 2018). Insights into interoceptive processing in dissociation may thus contribute to a better understanding of the underlying mechanisms implicated in dissociative symptoms.

1.3 Altered Interoceptive Integration as a Transdiagnostic Feature Across Mental Health Conditions

Interoception refers to the process by which the nervous system senses, interprets, and integrates signals originating from within the body, providing a moment-by-moment mapping of the body's internal landscape across conscious and unconscious levels (Khalsa et al. 2018). Neuronal, chemical, and mechanic signals are generated by all major organ systems involved in homeostasis, including the cardiac, respiratory, gastrointestinal, genitourinary, nociceptive, thermoregulatory, and immune system. Afferent signals from these organ systems are relayed primarily via spinal, vagal, and glossopharyngeal pathways and integrated in neuronal circuits at multiple levels within the spinal cord, brain stem, subcortical areas, and cortex. Within the brain interoception is distributed across a wide set of viscerosensory and visceromotor areas, including amongst others the nucleus tractus solitarius, thalamus, hypothalamus, hippocampus, amygdala, insula, anterior cingulate cortex, somatosensory cortex, motor cortex, and several frontal areas (Jänig 1996; Critchley et al. 2004; Khalsa et al. 2009; Berntson and Khalsa 2021). Interoception can thus be delineated across multiple hierarchical levels, including the nature of afferent signals, the accuracy with which they can be detected, and their neural representation. Interoceptive processing can also encompass higher-order measures pertaining to the awareness, attention, and interpretation of interoceptive bodily signals (Suksasilp and Garfinkel 2022).

Many of the brain areas involved in interoceptive processing overlap with principal affective and cognitive regions (Khalsa et al. 2018). Evidence for integration of interoception and cognitive processes is found in a number of studies showing modulatory influences of visceral cycles on aspects as diverse as self-representation, memory, emotion, pain, time perception, and resting state-brain dynamics (Edwards et al. 2001; Critchley and Harrison 2013; Quadt et al. 2018; Azzalini et al. 2019; Arslanova et al. 2023). Research has shown that alterations in interoceptive integration manifest across mental health conditions, affecting symptom areas ranging from attentional bias and anhedonia to cognitive bias and impaired metacognitive insight (Nord and Garfinkel 2022). For example, altered insula activity, a key hub for interoception (Critchley et al., 2004), has been observed in a wide range of mental health conditions, including anxiety, depression, schizophrenia, and addiction (Gehrlach et al. 2019). Altered anterior insula activity is also one of the most robust neurological findings in individuals with dissociation (Phillips et al. 2001; Sierra and David 2011; Medford et al. 2016; Nicholson et al. 2016; Harricharan et al. 2020; Lotfinia et al. 2020; Roydeva and Reinders 2021). Notably, the insula has been cited as both a key region for integration of interoceptive processing with other sensory modalities and a critical brain region for the emergence of human self-consciousness (Craig 2009). Other conditions in which abnormalities in interoceptive processing have been found include anxiety, depression, eating disorders, drug addiction, PTSD, and autism (see Khalsa et al. 2018 for a review).

Interoceptive dysfunction is, however, often not readily diagnosed by clinicians and has therefore not entered more widely into diagnostic criteria (Khalsa et al. 2018). This is at least partly because the mechanisms underlying interoceptive integration and their interaction with mental health are not fully explored. The extent of modular versus integrated interoceptive processing at each level of the neuroaxis remains unclear. Disruptions in interoceptive processing may, moreover, have diverse causes. On the one hand, physiological changes within the periphery, such as more reactive autonomic responses, may alter the nature of afferent interoceptive signals. On the other hand, the neural and sensory integration of interoceptive signals may be upregulated, downregulated, or impaired (Nord and Garfinkel 2022). From this perspective, and operationalised within the current manuscript, 'interoceptive integration' is defined as how the nature of afferent signals and their neural representation are combined and aligned with other types of sensory, cognitive, and emotional processing.

This article adopts a transdiagnostic approach to the study of dissociation with a focus on elucidating the role that changes in interoceptive integration may play across the four main symptom areas of dissociation, namely: (1) reduced self-representation, (2) impaired self-other/self-world differentiation, (3) reduced emotional experience, and (4) altered subjective recall.

2 Altered Interoceptive Integration Across the Main Symptom Areas of Dissociation

2.1 Interoception and Reduced Self-Representation

A core feature of dissociative states is a qualitative change in self-experience that can range from a sense of detachment to a breakdown of the integrity of the self. Insight into the cognitive and physiological processes that contribute to self-perception and their link to interoceptive integration are thus relevant for understanding dissociative symptoms.

Philosophers and psychologists have suggested various models of the self. Early philosophical work focused on the question whether the mind and the body are the same or separate entities. While Aristotle's monism treated mind and body as a unified whole, Plato's and Descartes' dualism stated that only the body has a physical reality, whereas the mind belongs to the spiritual realm and can therefore not be elucidated by investigation of the physical properties of the body (for a discussion, see, e.g., Stent 1998). This age-old conundrum has remained relevant throughout history. For example, William James's (1891) distinction between the 'I', i.e. the self as the subject of experience, and the 'Me', i.e. the self as an object of experience, can be considered a form of monism, as it suggests that both facets of the self can be understood through scientific enquiry. Yet, most popular theories of the self that followed James did not aim to unify body and mind, but focused on isolated aspects of the self, such as the conscious and unconscious aspects of the psyche in Psychoanalytical Theory (Freud 1923; Jung 1953), or the social contributions to identity in Social Identity Theory (Tajfel 1978; Tajfel and Turner 1979). As a result, at the beginning of the twenty-first century psychologists and philosophers continue to grapple with the 'hard problem' of consciousness (Chalmers 1995), which refers to the challenge of elucidating how raw sensory inputs can lead to subjective, first-person representations. Most properties of human mental experience – such as perception, memory, decision-making, and affect – can now be studied systematically across biopsychosocial levels. Yet, the findings from these studies remain largely insufficient to explain the mechanisms underlying the subjective qualities of 'mineness' (i.e. an immediately given, higher-order, non-conceptual sense of ownership), 'selfhood' (i.e. an experientially untranscendable feeling of being a self), and 'perspectivalness' (i.e. phenomenal first-person perspective) (Metzinger 2003).

To facilitate investigation of these phenomenal foundations of the self from a cognitive perspective, Gallagher (2000) suggested to focus on the 'minimal self' and the 'narrative self'. The minimal self refers to the pre-reflexive awareness of one's existence in the present moment. This concept is closely related to Neisser's (1988) 'ecological self' and Blanke and Metzinger's (2009) Minimal Phenomenal Selfhood (MPS). The narrative self refers to the temporal continuity of self-experience, linking the present moment to self-related memories and to beliefs about the future self.

The constituents of the minimal self are body ownership, i.e. the feeling of inhabiting our body, and sense of agency, i.e. the feeling that we are in control of our actions (Gallagher 2000). Body ownership has identifiable exteroceptive components, such as the visual perception of one's limbs and torso when looking down and when looking in the mirror. Simultaneously, one feels and senses one's movements via vestibular and proprioceptive signals. Notably, these multisensory precepts within one's body and in one's peripersonal space need to be integrated in space and time for a sense of bodily wholeness to persist (Blanke et al. 2015). On the other hand, perceptual incoherence may lead to disrupted self-experiences, including diminished sense of agency, depersonalisation, and schizophrenia (Postmes et al. 2014). Evidence for this can be found in studies that report induced experiences of disembodiment and reduced sense of agency due to experimental paradigms that create sensory mismatch between touch and vision (Ehrsson 2007; Roel Lesur et al. 2020).

Interoceptive signals such as heart, lung, and gut activity get integrated jointly with sensorimotor, proprioceptive, and exteroceptive information in the emergence of body ownership and the minimal self (Salomon 2017). The brain continuously monitors the internal state of the body via neural and humoral channels and triggers automatic responses in order to maintain homeostasis, as well as to prepare the body to adaptively respond to environmental demands (Critchley and Harrison 2013). With high regularity and constancy over time, the oscillating rhythms of the body's internal activity create a subtle, ever-present feeling space. Together with other interoceptive signals, their neural representation is presumed to provide a reference frame for the experience of the self, which interacts with exteroceptive and proprioceptive processing. Park and Tallon-Baudry (2014) termed this the 'neural subjective frame' and suggested that it likely lays the foundation for subjective experience, manifesting as a basic, pre-reflective consciousness of oneself - a feeling of presence in the current moment.

The more precisely visceral signals are sensed, the more they support the integrity of the self-image. Individual differences in interoceptive processing serve as an important moderator between interoceptive signals and self-perception.

Interoceptive accuracy is associated with the extent to which participants identify with signals from their own bodies, such as recordings of their own voice (Crow et al. 2021), and reduces the susceptibility to bodily illusions, such as the rubber hand illusion and the embreathment illusion (Tsakiris et al. 2011; Sel et al. 2017). Conversely, individuals with higher scores on the Cambridge Depersonalisation Scale are more susceptible to bodily illusions (Kanayama et al. 2009). This aligns with findings that participants with dissociative symptoms, specifically DPD with additional compartmentalisation symptoms, have been shown to have reduced interoceptive accuracy, such as impaired performance on a cardiac tracking task (Schäflein et al. 2018). However, deficits in interoceptive accuracy have not always been obtained in participants with DPD, which may indicate a discrepancy between their meta-cognitive beliefs and what they actually perceive (Michal et al. 2014).

While earlier work considered limbic areas, and in particular the insula and anterior cingulate, as the main interoceptive regions (Craig 2002; Critchley and Harrison 2013), recent research shows that links between foundations of the self and interoceptive integration are pervasive across the brain. Rhythmic visceral activity, including breathing patterns, gastric and cardiac cycles, is coupled to resting-state brain networks, the alpha rhythm, as well as activity in a distributed set of brain areas (Zelano et al. 2016; Richter et al. 2017; Rebollo et al. 2018; Azzalini et al. 2019). For example, the network of brain areas that show phase synchronisation with the stomach is significantly wider than autonomic control areas, and moreover includes somatosensory cortices, posteromedial areas, as well as visual association areas (Rebollo et al. 2018). It overlaps with areas of the default-mode network (DMN), which are consistently engaged during self-referential thought and self-recognition (Buckner et al. 2008; Qin and Northoff 2011). This may explain why greater gastric activity (measured in terms of pH level, temperature, and pressure) correlates with measures of body ownership and agency (Monti et al. 2022). In addition, there is strong human evidence showing that intracranial stimulation of the posteromedial cortex, and specifically the anterior precuneus (an area of the DMN related to self-related thought, the dorsal part of which is also implicated in the gastric network) consistently induces perceptions of disembodiment, distorted self-location, and altered sense of self (Parvizi et al. 2021; Lyu et al. 2023). Furthermore, a recent systematic review showed that, while only stimulation of parietal regions uniquely affects all areas of the bodily self, electrical stimulation of other key interoceptive regions, such as insula and cingulate cortex, can also induce disturbances in self-perception (Dary et al. 2023).

2.2 *Interoception and Impaired Self-Other/Self-World Differentiation*

While the sense of body ownership is probably the most foundational aspect of subjective experience, it has been argued that it provides only weak first-person perspective, offering mainly a visuospatial reference frame for sensory and mental processing. Strong first-person perspective, i.e. the ability to have a mental representation of oneself as the subject of experience, is proposed to only emerge when one experiences a feeling of control over one's actions (Blanke and Metzinger 2009). In other words, the minimal self relies on the joint presence of body ownership and sense of agency (Gallagher 2000). Sense of agency refers to feeling in control of one's actions, and through those actions, being able to influence events in the external world. It emerges when perceptual feedback, such as observing one's body execute intended movements, aligns with one's behavioural intentions (Haggard and Chambon 2012). In addition to exteroceptive feedback, most self-generated actions encompass somatosensory and proprioceptive components, including tactile feedback and a sense of spatial displacement (Balslev et al. 2007), which further aids to distinguish them from others' actions and external events. Notably, while interoceptive processing and interoceptive integration likely contribute to self-other/self-world distinction and sense of agency, they have received relatively limited scientific attention in this context.

Numerous studies have investigated agency from a motor perspective, leading to the application of comparator models to agency (Haggard 2017). These models suggest that to derive a feeling of agency over our own actions and differentiate them from external causes (e.g. distinguishing between being touched by someone else and touching ourselves), the motor cortex generates both a movement command to the relevant muscles and a mental prediction (termed 'efference copy') of the expected movement outcome to the somatosensory cortex and other cortical areas. This allows for suppression of sensory signals produced by our own actions (Blakemore et al. 1998). Multiple empirical studies support this theory, demonstrating attenuated activity in a large network of cortical areas, as well as altered neural processing at the level of the brainstem and spinal cord in response to self-generated actions (Whitford et al. 2017a; Boehme et al. 2019). The comparator model of agency helps to explain, for example, why most people cannot tickle themselves (Blakemore et al. 2000).

It has been suggested that interoceptive signals contribute to the experience of motor control and that agency involves not only the prediction of somatosensory, but also of interoceptive action consequences. This is supported, for example, by the fact that impairments in interoceptive accuracy and motor control often co-occur (Marshall et al. 2018). Moreover, a study investigating differences between self- and other-produced touch showed that self-touch correlated with deactivation of brain areas related to somatosensory, social cognitive, as well as interoceptive processing (Boehme et al. 2019). There is also evidence that individuals with deafferentation (i.e. selective loss of proprioception and cutaneous touch) experience reduced sense of agency, putatively because the absence of body internal sensory action feedback impairs their ability to predict the outcomes of their own actions (Balslev et al. 2007).

To date, only a few studies have specifically explored the relationship between interoception and the sense of agency. One example is a study employing virtual reality to induce sensorimotor conflict, which reported a positive correlation between interoceptive accuracy and self-reported sense of agency (Stern et al. 2020). In another study, the phase of the cardiac cycle was found to modulate implicit measures of agency (Herman and Tsakiris 2020). Cardiac phase effects have previously been shown in relation to many other domains of external stimulus processing, including tactile, auditory, and visual perception, pain, fear, and memory. These studies generally report reduced perception and detection of external stimuli at systole (i.e. at the moment of maximal cardiac ejection, when the baroreceptors in the aortic arch and carotid artery fire most intensely, thereby informing the brain of ongoing cardiac activity), compared to diastole (i.e. the period in between heartbeats when baroreceptors are silent) (Critchley and Garfinkel 2015).

Early explanations of cardiac effects on cognition treated interoceptive signals, such as the heartbeat, simply as distracting or interfering effects on cognitive and sensory functions (Lacey and Lacey 1978). Contemporary perspectives propose that heartbeats may entail spontaneous shifts between interoception and exteroception (Al et al. 2020) and that these periodic self-related signals from the body may not only interfere with, but can also enhance, certain types of stimulus processing (Critchley and Garfinkel 2018). While the mechanism underlying these modulatory effects remains unclear, possible explanations suggest that selective modulation of stimulus processing could occur through the entrainment of neural rhythms by visceral cycles. Such modulatory effects have been observed, for example, in relation to the respiratory cycle. Namely, activation of mechanoreceptors of the nasal mucosa that sense airflow appears to entrain neural oscillations in limbic areas and correlates with performance differences in emotion and memory processing (Zelano et al. 2016). Additionally, the dynamic interplay between body and brain function may involve central processing mechanisms that inhibit certain types of processing (e.g. nociception) while facilitating others (e.g. fear), depending on the nature of afferent signals and consequent pattern of neural activation (Critchley and Garfinkel 2016).

Overall, these explanations implicate different levels of interoceptive integration in shaping stimulus processing, whether through interference, attentional shifts, or neural dynamics, possibly with subsequent implications for the perceived self-relatedness of external stimuli. This three-way relationship between self-perception, visceral cycle modulation, and interoceptive integration was assessed in a temporal binding paradigm (Koreki et al. 2022). The temporal binding effect is an implicit measure of sense of agency, which relies on the phenomenon that individuals perceive the time between a self-initiated action and its consequent outcome as compressed if the action is voluntary (Haggard et al. 2002). The findings of this study revealed that individuals with greater interoceptive accuracy displayed a stronger sense of agency, as indicated by a greater temporal binding effect at systole (Koreki et al. 2022), which is in line with the notion that central processing of afferent signals can be registered as a periodic series of self-related events. In contrast, there is evidence that individuals with greater tendency to dissociation may have a more fragmented sense of self and elevated sensitivity to discrepancies in multisensory integration. A study using augmented-reality to introduce small delays between visual feedback and felt touch showed that temporal mismatch can induce dissociative symptoms, including disembodiment, deafference, and derealisation. In addition, participants with heightened trait dissociation exhibited greater sensitivity to visual delays during self-touch (Moffatt et al. 2024). These findings suggest that individuals with a greater propensity to 'bind' senses together during self-generated action are better able to form a smooth and integrated percept of the world and are less likely to display dissociative traits. Another study employing the temporal binding paradigm found no differences in implicit sense of agency (i.e. temporal binding windows) between participants with high and low occurrences of depersonalisation, but reported a trend-level effect for a more 'time-sensitive' explicit sense of agency in participants with higher occurrences of depersonalisation (i.e. their explicit sense of agency decreased more rapidly with increasing temporal mismatch) (Ciaunica et al. 2024). Notably, in this study this effect was absent for self-initiated actions, which aligns with findings that active self-touch restores bodily self-awareness after body-ownership illusions (which is one of the coping strategies often cited by individuals suffering from depersonalisation) (Crivelli et al. 2023).

Relatedly, it has been proposed that another reason why self-generated actions and self-related processes undergo perceptual suppression is to support the construction of multisensory body representation, avoid sensory conflict when moving one's body, and facilitate self-other distinction (Limanowski and Friston 2020). Notably, in healthy individuals many volitional actions, especially habitual, semi-automatic actions, and behavioural reflexes, lack perceptual salience (Haggard 2017). The attenuation of self-related action consequences may be impaired in DPD, and there is evidence that insufficient attenuation of self-related action consequences contributes to other self-related disorders, such as schizophrenia, where dissociative symptoms are often prominent (Frith 1992; Whitford et al. 2017b). Neuroimaging studies in individuals with DPD support this, indicating elevated neural activity in brain areas that are typically suppressed during self-perception (Ketay et al. 2014) and lack of differential neural activation for self-versus other-related events (Adler et al. 2016). Failures in attenuating self-generated action consequences may thus lead to dissociative symptoms by triggering hyper-reflexivity, i.e. paying excessive attention to one's bodily signals because they feel out of control, and self-related overthinking (Ciaunica et al. 2022). A study in healthy individuals showed that participants with heightened trait dissociation exhibited an equivalent decrease in heart rate when they stroked their own arm, as when the experimenter stroked their arm. Conversely, participants with low trait dissociation exhibited a greater decrease in heart rate in response to other-touch than to self-touch (Moffatt et al. 2024). This apparent relative 'failure' for autonomic signals to differentiate between self- and other-touch in individuals with high dissociative symptoms may contribute to, or reflect, a reduced sense of self and warrants future investigation in dissociative disorders.

An additional neurophysiological marker of interoception, the heartbeat-evoked potential (HEP), reflects the cortical processing of cardiac activity (Park and Blanke 2019) and can be conceptualised as cardiac-brain interoceptive integration. Various studies show a link between the HEP and self-perception. Specifically, the HEP amplitude increases with self-referential thought, imagination, and changes in bodily self-consciousness (Babo-Rebelo et al. 2016a, b, 2019; Park et al. 2016), whereas HEP amplitude decreases during self-referential cognitive conflict (Sel et al. 2017). Intracranial recordings in two epileptic patients suggest that self-related amplitude variations of the HEP can be measured in both the DMN and key interoceptive regions such as the insula, but amplitude variations in the right anterior insula may be specific to the 'I' dimension of thought, i.e. to being the protagonist or agent (Babo-Rebelo et al. 2016b). Furthermore, there is evidence that self-related HEP amplitude variations may be diminished in individuals with dissociation (Schulz et al. 2015). In line with these findings, reduced HEP amplitude and reactivity have been proposed as potential biomarkers of DPD (Gatus et al. 2022).

Self-other differentiation, moreover, requires understanding of other people's actions. It has been argued that this requires mental translocation of oneself into the location of the other, combined with mental simulation of their actions (Jeannerod and Anquetil 2008). Neuroimaging studies indicate substantial overlap between brain regions activated during first-person and third-person perspective-taking, but also key differences (Vogeley and Fink 2003; Rizzolatti et al. 2014). One region that appears to be selectively activated during third-person perspective-taking is the temporoparietal junction (TPJ), which is an area more generally associated with visuospatial transformation, perspective-taking, and Theory of Mind (Wurm and Schubotz 2018). Electrical stimulation of this area has been shown to evoke out-of-body experiences, similar to those seen in DPD (Blanke et al. 2002). Moreover, the TPJ has been suggested to be one of the brain areas where sensory processing from different sensory modalities, including interoception, is integrated (Blanke et al. 2015). Research in BPD, a condition with high prevalence of dissociative symptoms, points to structural and functional differences in the TPJ (De Meulemeester et al. 2021), combined with cortical interoceptive integration deficits – as evidenced, for example, by reduced HEP amplitudes (Back and Bertsch 2020).

Further evidence that abnormalities in interoceptive integration can affect self-other distinction can be found in the literature on bodily illusions. For example, in an extension of the classic rubber hand illusion paradigm, illumination of the virtual hand synchronously with the participant's heartbeat led to greater feelings of ownership, as well as greater proprioceptive drift (Suzuki et al. 2013). In a similar study, participants were more likely to identify with a photo of their own face, morphed with a stranger's face, when their cardiac signal was projected onto the photo (Sel et al. 2017). Conversely, participants with lower interoceptive accuracy were found to be more likely to identify with a stranger's face (Tajadura-Jiménez and Tsakiris 2014). Lastly, impairments in self-other differentiation may also have recursive impacts on the physiological state of the body. For example, illusory embodiment of an artificial limb leads to reduced skin temperature and increased histamine reactivity in the real limb (Moseley et al. 2008; Barnsley et al. 2011; van Stralen et al. 2014).

2.3 Interoception and Reduced Emotional Experience

Another key concept for the understanding of dissociation is reduced emotional experience. The mental experience of emotions is coupled with bodily activations and sensations (Wiens 2005; Kreibig 2010; Nummenmaa et al. 2018). There is substantial overlap between brain regions involved in emotion-processing, self-processing, and interoception. Beyond subcortical regions, the network involved in emotion-processing incorporates the insula, cingulate cortex, orbitofrontal cortex, amygdala, and thalamus (Saarimäki et al. 2022).

Nearly all major theories of emotion acknowledge an important physiological component of emotional experience. Classical theories, such as the James-Lange theory of emotion (Lange 1885; James 1948) posit that physiological changes are the precursor of emotional feeling states; i.e. the perception of an event triggers a physiological response, which then drives the emotion. Conversely, the Cannon-Bard Theory (Cannon 1927; Bard 1928) proposes that bodily arousal responses are too undifferentiated to account for the variety of distinct emotional feeling states. Relatedly, according to Schachter and Singer's Two-Factor Theory of Emotion (Schachter and Singer 1962) physical sensations experienced in response to stimuli are not inherently emotional; instead, it is one's cognitive appraisal of these sensations, informed by the context, that assigns them an emotional value. This idea was expanded by Scherer and Moors (2019), who suggested that emotions are dimensional constructs that result from cognitive appraisal of environmental stimuli, taking account of factors such as novelty and the individual's goals and coping potential. This appraisal then activates action tendencies, physiological changes, and motor expressions, the conjoint conscious experience of which can be interpreted as emotional feeling states. This conceptualisation aligns with other, newer emotions theories, such as the Theory of Constructed Emotion (Barrett 2016), which treats emotions as holistic brain-body phenomena, which emerge from the interpretation of exteroceptive and interoceptive signals in the context of past experiences, goals, cultural norms, and values. Empirical findings support a multi-dimensional structure of emotions, involving patterned activations across physiological, neural, cognitive, and behavioural levels (Nummenmaa and Saarimäki 2019).

Empirical results show that interoceptive signals can both change the experience of emotions and drive emotional states. Individuals with greater interoceptive accuracy report greater intensity of emotional experience (Wiens et al. 2000). Moreover, emotion experience is modulated by visceral cycles. At cardiac systole (an experimental proxy for states of arousal), threat detection is improved and perception of fear intensity is elevated, while factors such as interoceptive accuracy and differences in heart rate variability help to explain inter-individual differences in fear/threat response (Gray et al. 2009; Critchley and Garfinkel 2015; Garfinkel et al. 2021). Severe or chronic inflammation can trigger depression-like states (Harrison 2017) and there is growing evidence that the gut microbiome may be implicated in conditions such as mood disorders and anxiety (Mayer 2011; Cryan and Dinan 2012; Dinan and Cryan 2016). Compelling evidence in rodents shows that optogenetically increased heart rate provokes anxiety-like behaviours, but only in risky context; and that when simultaneously activity of the insula is optogenetically inhibited, this effect disappears (Hsueh et al. 2023). This finding underlines both that interoceptive states can drive emotions and that visceral changes need to be integrated in context to gain meaning.

Emotion dysregulation and affect intolerance are often linked to dissociation (Lynn et al. 2019). Individuals with DPD typically report marked reduction or even complete absence of the subjective dimension of affective experience (Ciaunica et al. 2021). At the same time, they retain the ability to cognitively appreciate someone else's emotional state and show normal behavioural reactions (Sierra and David 2011). This implies a disconnect between bodily and mental emotional experience (Phillips et al. 2001; Sierra and David 2011). Emotional dysregulation may also mediate the relationship between PTSD and dissociative symptoms (Powers et al. 2015). Moreover, numerous studies indicate a positive association between dissociation and alexithymia, i.e. an impaired ability to accurately identify and describe emotions (Lemche et al. 2013; Merckelbach et al. 2017; Černis et al. 2022). Emotion suppression in healthy fatigued participants leads to increased occurrence of depersonalisation and derealisation symptoms (Tibubos et al. 2018), whereas patients with spinal cord injury (who commonly experience autonomic dysfunction and reduced somatosensory sensations) display difficulties judging their own emotional response to fear and anger stimuli, while they maintain the ability to understand when others experience these emotions (Pistoia et al. 2015). This aligns with findings that dissociative symptoms are elevated among patients with spinal cord injury (Lenggenhager et al. 2012).

The available evidence on physiological response patterns in dissociation seems to suggest that a flattened reactivity curve in relation to emotional stimuli is paired with chronic autonomic dysregulation. For example, skin conductance responses to emotional stimuli, and in particular to aversive stimuli, are frequently reported to be blunted and show increased latency (Sierra et al. 2002, 2006; Medford et al. 2016; Dewe et al. 2016; Seligowski et al. 2019), while the startle response is reduced (Müller et al. 2015; Seligowski et al. 2019). Conversely, basal skin conductance levels are generally found to be elevated and noisier (Giesbrecht et al. 2010; Schoenberg et al. 2012; Seligowski et al. 2019), while skin conductance responses to non-specific stimuli exhibit lower latency (Sierra et al. 2002), which is reflected behaviourally in impaired ability to ignore irrelevant stimuli (Schabinger et al. 2018). Moreover, individuals with dissociation exhibit impaired autonomic arousal suppression under stress (Owens et al. 2015; Lemche et al. 2016). Findings in regard to heart rate and heart rate variability are mixed and require more research (Roydeva and Reinders 2021).

It thus seems possible that emotional dysregulation and emotional numbing in dissociation could arise though reduced ability to detect interoceptive signals. This could arise through impairments in interoceptive accuracy directly or could be driven by noisier baseline autonomic signals as well as reduced autonomic reactivity, rendering the signals themselves harder to detect with precision. Future research should probe these alternative hypotheses.

2.4 Interoception and Altered Subjective Recall

Altered subjective recall is the fourth major factor of the dissociation symptomatology, next to reduced self-representation, impaired self-other/self-world differentiation, and reduced emotional experience. Dissociative identity disorder and dissociative amnesia are characterised by memory deficiencies and fragmentation of memory and identity (Lynn et al. 2022). Changes in subjective recall in other forms of dissociation, such as depersonalisation and derealisation disorder, can better be described as a sense of detachment from memories: 'I can remember things, but it seems as if what I remember did not really happen to me' (Sierra and David 2011, pp. 101, 102).

It is well-established that emotional material is better remembered than non-emotional material and that autonomic arousal facilitates memory encoding and retrieval (for a review, see: Tyng et al. 2017). For example, familiar words evoke greater autonomic arousal than unfamiliar words, even if they are not actively recognised (Morris et al. 2008). Participants are more likely to endorse a face as familiar when the face is presented at systole (i.e. the point in the cardiac cycle where cardiovascular arousal is represented in brain) (Fiacconi et al. 2016), thus implicating interoceptive integration in the subjective experience of familiarity. Moreover, interoceptive integration appears to modulate memory processes. For example, memory encoding and retrieval performance for words and digits differs based on when in the cardiac cycle these stimuli are processed (Garfinkel et al. 2013; Quelhas Martins et al. 2014). Similar effects are observed for respiration cycles, with improved memory retrieval for pictures encoded during nasal inspiration, compared to exhalation (Zelano et al. 2016). Changes in interoceptive integration in individuals with dissociation, such as apparent flattening of physiological reactivity to emotional stimuli, are thus likely to have implications for memory.

Research specifically targeting the link between impaired interoceptive integration and altered subjective recall in dissociation stems mainly from studies in participants with DPD. For example, impairment in bodily emotion reactivity in DPD (operationalised as lower skin conductance levels when viewing emotional film clips) was found to be associated with increased memory fragmentation (Giesbrecht et al. 2010). Moreover, during memory encoding of aversive stimuli, individuals with DPD show reduced neural activity in brain areas related to emotion-processing, such as posterior occipital-temporal cortex and insula, paired with increased activity in areas associated with emotion regulation such as inferior and lateral frontal cortex (Medford et al. 2006). It has been suggested that this is evidence for emotionally avoidant information processing in DPD, as further evidenced by a study showing better explicit memory for neutral than for negative words in a mixed sample of participants with DPD and participants with dissociative disorder not-otherwise-specified. Notably, this pattern was found to be reversed after stress induction (Simeon et al. 2023). The relationship between dissociative experiences and memory impairments is corroborated by research in healthy participants, showing that sleep deprivation not only leads to an increase in dissociative experiences, but simultaneously results in memory impairments for emotional material, especially in individuals with greater tendency to dissociative experiences (van Heugten-van der Kloet et al. 2015).

It is conceivable that the above-described dissociation-related changes in self-representation self-other/self-world differentiation, emotion-processing, and subjective recall (Fig. 1) may contribute to more severe memory alterations, such as seen in dissociative amnesia and dissociative identity disorder. As stated by Marc Jeannerod, "the conscious narrative self tends to build cause-and-effect stories about the origin of the action" (Jeannerod 2007, p. 162). It has been suggested that an essential condition for an intact 'narrative self' is proper functioning of episodic memory, which in turn depends on fronto-limbic structures (Gallagher 2000). Notably, fronto-limbic structures are also involved in interoception and

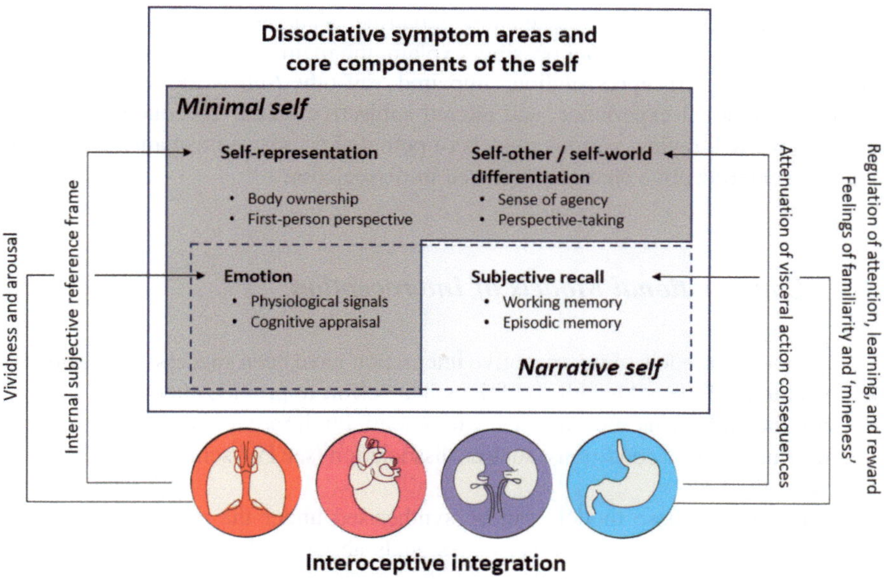

Fig. 1 Graphical illustration of the relevance of interoceptive integration for integrity of the core components of the self, in relation to the major symptom areas of dissociation. The minimal self refers to pre-reflexive awareness of one's existence in the present moment, whereas the narrative self refers to continuity of the self in memory and imagination. Interoceptive signals contribute self-related internal sensations to feelings of body ownership, supporting both a first-person perspective and a mental subjective reference frame. Sense of agency and self-other/self-world differentiation require successful suppression of self-generated actions and action consequences, including visceral perceptions. Emotions have important physiological components that contribute to their perceived intensity. Moreover, emotions and memories interact in such a way that emotionally charged content is remembered more easily, and emotions can alter memory encoding and retrieval. Impairments in interoceptive integration may therefore plausibly be implicated in reduced self-representation, impaired self-other/self-world differentiation, reduced emotional experience, and altered subjective recall

visceromotor control (Öngür et al. 2003; Khalsa et al. 2018) and may be integral for the prediction of future interoceptive states (Barrett and Simmons 2015) (see also Sect. 3.2 for a detailed discussion of interoceptive predictions). We can hypothesise that a marked reduction, or even complete absence, of the bodily basis of subjective experience due to impairments in interoceptive integration may, over time, result in memories being attributed to something or someone other than the self. However, research in this area is lacking and it remains unclear whether the various dissociative disorders are truly comparable in aetiology and underlying mechanisms.

3 Models of Altered Interoceptive Integration in Dissociation

In the previous section, we detailed how impairments and alterations in interoceptive integration relate to, and may possibly explain, the main symptom areas of dissociation: reduced self-representation, impaired self-other/self-world differentiation, reduced emotional experience, and altered subjective recall. To conclude our discussion, we will review various models of potential causative mechanisms underlying these interoceptive changes observed in dissociation.

3.1 Computational Models of Interoception

Computational models of interoceptive integration have been successfully applied to a broad range of mental health conditions in an effort to provide alternative perspectives on the underlying mechanisms with potential to improve our understanding of less well-understood conditions, such as dissociation (see Petzschner et al. (2021) for a review).

Conceptually, these models can be synthesised under the notion of a sensory control loop, in which afferent sensory signals about the body's internal state are translated by the organism into appropriate responses to maintain homeostsis. The various computational models of interoceptive integration take different stances on how the nervous system possibly implements this sensory control loop (Petzschner et al. 2021). Building on Bayesian inference, Interoceptive Predictive Coding (Seth et al. 2012; Seth 2013) suggests that the brain maintains a 'generative model' that allows it to estimate the internal state of the body, based on the coarse sensory signals that it receives via various interoceptive and exteroceptive channels. Differences between expected internal states (prior) and actual sensory perceptions (likelihood) are computed in specific neuronal populations, termed 'prediction error units' at various hierarchical layers in the brain. This computation results in an estimate of the most-likely internal state (posterior), as well as a measure of the discrepancy between prior and likelihood (i.e. the prediction error), which is used to update the generative model of the body and thereby improve future predictions of internal states. Interoceptive Active Inference (Seth and Friston 2016) expands this model by suggesting that priors actually reflect the organism's desired state, most congruent with survival, and that it can actively bring about this desired state by action selection, such as in allostatic control (i.e. adjusting homeostatic setpoints through neural, physiological, or behavioural change).

The Embodied Predictive Interoceptive Coding (EPIC) model (Barrett and Simmons 2015) provides a neuroanatomical framework for how Interoceptive Active Inference might be implemented in the brain. It proposes that agranular visceromotor cortices located in cingulate cortex, posterior ventral medial prefrontal cortex, posterior orbitofrontal cortex, and ventral anterior insula issue allostatic control commands (to hypothalamus, brainstem, and spinal cord) and interoceptive predictions (to mid- and posterior insula), which describe the interoceptive signals that are expected to arise due to these allostatic control commands. It moreover suggests that primary interoceptive regions in the granular cortex of the mid- and posterior insula are anatomically well suited to compute interoceptive prediction errors. Recent evidence supports that neurons in the insula are indeed involved in prediction error computation and encode both current and future physiological states (Gehrlach et al. 2019; Livneh et al. 2020; Harrison et al. 2021).

The authors of the EPIC model moreover suggest that given the coarseness of interoceptive signals, interoceptive predictions should generally be rather insensitive to prediction errors (Barrett and Simmons 2015). In the case that visceral signals deviate substantially from predictions, the brain can theoretically do any of the following: attempt to adjust the internal state of the body via allostatic control, direct attention away from interoceptive signals, or downregulate the sampling of interoceptive signals (putatively via limiting the influx of interoceptive signals via the thalamus) (Barrett and Simmons 2015). The latter two cases may plausibly lead to impairments in self-representation and reduction in feelings of agency, as seen in dissociation, because the generative model deviates progressively more from the true state of the body and becomes increasingly reliant on only exteroceptive signals (Limanowski and Friston 2020).

Downregulation of interoceptive integration in dissociation is supported by a recent computational model of DPD within the active inference framework. This model simulates an artificial agent affected by chronically elevated levels of interoceptive arousal in the absence of an identifiable external threat. The model results suggest that when threat is perceived as inescapable, interoceptive signals will be downregulated, while exteroceptive processing will be left intact. Repeated stimulations indicate that this would ultimately lead to a disengagement from bodily responses, as seen in DPD (Saini et al. 2022). These model outcomes align with findings from clinical practice that suggest that individuals with dissociation have learned to distrust their bodily signals, due to early life adversity, chronic stress, or trauma, as for example often seen in BPD (Back and Bertsch 2020).

3.2 Models of Underlying Neurophysiological Mechanisms

Various models of the possible neurophysiological mechanisms underlying reduced interoceptive integration in dissociation have been proposed.

The fronto-limbic inhibitory model proposes that in states of depersonalisation and derealisation the prefrontal cortex may selectively suppress activity in response to negative stimuli in limbic areas, including key interoceptive regions such as the ACC, and anterior insula (Sierra and Berrios 1998; see also Sierra and David 2011). Enhanced ventromedial prefrontal cortex activity in response to consciously perceived fearful faces (but not in response to backward masked fearful faces) has been observed in individuals with PTSD and dissociative symptoms, when compared to PTSD without dissociative symptoms (Felmingham et al. 2008). A recent systematic review of potential biomarkers of dissociation (Roydeva and Reinders 2021) found strong evidence for increased prefrontal cortex activation in response to emotional stimuli. Moreover, multiple studies using repetitive transcranial magnetic stimulation to inhibit the prefrontal cortex report temporary reduction of dissociative symptoms (Jay et al. 2014, 2016). It has been suggested that prefrontal suppression may possibly also silence interoceptive signals and thereby lead to a disembodied perception of the self (Saini et al. 2022).

In an alternative account, it has been proposed that dissociative states may possibly be explained within the framework of the defence cascade model (Kozlowska et al. 2015). Based on animal research, the defence cascade model shows that tonic immobility can be a natural defensive response in situations of inescapable danger, mediated via the release of endogenous opioids which can cause functional deafferentation at the level of the thalamus. In humans, elevated levels of endogenous opioids have been documented in individuals with PTSD and it has been suggested that specifically the kappa-opioid system may be causing dissociative symptoms, including alterations in consciousness and emotional numbing (see Lanius et al. 2018 for a review).

Furthermore, stress may be implicated in the genesis of dissociative disorders. Schulz and Vögele (2015) put forward a positive feedback model, wherein early life adversity, chronic stress, and traumatic events can lead to dysregulation of the physiological stress axis. This leads to altered interoceptive integration, and as a result triggers physical symptoms such as chest pain, palpitations, and breathlessness (it should be noted that it is precisely these uncontrollable physical symptoms that would be downregulated in dissociation, according to the computational account discussed above). There is evidence for impaired interoceptive integration due to stress hormones, such as cortisol, as a study showed that venous infusion of 4 mg of cortisol was associated with reduced HEP amplitudes (notably only when participants had their eyes closed) (Schulz et al. 2013). However, while a relatively larger number of studies have measured cortisol in relation to pathological dissociation, most of them did not find a change in cortisol levels (see Roydeva and Reinders 2021 for a review).

Vestibular information plays a distinct role in multisensory integration as it contextualises other sensory modalities in relation to space and movement (Kearney and Lanius 2022). It has been proposed that coherent body representations may crucially rely on successful integration of vestibular signals with signals from other sensory modalities, given that vestibular conditions as well as vestibular stimulation have also been linked to misperception of one's body parts, feelings of disembodiment, as well as autoscopic phenomena, such as out-of-body experiences (for a review, see Ferrè and Haggard 2016). Moreover, dizziness, a condition related to altered proprioceptive signalling from the vestibular system, has been related to increased levels of depersonalisation, derealisation, and out-of-body experiences (Lopez and Elzière 2018). A recent study in individuals with the dissociative subtype of PTSD found reduced functional connectivity between vestibular nuclei and the insula, a key interoceptive region (Harricharan et al. 2017).

Future research is required to delineate the neurophysiological mechanisms underlying dissociation in more detail and test the implications of the various models for disrupted interoceptive integration.

4 Future Directions

Converging evidence suggests the relevance of impaired interoceptive integration for the main symptom areas of dissociation. However, much of the previous research that links interoception to the integrity of the self has been carried out in healthy individuals. While studies are now providing converging and complementary evidence of impaired interoceptive integration in individuals with dissociative disorders, further research is needed to fully delineate the specific nature of interoceptive impairments and the levels at which they operate in clinical groups defined by high levels of dissociation.

Based on the available data, a number of hypotheses about a possible causal role for impaired interoceptive integration in dissociation can be derived. Namely, impaired interoception and reduced interoceptive integration could contribute to dissociative experiences at various hierarchical levels. At the physiological level, chronic autonomic dysregulation (Giesbrecht et al. 2010; Schoenberg et al. 2012; Owens et al. 2015; Lemche et al. 2016; Seligowski et al. 2019) coupled with reduced bodily reactivity to emotional stimuli (Sierra et al. 2002; Medford et al. 2016; Dewe et al. 2016) may render afferent signals less informative, dampening this crucial channel of physiological information. This, in turn, may lead to reductions in multisensory integration of interoceptive information and overreliance on other sensory modalities (Barrett and Simmons 2015; Allen et al. 2022; Saini et al. 2022). At the neural level, impaired interoceptive integration is reflected in attenuation of neural representations of autonomic signals, such as the HEP, and altered neural activity in key interoceptive regions, particularly in the insula (Schulz et al. 2015; Roydeva and Reinders 2021). These changes may ultimately diminish the body-to-brain axis and the neural subjective reference frame (Park and Tallon-

Baudry 2014), thereby undermining the feelings of presence and minimal selfhood. At the perceptual level, diminished sensation of the internal state of the body may trigger feelings of detachment from oneself and the surrounding world. Moreover, due to decreased interoceptive integration, affective colouring of subjective experience may be reduced (Phillips et al. 2001; Medford et al. 2006, 2016; Lemche et al. 2008, 2013), with deleterious effects for emotion, memory, and other mental processes.

The following areas appear promising for future investigation: (1) Comprehensive detailing of afferent signalling change, including characterising ‘noisy’ baseline levels and/or reduced reactivity. There are many interoceptive axes that could be studied in relation to dissociation to build a more comprehensive understanding of afferent bodily change at rest (Rebollo et al. 2018), during social interaction (Crucianelli et al. 2018), under stress (Schulz and Vögele 2015), and during tasks pertinent to dissociation (e.g. tests of episodic memory (Zelano et al. 2016), agency (Ciaunica et al. 2022), and self-related thought (Babo-Rebelo et al. 2019)); (2) Testing at which level of the neuroaxis impairments in interoceptive sensing and integration may occur. For instance, comprehensive imaging techniques that monitor the spinal cord and the brainstem concurrent with the cortex could help identify the level of neuronal impairment (Mardell et al. 2022); (3) Systematic research detailing interoceptive alterations in dissociation in relation to changes in agency, memory, and emotion in clinical dissociation to complement the growing body of work detailing the importance of interoception for shaping cognition and emotion-processing. Here, tasks that detail how afferent signals influence memory, emotion, and agency (Gray et al. 2009; Park et al. 2016; Moffatt et al. 2024) could be applied to dissociation cohorts with specific predictions that interoceptive signals will have reduced potential to influence cognition (e.g. memory) in individuals with dissociation. Finally, different clinical populations with dissociation may have distinct underlying aetiologies. For example, future research should investigate how childhood adversity in individuals with PTSD and dissociation might influence different interoceptive profiles relative to individuals with DPD.

5 Conclusion

Dissociative symptoms manifest across the main symptom areas of reduced self-representation, impaired self-other/self-world differentiation, reduced emotional experience, and altered subjective recall. They entail a breakdown in the normal integration of consciousness, memory, emotion, and self-related processing.

Dissociative symptoms commonly manifest across multiple domains of neurobehavioral functioning. A key feature shared by most, if not all, dissociative symptoms is altered interoceptive processing, with increasing evidence suggesting impaired integration of interoceptive signals with other sensory modalities. The relevance of interoceptive integration for a better understanding of dissociation is further supported by an extensive body of literature detailing the importance of intact interoceptive integration for supporting core elements of self-representation, contributing to the sense of body ownership, agency, shaping emotional feeling states, and memory. Aberrant interoceptive processing may thus serve as a potential transdiagnostic mechanism underlying a range of dissociative phenomena.

Interoceptive mechanisms are increasingly implicated in the understanding and treatment of mental health conditions (Khalsa et al. 2018; Nord and Garfinkel 2022). Interoceptive mechanisms are modifiable through a range of pharmacological and behavioural interventions, offering an accessible way to exert mechanistic change. Established pharmacological agents (e.g. selective serotonin reuptake inhibitors (SSRIs)) may work through interoceptive change, particularly in the domain of interoceptive awareness (Livermore et al. 2022), and interoceptive accuracy is amenable to targeted training in a therapeutic context (Quadt et al. 2021). If interoceptive disturbances drive alterations in emotional, cognitive, and self-related processes associated with dissociation, then interoceptive mechanisms may prove a promising direction for future treatment targets.

References

Adler J, Schabinger N, Michal M et al (2016) Is that me in the mirror? Depersonalisation modulates tactile mirroring mechanisms. Neuropsychologia 85:148–158. https://doi.org/10.1016/j.neuropsychologia.2016.03.009

Al E, Iliopoulos F, Forschack N et al (2020) Heart–brain interactions shape somatosensory perception and evoked potentials. Proc Natl Acad Sci 117:10575–10584. https://doi.org/10.1073/pnas.1915629117

Allen M, Levy A, Parr T, Friston KJ (2022) In the body's eye: the computational anatomy of interoceptive inference. PLoS Comput Biol 18:e1010490. https://doi.org/10.1371/journal.pcbi.1010490

American Psychiatric Association (2022) Dissociative disorders. In: Diagnostic and statistical manual of mental disorders. American Psychiatric Association Publishing

Arslanova I, Kotsaris V, Tsakiris M (2023) Perceived time expands and contracts within each heartbeat. Curr Biol 33:1389–1395.e4. https://doi.org/10.1016/j.cub.2023.02.034

Azzalini D, Rebollo I, Tallon-Baudry C (2019) Visceral signals shape brain dynamics and cognition. Trends Cogn Sci 23:488–509. https://doi.org/10.1016/j.tics.2019.03.007

Babo-Rebelo M, Richter CG, Tallon-Baudry C (2016a) Neural responses to heartbeats in the default network encode the self in spontaneous thoughts. J Neurosci 36:7829–7840. https://doi.org/10.1523/JNEUROSCI.0262-16.2016

Babo-Rebelo M, Wolpert N, Adam C et al (2016b) Is the cardiac monitoring function related to the self in both the default network and right anterior insula? Philos Trans R Soc B Biol Sci 371:20160004. https://doi.org/10.1098/rstb.2016.0004

Babo-Rebelo M, Buot A, Tallon-Baudry C (2019) Neural responses to heartbeats distinguish self from other during imagination. Neuroimage 191:10–20. https://doi.org/10.1016/j.neuroimage.2019.02.012

Back SN, Bertsch K (2020) Interoceptive processing in borderline personality pathology: a review on neurophysiological mechanisms. Curr Behav Neurosci Rep 7:232–238. https://doi.org/10.1007/s40473-020-00217-2

Balslev D, Cole J, Miall RC (2007) Proprioception contributes to the sense of agency during visual observation of hand movements: evidence from temporal judgments of action. J Cogn Neurosci 19:1535–1541. https://doi.org/10.1162/jocn.2007.19.9.1535

Bard P (1928) A diencephalic mechanism for the expression of rage with special reference to the sympathetic nervous system. Am J Physiol-Leg Content 84:490–515. https://doi.org/10.1152/ajplegacy.1928.84.3.490
Barnsley N, McAuley JH, Mohan R et al (2011) The rubber hand illusion increases histamine reactivity in the real arm. Curr Biol 21:R945–R946. https://doi.org/10.1016/j.cub.2011.10.039
Barrett LF (2016) The theory of constructed emotion: an active inference account of interoception and categorization. Soc Cogn Affect Neurosci:nsw154. https://doi.org/10.1093/scan/nsw154
Barrett LF, Simmons WK (2015) Interoceptive predictions in the brain. Nat Rev Neurosci 16:419–429. https://doi.org/10.1038/nrn3950
Bernstein EM, Putnam FW (1986) Development, reliability, and validity of a dissociation scale. J Nerv Ment Dis 174:727
Berntson GG, Khalsa SS (2021) Neural circuits of interoception. Trends Neurosci 44:17–28. https://doi.org/10.1016/j.tins.2020.09.011
Blakemore SJ, Wolpert DM, Frith CD (1998) Central cancellation of self-produced tickle sensation. Nat Neurosci 1:635–640
Blakemore S-J, Wolpert D, Frith C (2000) Why can't you tickle yourself? Neuroreport 11:R11–R16
Blanke O, Metzinger T (2009) Full-body illusions and minimal phenomenal selfhood. Trends Cogn Sci 13:7–13. https://doi.org/10.1016/j.tics.2008.10.003
Blanke O, Ortigue S, Landis T, Seeck M (2002) Stimulating illusory own-body perceptions. Nature 419:269–270. https://doi.org/10.1038/419269a
Blanke O, Slater M, Serino A (2015) Behavioral, neural, and computational principles of bodily self-consciousness. Neuron 88:145–166. https://doi.org/10.1016/j.neuron.2015.09.029
Boehme R, Hauser S, Gerling GJ et al (2019) Distinction of self-produced touch and social touch at cortical and spinal cord levels. Proc Natl Acad Sci 116:2290–2299. https://doi.org/10.1073/pnas.1816278116
Briere J, Weathers FW, Runtz M (2005) Is dissociation a multidimensional construct? Data from the multiscale dissociation inventory. J Trauma Stress 18:221–231. https://doi.org/10.1002/jts.20024
Buchnik-Daniely Y, Vannikov-Lugassi M, Shalev H, Soffer-Dudek N (2021) The path to dissociative experiences: a direct comparison of different etiological models. Clin Psychol Psychother 28:1091–1102. https://doi.org/10.1002/cpp.2559
Buckner RL, Andrews-Hanna JR, Schacter DL (2008) The brain's default network. Ann N Y Acad Sci 1124:1–38. https://doi.org/10.1196/annals.1440.011
Cannon WB (1927) The James-Lange theory of emotions: a critical examination and an alternative theory. Am J Psychol 39:106–124. https://doi.org/10.2307/1415404
Černis E, Evans R, Ehlers A, Freeman D (2021) Dissociation in relation to other mental health conditions: an exploration using network analysis. J Psychiatr Res 136:460–467. https://doi.org/10.1016/j.jpsychires.2020.08.023
Černis E, Ehlers A, Freeman D (2022) Psychological mechanisms connected to dissociation: generating hypotheses using network analyses. J Psychiatr Res 148:165–173. https://doi.org/10.1016/j.jpsychires.2022.01.049
Chalmers D (1995) Facing up to the problem of consciousness. J Conscious Stud 2:200–219
Ciaunica A, Charlton J, Farmer H (2021) When the window cracks: transparency and the fractured self in depersonalisation. Phenomenol Cogn Sci 20:1–19. https://doi.org/10.1007/s11097-020-09677-z
Ciaunica A, Seth A, Limanowski J et al (2022) I overthink – therefore I am not: an active inference account of altered sense of self and agency in depersonalisation disorder. Conscious Cogn 101:103320. https://doi.org/10.1016/j.concog.2022.103320
Ciaunica A, Ayache J, Haggard P et al (2024) Am I in control? The effect of depersonalization on the sense of agency. PsyArvix Prepr. https://doi.org/10.31234/osf.io/3knzd
Craig AD (2002) How do you feel? Interoception: the sense of the physiological condition of the body. Nat Rev Neurosci 3:655–666. https://doi.org/10.1038/nrn894

Craig (2009) How do you feel – now? The anterior insula and human awareness. Nat Rev Neurosci 10:59–70. https://doi.org/10.1038/nrn2555

Critchley HD, Garfinkel SN (2015) Interactions between visceral afferent signaling and stimulus processing. Front Neurosci 9. https://doi.org/10.3389/fnins.2015.00286

Critchley HD, Garfinkel SN (2016) Bodily arousal differentially impacts stimulus processing and memory: norepinephrine in interoception. Behav Brain Sci 39:e205. https://doi.org/10.1017/S0140525X15001764

Critchley HD, Garfinkel SN (2018) The influence of physiological signals on cognition. Curr Opin Behav Sci 19:13–18. https://doi.org/10.1016/j.cobeha.2017.08.014

Critchley HD, Harrison NA (2013) Visceral influences on brain and behavior. Neuron 77:624–638. https://doi.org/10.1016/j.neuron.2013.02.008

Critchley HD, Wiens S, Rotshtein P et al (2004) Neural systems supporting interoceptive awareness. Nat Neurosci 7:189–195. https://doi.org/10.1038/nn1176

Crivelli D, Cataldo A, Bottini G et al (2023) Active self-touch restores bodily self-awareness following disruption by "rubber hand illusion." 2023.03.06.531301

Crow KM, van Mersbergen M, Payne AE (2021) Vocal congruence: the voice and the self measured by interoceptive awareness. J Voice 35:324.e15–324.e28. https://doi.org/10.1016/j.jvoice.2019.08.027

Crucianelli L, Krahé C, Jenkinson PM, Fotopoulou A, (Katerina) (2018) Interoceptive ingredients of body ownership: affective touch and cardiac awareness in the rubber hand illusion. Cortex 104:180–192. https://doi.org/10.1016/j.cortex.2017.04.018

Cryan JF, Dinan TG (2012) Mind-altering microorganisms: the impact of the gut microbiota on brain and behaviour. Nat Rev Neurosci 13:701–712. https://doi.org/10.1038/nrn3346

Cuthbert BN, Insel TR (2013) Toward the future of psychiatric diagnosis: the seven pillars of RDoC. BMC Med 11:126. https://doi.org/10.1186/1741-7015-11-126

Dalenberg CJ, Brand BL, Gleaves DH et al (2012) Evaluation of the evidence for the trauma and fantasy models of dissociation. Psychol Bull 138:550–588. https://doi.org/10.1037/a0027447

Dary Z, Lenggenhager B, Lagarde S et al (2023) Neural bases of the bodily self as revealed by electrical brain stimulation: a systematic review. Hum Brain Mapp 44:2936–2959. https://doi.org/10.1002/hbm.26253

De Meulemeester C, Lowyck B, Luyten P (2021) The role of impairments in self–other distinction in borderline personality disorder: a narrative review of recent evidence. Neurosci Biobehav Rev 127:242–254. https://doi.org/10.1016/j.neubiorev.2021.04.022

Dell PF (2006) The multidimensional inventory of dissociation (MID): a comprehensive measure of pathological dissociation. J Trauma Dissociation 7:77–106. https://doi.org/10.1300/J229v07n02_06

Dewe H, Watson DG, Braithwaite JJ (2016) Uncomfortably numb: new evidence for suppressed emotional reactivity in response to body-threats in those predisposed to sub-clinical dissociative experiences. Cogn Neuropsychiatry 21:377–401. https://doi.org/10.1080/13546805.2016.1212703

Dinan TG, Cryan JF (2016) Mood by microbe: towards clinical translation. Genome Med 8:36. https://doi.org/10.1186/s13073-016-0292-1

Duerler P, Brem S, Fraga-González G et al (2022) Psilocybin induces aberrant prediction error processing of tactile mismatch responses – a simultaneous EEG–FMRI study. Cereb Cortex 32:186–196. https://doi.org/10.1093/cercor/bhab202

Edwards L, Ring C, McIntyre D, Carroll D (2001) Modulation of the human nociceptive flexion reflex across the cardiac cycle. Psychophysiology 38:712–718. https://doi.org/10.1111/1469-8986.3840712

Ehrsson, HH (2007) The experimental induction of out-of-body experiences. Science, 317:1048

Ellickson-Larew S, Stasik-O'Brien SM, Stanton K, Watson D (2020) Dissociation as a multidimensional transdiagnostic symptom. Psychol Conscious Theory Res Pract 7:126. https://doi.org/10.1037/cns0000218

Felmingham K, Kemp AH, Williams L et al (2008) Dissociative responses to conscious and non-conscious fear impact underlying brain function in post-traumatic stress disorder. Psychol Med 38:1771–1780. https://doi.org/10.1017/S0033291708002742

Ferrè ER, Haggard P (2016) The vestibular body: vestibular contributions to bodily representations. Cogn Neuropsychol 33:67–81. https://doi.org/10.1080/02643294.2016.1168390

Fiacconi CM, Peter EL, Owais S, Köhler S (2016) Knowing by heart: visceral feedback shapes recognition memory judgments. J Exp Psychol Gen 145:559–572. https://doi.org/10.1037/xge0000164

Freud S (1923) The Ego and the Id. In: Strachey J (ed) The standard edition of the complete psychological works of Sigmund Freud. Hogarth Press, London, pp 1–66

Frith CD (1992) The cognitive neuropsychology of schizophrenia. Lawrence Erlbaum Associates, Inc, Hillsdale

Gallagher S (2000) Philosophical conceptions of the self: implications for cognitive science. Trends Cogn Sci 4:14–21. https://doi.org/10.1016/S1364-6613(99)01417-5

Ganslev CA, Storebø OJ, Callesen HE et al (2020) Psychosocial interventions for conversion and dissociative disorders in adults. Cochrane Database Syst Rev 7:CD005331. https://doi.org/10.1002/14651858.CD005331.pub3

Garfinkel SN, Barrett AB, Minati L et al (2013) What the heart forgets: cardiac timing influences memory for words and is modulated by metacognition and interoceptive sensitivity. Psychophysiology 50:505–512. https://doi.org/10.1111/psyp.12039

Garfinkel SN, Gould van Praag CD, Engels M et al (2021) Interoceptive cardiac signals selectively enhance fear memories. J Exp Psychol Gen 150:1165–1176. https://doi.org/10.1037/xge0000967

Gatus A, Jamieson G, Stevenson B (2022) Past and future explanations for depersonalization and derealization disorder: a role for predictive coding. Front Hum Neurosci 16. https://doi.org/10.3389/fnhum.2022.744487

Gehrlach DA, Dolensek N, Klein AS et al (2019) Aversive state processing in the posterior insular cortex. Nat Neurosci 22:1424–1437. https://doi.org/10.1038/s41593-019-0469-1

Giesbrecht T, Merckelbach H, van Oorsouw K, Simeon D (2010) Skin conductance and memory fragmentation after exposure to an emotional film clip in depersonalization disorder. Psychiatry Res 177:342–349. https://doi.org/10.1016/j.psychres.2010.03.010

Gray MA, Rylander K, Harrison NA et al (2009) Following one's heart: cardiac rhythms gate central initiation of sympathetic reflexes. J Neurosci 29:1817–1825. https://doi.org/10.1523/JNEUROSCI.3363-08.2009

Haggard P (2017) Sense of agency in the human brain. Nat Rev Neurosci 18:196–207. https://doi.org/10.1038/nrn.2017.14

Haggard P, Chambon V (2012) Sense of agency. Curr Biol 22:R390–R392. https://doi.org/10.1016/j.cub.2012.02.040

Haggard P, Clark S, Kalogeras J (2002) Voluntary action and conscious awareness. Nat Neurosci 5:382–385. https://doi.org/10.1038/nn827

Harricharan S, Nicholson AA, Densmore M et al (2017) Sensory overload and imbalance: resting-state vestibular connectivity in PTSD and its dissociative subtype. Neuropsychologia 106:169–178. https://doi.org/10.1016/j.neuropsychologia.2017.09.010

Harricharan S, Nicholson AA, Thome J et al (2020) PTSD and its dissociative subtype through the lens of the insula: anterior and posterior insula resting-state functional connectivity and its predictive validity using machine learning. Psychophysiology 57:e13472. https://doi.org/10.1111/psyp.13472

Harricharan S, McKinnon MC, Lanius RA (2021) How processing of sensory information from the internal and external worlds shape the perception and engagement with the world in the aftermath of trauma: implications for PTSD. Front Neurosci 15

Harrison NA (2017) Brain structures implicated in inflammation-associated depression. In: Dantzer R, Capuron L (eds) Inflammation-associated depression: evidence, mechanisms and implications. Springer, Cham, pp 221–248

Harrison OK, Köchli L, Marino S et al (2021) Interoception of breathing and its relationship with anxiety. Neuron 109:4080–4093.e8. https://doi.org/10.1016/j.neuron.2021.09.045

Herman AM, Tsakiris M (2020) Feeling in control: the role of cardiac timing in the sense of agency. Affect Sci 1:155–171. https://doi.org/10.1007/s42761-020-00013-x

van Heugten-van der Kloet D, Giesbrecht T, Merckelbach H (2015) Sleep loss increases dissociation and affects memory for emotional stimuli. J Behav Ther Exp Psychiatry 47:9–17. https://doi.org/10.1016/j.jbtep.2014.11.002

Hsueh B, Chen R, Jo Y et al (2023) Cardiogenic control of affective behavioural state. Nature 615: 292–299. https://doi.org/10.1038/s41586-023-05748-8

Hunter ECM, Sierra M, David AS (2004) The epidemiology of depersonalisation and derealisation. Soc Psychiatry Psychiatr Epidemiol 39:9–18. https://doi.org/10.1007/s00127-004-0701-4

Indelli P, Landeira-Fernandez J, Mograbi DC (2018) In search of connection: towards a transdiagnostic view of dissociative phenomena through research domain criteria (RDoC) framework. Psicol Clínica 30:509–540. https://doi.org/10.33208/PC1980-5438v0030n03A06

Insel T, Cuthbert B, Garvey M et al (2010) Research domain criteria (RDoC): toward a new classification framework for research on mental disorders. Am J Psychiatry 167:748–751. https://doi.org/10.1176/appi.ajp.2010.09091379

James W (1891) The principles of psychology. Harvard University Press, Cambridge

James W (1948) What is emotion? 1884. In: Readings in the history of psychology. Appleton-Century-Crofts, East Norwalk, pp 290–303

Jänig W (1996) Neurobiology of visceral afferent neurons: neuroanatomy, functions, organ regulations and sensations. Biol Psychol 42:29–51. https://doi.org/10.1016/0301-0511(95)05145-7

Jay E-L, Sierra M, Van den Eynde F et al (2014) Testing a neurobiological model of depersonalization disorder using repetitive transcranial magnetic stimulation. Brain Stimul 7:252–259. https://doi.org/10.1016/j.brs.2013.12.002

Jay E-L, Nestler S, Sierra M et al (2016) Ventrolateral prefrontal cortex repetitive transcranial magnetic stimulation in the treatment of depersonalization disorder: a consecutive case series. Psychiatry Res 240:118–122. https://doi.org/10.1016/j.psychres.2016.04.027

Jeannerod M (2007) Being oneself J Physiol-Paris 101:161–168. https://doi.org/10.1016/j.jphysparis.2007.11.005

Jeannerod M, Anquetil T (2008) Putting oneself in the perspective of the other: a framework for self–other differentiation. Soc Neurosci 3:356–367. https://doi.org/10.1080/17470910701563715

Jung CG (1953) Collected works, Psychology and alchemy, vol 12. Pantheon Books, New York

Kanayama N, Sato A, Ohira H (2009) The role of gamma band oscillations and synchrony on rubber hand illusion and crossmodal integration. Brain Cogn 69:19–29. https://doi.org/10.1016/j.bandc.2008.05.001

Kearney BE, Lanius RA (2022) The brain-body disconnect: a somatic sensory basis for trauma-related disorders. Front Neurosci 16

Ketay S, Hamilton HK, Haas BW, Simeon D (2014) Face processing in depersonalization: an fMRI study of the unfamiliar self. Psychiatry Res Neuroimaging 222:107–110. https://doi.org/10.1016/j.pscychresns.2014.02.003

Khalsa SS, Rudrauf D, Feinstein JS, Tranel D (2009) The pathways of interoceptive awareness. Nat Neurosci 12:1494–1496. https://doi.org/10.1038/nn.2411

Khalsa SS, Adolphs R, Cameron OG et al (2018) Interoception and mental health: a roadmap. Biol Psychiatry Cogn Neurosci Neuroimaging 3:501–513. https://doi.org/10.1016/j.bpsc.2017.12.004

Koreki A, Goeta D, Ricciardi L et al (2022) The relationship between interoception and agency and its modulation by heartbeats: an exploratory study. Sci Rep 12:13624. https://doi.org/10.1038/s41598-022-16569-6

Kozlowska K, Walker P, McLean L, Carrive P (2015) Fear and the defense cascade: clinical implications and management. Harv Rev Psychiatry 23:263–287. https://doi.org/10.1097/HRP.0000000000000065

Kreibig SD (2010) Autonomic nervous system activity in emotion: a review. Biol Psychol 84:394–421. https://doi.org/10.1016/j.biopsycho.2010.03.010

Lacey BC, Lacey JI (1978) Two-way communication between the heart and the brain: significance of time within the cardiac cycle. Am Psychol 33:99–113. https://doi.org/10.1037/0003-066X.33.2.99

Lange CG (1885) The mechanism of the emotions. In: Rand B (ed) The classical psychologists. Copenhagen (Original work published 1885, Om Sindsbevaegelser et Psyko-Fysiologisk Studie), pp 672–684

Langeland W, Jepsen EKK, Brand BL et al (2020) The economic burden of dissociative disorders: a qualitative systematic review of empirical studies. Psychol Trauma Theory Res Pract Policy 12:730. https://doi.org/10.1037/tra0000556

Lanius RA, Brand B, Vermetten E et al (2012) The dissociative subtype of posttraumatic stress disorder: rationale, clinical and neurobiological evidence, and implications. Depress Anxiety 29:701–708. https://doi.org/10.1002/da.21889

Lanius RA, Boyd JE, McKinnon MC et al (2018) A review of the neurobiological basis of trauma-related dissociation and its relation to cannabinoid- and opioid-mediated stress response: a transdiagnostic, translational approach. Curr Psychiatry Rep 20:118. https://doi.org/10.1007/s11920-018-0983-y

Lee WE, Kwok CHT, Hunter ECM et al (2012) Prevalence and childhood antecedents of depersonalization syndrome in a UK birth cohort. Soc Psychiatry Psychiatr Epidemiol 47:253–261. https://doi.org/10.1007/s00127-010-0327-7

Lemche E, Anilkumar A, Giampietro VP et al (2008) Cerebral and autonomic responses to emotional facial expressions in depersonalisation disorder. Br J Psychiatry 193:222–228. https://doi.org/10.1192/bjp.bp.107.044263

Lemche E, Brammer MJ, David AS et al (2013) Interoceptive-reflective regions differentiate alexithymia traits in depersonalization disorder. Psychiatry Res 214:66–72. https://doi.org/10.1016/j.pscychresns.2013.05.006

Lemche E, Sierra-Siegert M, David AS et al (2016) Cognitive load and autonomic response patterns under negative priming demand in depersonalization-derealization disorder. Eur J Neurosci 43:971–978. https://doi.org/10.1111/ejn.13183

Lenggenhager B, Pazzaglia M, Scivoletto G et al (2012) The sense of the body in individuals with spinal cord injury. PLoS One 7:e50757. https://doi.org/10.1371/journal.pone.0050757

Limanowski J, Friston K (2020) Attenuating oneself: an active inference perspective on "selfless" experiences. Philos Mind Sci 1:1–16. https://doi.org/10.33735/phimisci.2020.I.35

Livermore JJA, Holmes CL, Moga G et al (2022) A single oral dose of citalopram increases interoceptive insight in healthy volunteers. Psychopharmacology (Berl) 239:2289–2298. https://doi.org/10.1007/s00213-022-06115-7

Livneh Y, Sugden AU, Madara JC et al (2020) Estimation of current and future physiological states in insular cortex. Neuron 105:1094–1111.e10. https://doi.org/10.1016/j.neuron.2019.12.027

Lopez C, Elzière M (2018) Out-of-body experience in vestibular disorders – a prospective study of 210 patients with dizziness. Cortex 104:193–206. https://doi.org/10.1016/j.cortex.2017.05.026

Lotfinia S, Soorgi Z, Mertens Y, Daniels J (2020) Structural and functional brain alterations in psychiatric patients with dissociative experiences: a systematic review of magnetic resonance imaging studies. J Psychiatr Res 128:5–15. https://doi.org/10.1016/j.jpsychires.2020.05.006

Luckenbaugh DA, Niciu MJ, Ionescu DF et al (2014) Do the dissociative side effects of ketamine mediate its antidepressant effects? J Affect Disord 159:56–61. https://doi.org/10.1016/j.jad.2014.02.017

Lynn SJ, Maxwell R, Merckelbach H et al (2019) Dissociation and its disorders: competing models, future directions, and a way forward. Clin Psychol Rev 73:101755. https://doi.org/10.1016/j.cpr.2019.101755

Lynn SJ, Polizzi C, Merckelbach H et al (2022) Dissociation and dissociative disorders reconsidered: beyond sociocognitive and trauma models toward a transtheoretical framework. Annu Rev Clin Psychol 18:259–289. https://doi.org/10.1146/annurev-clinpsy-081219-102424

Lyssenko L, Schmahl C, Bockhacker L et al (2018) Dissociation in psychiatric disorders: a meta-analysis of studies using the dissociative experiences scale. Am J Psychiatry 175:37–46. https://doi.org/10.1176/appi.ajp.2017.17010025

Lyu D, Stieger JR, Xin C et al (2023) Causal evidence for the processing of bodily self in the anterior precuneus. Neuron 111:2502–2512.e4. https://doi.org/10.1016/j.neuron.2023.05.013

Mardell LC, O'Neill GC, Tierney TM et al (2022) Concurrent spinal and brain imaging with optically pumped magnetometers. 2022.05.12.491623

Marshall AC, Gentsch A, Schütz-Bosbach S (2018) The interaction between interoceptive and action states within a framework of predictive coding. Front Psychol 9

Mayer EA (2011) Gut feelings: the emerging biology of gut–brain communication. Nat Rev Neurosci 12. https://doi.org/10.1038/nrn3071

Medford N, Brierley B, Brammer M et al (2006) Emotional memory in depersonalization disorder: a functional MRI study. Psychiatry Res Neuroimaging 148:93–102. https://doi.org/10.1016/j.pscychresns.2006.05.007

Medford N, Sierra M, Stringaris A et al (2016) Emotional experience and awareness of self: functional MRI studies of depersonalization disorder. Front Psychol 7:432. https://doi.org/10.3389/fpsyg.2016.00432

Merckelbach H, Boskovic I, Pesy D et al (2017) Symptom overreporting and dissociative experiences: a qualitative review. Conscious Cogn 49:132–144. https://doi.org/10.1016/j.concog.2017.01.007

Metzinger T (2003) Phenomenal transparency and cognitive self-reference. Phenomenol Cogn Sci 2:353–393. https://doi.org/10.1023/B:PHEN.0000007366.42918.eb

Michal M, Reuchlein B, Adler J et al (2014) Striking discrepancy of anomalous body experiences with normal interoceptive accuracy in depersonalization-derealization disorder. PLoS One 9: e89823. https://doi.org/10.1371/journal.pone.0089823

Moffatt JA, Lesur MR, Lenggenhager B et al (2024) Dissociative experiences arise from disrupted multi-sensory integration

Monti A, Porciello G, Panasiti MS, Aglioti SM (2022) Gut markers of bodily self-consciousness in men. iScience 25:105061. https://doi.org/10.1016/j.isci.2022.105061

Morris AL, Cleary AM, Still ML (2008) The role of autonomic arousal in feelings of familiarity. Conscious Cogn 17:1378–1385. https://doi.org/10.1016/j.concog.2008.04.005

Morris SE, Sanislow CA, Pacheco J et al (2022) Revisiting the seven pillars of RDoC. BMC Med 20:220. https://doi.org/10.1186/s12916-022-02414-0

Moseley GL, Olthof N, Venema A et al (2008) Psychologically induced cooling of a specific body part caused by the illusory ownership of an artificial counterpart. Proc Natl Acad Sci 105: 13169–13173. https://doi.org/10.1073/pnas.0803768105

Müller LE, Schulz A, Andermann M et al (2015) Cortical representation of afferent bodily signals in borderline personality disorder: neural correlates and relationship to emotional dysregulation. JAMA Psychiatry 72:1077. https://doi.org/10.1001/jamapsychiatry.2015.1252

Neisser U (1988) Five kinds of self-knowledge. Philos Psychol 1:35–59. https://doi.org/10.1080/09515088808572924

Nicholson AA, Sapru I, Densmore M et al (2016) Unique insula subregion resting-state functional connectivity with amygdala complexes in posttraumatic stress disorder and its dissociative subtype. Psychiatry Res Neuroimaging 250:61–72. https://doi.org/10.1016/j.pscychresns.2016.02.002

Nord CL, Garfinkel SN (2022) Interoceptive pathways to understand and treat mental health conditions. Trends Cogn Sci 26:499–513. https://doi.org/10.1016/j.tics.2022.03.004

Nummenmaa L, Saarimäki H (2019) Emotions as discrete patterns of systemic activity. Neurosci Lett 693:3–8. https://doi.org/10.1016/j.neulet.2017.07.012

Nummenmaa L, Hari R, Hietanen JK, Glerean E (2018) Maps of subjective feelings. Proc Natl Acad Sci 115:9198–9203. https://doi.org/10.1073/pnas.1807390115

Öngür D, Ferry AT, Price JL (2003) Architectonic subdivision of the human orbital and medial prefrontal cortex. J Comp Neurol 460:425–449. https://doi.org/10.1002/cne.10609

Owens AP, David AS, Low DA et al (2015) Abnormal cardiovascular sympathetic and parasympathetic responses to physical and emotional stimuli in depersonalization disorder. Front Neurosci 9

Park H-D, Blanke O (2019) Heartbeat-evoked cortical responses: underlying mechanisms, functional roles, and methodological considerations. Neuroimage 197:502–511. https://doi.org/10.1016/j.neuroimage.2019.04.081

Park H-D, Tallon-Baudry C (2014) The neural subjective frame: from bodily signals to perceptual consciousness. Philos Trans R Soc B Biol Sci 369:20130208. https://doi.org/10.1098/rstb.2013.0208

Park H-D, Bernasconi F, Bello-Ruiz J et al (2016) Transient modulations of neural responses to heartbeats covary with bodily self-consciousness. J Neurosci 36:8453–8460. https://doi.org/10.1523/JNEUROSCI.0311-16.2016

Parvizi J, Braga RM, Kucyi A et al (2021) Altered sense of self during seizures in the posteromedial cortex. Proc Natl Acad Sci 118:e2100522118. https://doi.org/10.1073/pnas.2100522118

Patihis L, Lynn SJ (2017) Psychometric comparison of dissociative experiences scales II and C: a weak trauma-dissociation link. Appl Cogn Psychol 31:392–403. https://doi.org/10.1002/acp.3337

Petzschner FH, Garfinkel SN, Paulus MP et al (2021) Computational models of interoception and body regulation. Trends Neurosci 44:63–76. https://doi.org/10.1016/j.tins.2020.09.012

Phillips ML, Medford N, Senior C et al (2001) Depersonalization disorder: thinking without feeling. Psychiatry Res Neuroimaging 108:145–160. https://doi.org/10.1016/S0925-4927(01)00119-6

Pistoia F, Carolei A, Sacco S et al (2015) Contribution of interoceptive information to emotional processing: evidence from individuals with spinal cord injury. J Neurotrauma 32:1981–1986. https://doi.org/10.1089/neu.2015.3897

Postmes L, Sno HN, Goedhart S et al (2014) Schizophrenia as a self-disorder due to perceptual incoherence. Schizophr Res 152:41–50. https://doi.org/10.1016/j.schres.2013.07.027

Powers A, Cross D, Fani N, Bradley B (2015) PTSD, emotion dysregulation, and dissociative symptoms in a highly traumatized sample. J Psychiatr Res 61:174–179. https://doi.org/10.1016/j.jpsychires.2014.12.011

Qin P, Northoff G (2011) How is our self related to midline regions and the default-mode network? Neuroimage 57:1221–1233. https://doi.org/10.1016/j.neuroimage.2011.05.028

Quadt L, Critchley HD, Garfinkel SN (2018) The neurobiology of interoception in health and disease: neuroscience of interoception. Ann N Y Acad Sci 1428:112–128. https://doi.org/10.1111/nyas.13915

Quadt L, Garfinkel SN, Mulcahy JS et al (2021) Interoceptive training to target anxiety in autistic adults (ADIE): a single-center, superiority randomized controlled trial. eClinicalMedicine 39. https://doi.org/10.1016/j.eclinm.2021.101042

Quelhas Martins A, McIntyre D, Ring C (2014) Effects of baroreceptor stimulation on performance of the Sternberg short-term memory task: a cardiac cycle time study. Biol Psychol 103:262–266. https://doi.org/10.1016/j.biopsycho.2014.10.001

Rebollo I, Devauchelle A-D, Béranger B, Tallon-Baudry C (2018) Stomach-brain synchrony reveals a novel, delayed-connectivity resting-state network in humans. eLife 7:e33321. https://doi.org/10.7554/eLife.33321

Reinders AATS, Marquand AF, Schlumpf YR et al (2019) Aiding the diagnosis of dissociative identity disorder: pattern recognition study of brain biomarkers. Br J Psychiatry 215:536–544. https://doi.org/10.1192/bjp.2018.255

Richter CG, Babo-Rebelo M, Schwartz D, Tallon-Baudry C (2017) Phase-amplitude coupling at the organism level: the amplitude of spontaneous alpha rhythm fluctuations varies with the phase of the infra-slow gastric basal rhythm. Neuroimage 146:951–958. https://doi.org/10.1016/j.neuroimage.2016.08.043

Rizzolatti G, Cattaneo L, Fabbri-Destro M, Rozzi S (2014) Cortical mechanisms underlying the organization of goal-directed actions and mirror neuron-based action understanding. Physiol Rev 94:655–706. https://doi.org/10.1152/physrev.00009.2013

Roel Lesur M, Weijs ML, Simon C et al (2020) Psychometrics of disembodiment and its differential modulation by visuomotor and visuotactile mismatches. iScience 23:100901. https://doi.org/10.1016/j.isci.2020.100901

Roydeva MI, Reinders AATS (2021) Biomarkers of pathological dissociation: a systematic review. Neurosci Biobehav Rev 123:120–202. https://doi.org/10.1016/j.neubiorev.2020.11.019

Saarimäki H, Glerean E, Smirnov D et al (2022) Classification of emotion categories based on functional connectivity patterns of the human brain. Neuroimage 247:118800. https://doi.org/10.1016/j.neuroimage.2021.118800

Saini F, Ponzo S, Silvestrin F et al (2022) Depersonalization disorder as a systematic downregulation of interoceptive signals. Sci Rep 12:22123. https://doi.org/10.1038/s41598-022-22277-y

Salomon R (2017) The assembly of the self from sensory and motor foundations. Soc Cogn 35:87–106. https://doi.org/10.1521/soco.2017.35.2.87

Schabinger N, Gillmeister H, Berti S et al (2018) Detached and distracted: ERP correlates of altered attentional function in depersonalisation. Biol Psychol 134:64–71. https://doi.org/10.1016/j.biopsycho.2018.02.014

Schachter S, Singer J (1962) Cognitive, social, and physiological determinants of emotional state. Psychol Rev 69:379–399. https://doi.org/10.1037/h0046234

Schäflein E, Sattel HC, Pollatos O, Sack M (2018) Disconnected – impaired interoceptive accuracy and its association with self-perception and cardiac vagal tone in patients with dissociative disorder. Front Psychol:9

Scherer KR, Moors A (2019) The emotion process: event appraisal and component differentiation. Annu Rev Psychol 70:719–745. https://doi.org/10.1146/annurev-psych-122216-011854

Schmid Y, Enzler F, Gasser P et al (2015) Acute effects of lysergic acid diethylamide in healthy subjects. Biol Psychiatry 78:544–553. https://doi.org/10.1016/j.biopsych.2014.11.015

Schoenberg PLA, Sierra M, David AS (2012) Psychophysiological investigations in depersonalization disorder and effects of electrodermal biofeedback. J Trauma Dissociation 13:311–329. https://doi.org/10.1080/15299732.2011.606742

Schulz A, Vögele C (2015) Interoception and stress. Front Psychol 6:993. https://doi.org/10.3389/fpsyg.2015.00993

Schulz A, Strelzyk F, Ferreira de Sá DS et al (2013) Cortisol rapidly affects amplitudes of heartbeat-evoked brain potentials – implications for the contribution of stress to an altered perception of physical sensations? Psychoneuroendocrinology 38:2686–2693. https://doi.org/10.1016/j.psyneuen.2013.06.027

Schulz A, Köster S, Beutel ME et al (2015) Altered patterns of heartbeat-evoked potentials in depersonalization/derealization disorder: neurophysiological evidence for impaired cortical representation of bodily signals. Psychosom Med 77:506. https://doi.org/10.1097/PSY.0000000000000195

Sel A, Azevedo RT, Tsakiris M (2017) Heartfelt self: cardio-visual integration affects self-face recognition and interoceptive cortical processing. Cereb Cortex 27:5144–5155. https://doi.org/10.1093/cercor/bhw296

Seligowski AV, Lebois LAM, Hill SB et al (2019) Autonomic responses to fear conditioning among women with PTSD and dissociation. Depress Anxiety 36:625–634. https://doi.org/10.1002/da.22903

Seth AK (2013) Interoceptive inference, emotion, and the embodied self. Trends Cogn Sci 17:565–573. https://doi.org/10.1016/j.tics.2013.09.007

Seth AK, Friston KJ (2016) Active interoceptive inference and the emotional brain. Philos Trans R Soc B Biol Sci 371:20160007. https://doi.org/10.1098/rstb.2016.0007

Seth A, Suzuki K, Critchley H (2012) An interoceptive predictive coding model of conscious presence. Front Psychol 2

Sierra M, Berrios GE (1998) Depersonalization: neurobiological perspectives. Biol Psychiatry 44: 898–908. https://doi.org/10.1016/S0006-3223(98)00015-8

Sierra M, Berrios GE (2000) The Cambridge depersonalisation scale: a new instrument for the measurement of depersonalisation. Psychiatry Res 93:153–164. https://doi.org/10.1016/S0165-1781(00)00100-1
Sierra M, David AS (2011) Depersonalization: a selective impairment of self-awareness. Conscious Cogn 20:99–108. https://doi.org/10.1016/j.concog.2010.10.018
Sierra M, Senior C, Dalton J et al (2002) Autonomic response in depersonalization disorder. Arch Gen Psychiatry 59:833–838. https://doi.org/10.1001/archpsyc.59.9.833
Sierra M, Baker D, Medford N, David AS (2005) Unpacking the depersonalization syndrome: an exploratory factor analysis on the Cambridge Depersonalization Scale. Psychol Med 35:1523–1532. https://doi.org/10.1017/S0033291705005325
Sierra M, Senior C, Phillips ML, David AS (2006) Autonomic response in the perception of disgust and happiness in depersonalization disorder. Psychiatry Res 145:225–231. https://doi.org/10.1016/j.psychres.2005.05.022
Simeon D, Kozin DS, Segal K et al (2008) De-constructing depersonalization: further evidence for symptom clusters. Psychiatry Res 157:303–306. https://doi.org/10.1016/j.psychres.2007.07.007
Simeon D, Knutelska M, Putnam FW et al (2023) Attention and memory in depersonalization-spectrum dissociative disorders: impact of selective-divided attentional condition, stimulus emotionality, and stress. J Trauma Dissociation 24:42–62. https://doi.org/10.1080/15299732.2022.2079798
Stein DJ, Koenen KC, Friedman MJ et al (2013) Dissociation in posttraumatic stress disorder: evidence from the world mental health surveys. Biol Psychiatry 73:302–312. https://doi.org/10.1016/j.biopsych.2012.08.022
Stent GS (1998) Epistemic dualism of mind and body. Proc Am Philos Soc 142:578–588
Sterlini GL, Bryant RA (2002) Hyperarousal and dissociation: a study of novice skydivers. Behav Res Ther 40:431–437. https://doi.org/10.1016/S0005-7967(01)00021-3
Stern Y, Koren D, Moebus R et al (2020) Assessing the relationship between sense of agency, the bodily-self and stress: four virtual-reality experiments in healthy individuals. J Clin Med 9:2931. https://doi.org/10.3390/jcm9092931
Suksasilp C, Garfinkel SN (2022) Towards a comprehensive assessment of interoception in a multi-dimensional framework. Biol Psychol 168:108262. https://doi.org/10.1016/j.biopsycho.2022.108262
Suzuki K, Garfinkel SN, Critchley HD, Seth AK (2013) Multisensory integration across exteroceptive and interoceptive domains modulates self-experience in the rubber-hand illusion. Neuropsychologia 51:2909–2917. https://doi.org/10.1016/j.neuropsychologia.2013.08.014
Tajadura-Jiménez A, Tsakiris M (2014) Balancing the "inner" and the "outer" self: interoceptive sensitivity modulates self–other boundaries. J Exp Psychol Gen 143:736–744. https://doi.org/10.1037/a0033171
Tajfel H (1978) The achievement of inter-group differentiation. In: Tajfel H (ed) Differentiation between social groups. Academic Press, London, pp 77–100
Tajfel H, Turner JC (1979) An integrative theory of inter-group conflict. In: Austin WG, Worchel S (eds) The social psychology of inter-group relations. Brooks/Cole, Monterey, pp 33–47
Terpou BA, Harricharan S, McKinnon MC et al (2019) The effects of trauma on brain and body: a unifying role for the midbrain periaqueductal gray. J Neurosci Res 97:1110–1140. https://doi.org/10.1002/jnr.24447
Tibubos AN, Grammes J, Beutel ME et al (2018) Emotion regulation strategies moderate the relationship of fatigue with depersonalization and derealization symptoms. J Affect Disord 227:571–579. https://doi.org/10.1016/j.jad.2017.11.079
Tsakiris M, Jiménez AT, Costantini M (2011) Just a heartbeat away from one's body: interoceptive sensitivity predicts malleability of body-representations. Proc R Soc B Biol Sci 278:2470–2476. https://doi.org/10.1098/rspb.2010.2547
Tyng CM, Amin HU, Saad MNM, Malik AS (2017) The influences of emotion on learning and memory. Front Psychol 8

van Stralen HE, van Zandvoort MJE, Hoppenbrouwers SS et al (2014) Affective touch modulates the rubber hand illusion. Cognition 131:147–158. https://doi.org/10.1016/j.cognition.2013.11.020

Vogeley K, Fink GR (2003) Neural correlates of the first-person-perspective. Trends Cogn Sci 7: 38–42. https://doi.org/10.1016/S1364-6613(02)00003-7

Whitford TJ, Jack BN, Pearson D et al (2017a) Neurophysiological evidence of efference copies to inner speech. eLife 6:e28197. https://doi.org/10.7554/eLife.28197

Whitford TJ, Mitchell AM, Mannion DJ (2017b) The ability to tickle oneself is associated with level of psychometric schizotypy in non-clinical individuals. Conscious Cogn 52:93–103. https://doi.org/10.1016/j.concog.2017.04.017

Wiens S (2005) Interoception in emotional experience. Curr Opin Neurol 18:442. https://doi.org/10.1097/01.wco.0000168079.92106.99

Wiens S, Mezzacappa ES, Katkin ES (2000) Heartbeat detection and the experience of emotions. Cognit Emot 14:417–427. https://doi.org/10.1080/026999300378905

World Health Organization (2019) ICD-11: international classification of diseases (11th revision). https://icd.who.int/. Accessed 27 Jan 2023

Wurm MF, Schubotz RI (2018) The role of the temporoparietal junction (TPJ) in action observation: agent detection rather than visuospatial transformation. Neuroimage 165:48–55. https://doi.org/10.1016/j.neuroimage.2017.09.064

Yang J, Millman LSM, David AS, Hunter ECM (2022) The prevalence of depersonalization-derealization disorder: a systematic review. J Trauma Dissociation 0:1–34. https://doi.org/10.1080/15299732.2022.2079796

Zelano C, Jiang H, Zhou G et al (2016) Nasal respiration entrains human limbic oscillations and modulates cognitive function. J Neurosci 36:12448–12467. https://doi.org/10.1523/JNEUROSCI.2586-16.2016

Predictive Processing and the Pathophysiology of Functional Neurological Disorder

Johannes Jungilligens and David L. Perez

Contents

Abstract The contemporary neuroscience understanding of the brain as an active inference organ supports that our conscious experiences, including sensorimotor perceptions, depend on the integration of probabilistic predictions with incoming sensory input across hierarchically organized levels. As in other systems, these

J. Jungilligens (✉)
Behavioral Neurology Research Group, Department of Neurology, University Hospital Knappschaftskrankenhaus Bochum, Ruhr University Bochum, Bochum, Germany

D. L. Perez
Division of Behavioral Neurology and Integrated Brain Medicine, Department of Neurology, Functional Neurological Disorder Unit, Massachusetts General Hospital, Harvard Medical School, Boston, MA, USA

Athinoula A. Martinos Center for Biomedical Imaging, Massachusetts General Hospital, Harvard Medical School, Boston, MA, USA

Division of Neuropsychiatry, Department of Psychiatry, Massachusetts General Hospital, Harvard Medical School, Boston, MA, USA

Curr Topics Behav Neurosci (2025) 74: 279–305
https://doi.org/10.1007/7854_2024_473
Published Online: 17 May 2024

complex processes are prone to error under certain circumstances, which may lead to alterations in their outcomes (i.e., variations in sensations and movements). Such variations are an important aspect of functional neurological disorder, a complex disorder at the interface of brain–mind–body interactions. Thus, predictive processing frameworks offer fundamental mechanistic insights into the pathophysiology of functional neurological disorder. In recent years, many of the aspects relevant to the neurobiology of functional neurological disorder – e.g., aberrant motor and sensory processes, symptom expectation, self-agency, and illness beliefs, as well as interoception, allostasis, and emotion – have been investigated through the lens of predictive processing frameworks. Here, we provide an overview of the current state of research on predictive processing and the pathophysiology of functional neurological disorder.

Keywords Functional neurological disorder · Predictive processing · Pathophysiology · Neurobiology · Active inference · Neuroimaging

1 Introduction: What Is a Functional Neurological Disorder and What Are the Major Themes in the Emerging Pathophysiology of This Condition?

> *"Functional neurological disorder is arguably what might be expected to happen when predictive processing in the brain goes awry."* (Hallett et al. 2022, p. 547)

Functional neurological disorder (FND) is a common, costly, and potentially disabling neuropsychiatric disorder, affecting about 50 in every 100,000 persons and incurring healthcare costs of 1.2 billion USD for emergency department and inpatient care in the USA alone (Stephen et al. 2021; Stone et al. 2009b; Carson and Lehn 2016). Patients with FND report reduced health-related quality of life comparable to multiple sclerosis and Parkinson's disease and are frequently unemployed due to health concerns (Anderson et al. 2007; Stone et al. 2003; Carson et al. 2011). FND is characterized by a range of physical symptoms (e.g., abnormal movements, seizures, difficulty speaking, dizziness, cognitive trouble), also commonly including pain and fatigue (Maggio et al. 2020; Butler et al. 2021; Hallett et al. 2022). As such, FND can resemble other neurological disorders, like epilepsy, stroke, or Parkinson's disease. However, FND is diagnosed using positive clinical signs that are specific for the condition (Aybek and Perez 2022). As currently described, FND is a condition that encompasses several subtypes – i.e., functional seizures, functional movement disorder (such as functional limb weakness or functional tremor), functional speech/voice disorder, persistent postural perceptual dizziness, functional cognitive disorder, and functional sensory disorders – with a subset of individuals presenting with mixed symptoms (Aybek and Perez 2022). FND received little attention during the second half of the 20th century despite driving the development of theories that

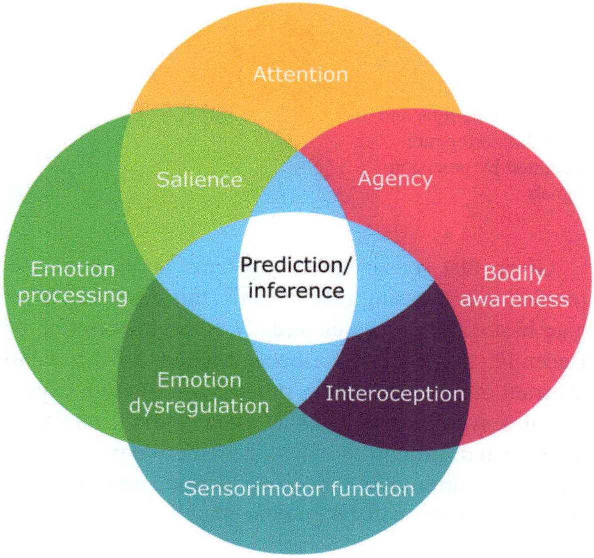

Fig. 1 Aspects and concepts relevant to functional neurological disorder pathophysiology and their relationship to predictive processing. Figure from Drane et al. (2020) and Aybek and Perez (2022) – reproduced with permission

were central to the psychological landscape of the late 19th and early 20th centuries, such as Sigmund Freud's "conversion model" and the concept of dissociation introduced by Pierre Janet (Stone 2016). Clinical and research interest in FND resurfaced in the late 1990s and early 2000s, simultaneous with and in part driven by emerging technologies such as functional magnetic resonance imaging (MRI), and advances in understanding computational aspects of brain function relevant to FND, e.g., the difference between volitional and involuntary movement (Voon et al. 2010a).

In recent years, research has contributed to a growing understanding of the multifaceted pathophysiology underlying FND. Commonly identified themes include the role of attention, salience, self-agency, interoception, and affective regulation, which contribute to alterations in bodily awareness and sensorimotor abilities (Fig. 1) (Drane et al. 2020). In addition, patients with FND often report a substantial psychiatric burden (e.g., trauma-related and mood/anxiety disorders), which may additionally influence an individual's clinical presentation and underlying pathophysiological circuits (Ducroizet et al. 2023). Finally, a significant number of patients develop FND following physical injuries such as accidents or surgery, or illnesses such as other neurological disorders or medication/vaccination side effects (objectively verified or subjectively experienced) (Stone et al. 2009a). The association with physical illness is thought to draw attention to the body and also potentially reflects the role of bodily experiences in shaping explicit and implicit expectations about, for example, the (in)ability to move a particular limb at will, or the bodily consequences of a state of heightened arousal and negative affect (Pick et al. 2016; Stone et al. 2012).

Fig. 2 Predictions are based on life experiences, which might be updated in case a prediction error occurs. Predictions and prediction errors are weighted by precision signals

Given that sensory-motor systems implicated in core FND symptoms rely on predictive processing mechanisms, this condition offers the opportunity to investigate biological processes directly at the intersection of physical and mental health. Recently, many of the aspects relevant to the pathophysiology of FND – e.g., aberrant motor and sensory processes, symptom expectation, self-agency, attention, and illness beliefs as well as interoception, allostasis, and emotion – have been investigated through the lens of predictive processing frameworks. Additionally, a range of neuroimaging, neuroendocrine, autonomic, and behavioral findings found in the literature can be re-conceptualized and integrated within these predictive processing frameworks. Here, we provide an overview of the current state of research on predictive processing and the pathophysiology of FND (for a brief introduction to predictive processing, see Box 1).

Box 1. Predictive Processing in the Brain: An Overview

Within theoretical and computational neuroscience, predictive processing has emerged as a powerful framework to describe how anticipatory calculations are generated and refined within and across neural networks to explain certain brain functions (like motor action generation, belief updating, and sensory processing). This family of approaches includes several variations, all generally based on the idea of Bayesian inference, which is a formal method of combining prior knowledge with new evidence to update beliefs about uncertain variables. These approaches share three key principles (Clark 2013) (Fig. 2):

1. Based on prior life experiences, the brain continually predicts the immediate future through an array of parallel predictive simulations that all come with a certain probability (weight). The (top-down) predictions with the highest probability are compared to incoming (bottom-up) sensory input. If the top-down prediction and the bottom-up sensory input match, it becomes a perception.
2. In case of a mismatch between prediction and sensory input, a prediction error occurs. The prediction error tracks the difference between anticipated and actual sensory input, which can then be used to update and improve subsequent predictions (i.e., learning).

(continued)

Box 1 (continued)

3. The strength or magnitude of predictions and prediction errors are weighted by their modeled importance through precision signals. In cases where the prediction outweighs the sensory evidence, the prediction error is discarded, and the subsequent predictions remain unchanged.

This general idea has proven powerful in explaining a range of brain functions and neuropsychological phenomena, while also integrating well with fundamental neurobiological principles of brain organization (Paquola et al. 2022). Predictive processing models have been developed for many perceptual and cognitive features, such as different sensory modalities, motor function, and cognitive abilities such as memory (Hohwy 2020).

2 Overview of Predictive Processing Theories for Functional Neurological and Functional Somatic Symptoms

Predictive processing frameworks have been used to explain aspects of functional neurological and functional somatic symptoms. We briefly review several such models here, highlighting their contributions to understanding the role of predictive processing in the pathophysiology of FND (also see Table 1 for an overview).

Edwards, Adams, and colleagues were first to use a Bayesian framework to explain functional motor and sensory symptoms (Edwards et al. 2012). Their approach is based on the conceptualization that all sensory perceptions result from comparing top-down *predictions* of sensations (generated based on past experiences)

Table 1 Overview of major predictive processing frameworks for functional neurological disorder

Model	Description
Bayesian framework (Edwards et al. 2012)	Focuses on the role of attentional and belief-driven processes including pathologically precise prior beliefs in the emergence of functional motor and sensory symptoms
Integrative cognitive model (Brown and Reuber, 2016)	Explains how functional seizures arise through a seizure scaffold as a mental representation of the semiology and phenomenology of a seizure
Inferential leap model (Van den Bergh et al. 2017)	Proposes that symptom perception is subject to belief-driven hypothesis testing, which may result in false percepts, and ranges from symptoms resulting from physical dysfunction to functional symptoms that have no direct relationship to physiological insults
Theory of constructed emotion (Jungilligens et al. 2022a)	Based on the notion that the brain regulates the body through allostasis, which relies on a predictive interoceptive model of the body's physiological state; functional neurological and functional somatic symptoms may emerge due to aberrant emotion construction

to the incoming or bottom-up sensory input; the difference between those is tracked by a *prediction error*. These prior beliefs/predictions are probabilistically represented by neuronal activity, and their influence in the predictive processing cascade is weighed through attentional processes, in part via neurotransmitter systems such as the dopaminergic system (Parr and Friston 2017; Diederen and Fletcher 2021). As the framework is based on the free-energy principle (Friston 2010), it is tied to the notion that the brain strives to self-organize in a way that minimizes prediction error, in this case unexpected sensations. This can be achieved either by adjusting the top-down predictions to fit the bottom-up sensory input (i.e., learning and updating prior beliefs) or by giving greater weight to the top-down predictions vs. bottom-up input (which results in altered perceptions). This means that, in the case of functional sensory symptoms, prior beliefs with a high probabilistic weight can override incoming sensory input to create evidence for their own existence; thus, resulting in sensation without a corresponding physical input. The same idea holds for movement. In case of a strongly weighted prior belief about a movement, the prediction of the sensory input related to the movement (for example, the proprioceptive information about the position of a certain limb) can be fulfilled by actually moving the limb as predicted: a movement such as a tremor is produced to fulfill the prediction of the proprioceptive sensation tied to this tremor. Therefore, the authors state that "all [functional motor and sensory symptoms] are created by attentional and belief-driven processes" (Edwards et al. 2012) within the functional architecture of normal perceptions and voluntary movements. Taken together, functional sensorimotor symptoms are understood to emerge due to pathologically precise prior beliefs in the process of minimizing prediction error (and/or aberrant prediction error learning), assigning the expectation a higher relative weight than the bottom-up sensory evidence. This approach is strictly focused on functional motor and sensory symptoms; thus, it does not focus on other related phenomena such as altered autonomic arousal or misinterpretation of normal bodily sensations, and the broader clinical presentation including the affective disturbances often associated with FND.

Recently, our group has developed a framework based on the Theory of Constructed Emotion to explain the role of emotion and affect in the pathophysiology of FND (Jungilligens et al. 2022a; Barrett 2017). Fundamental to this approach is the notion that the brain regulates the body through the predictive process of allostasis, which relies in part on an interoceptive model of the body's physiological state. A low-dimensional summary of this state – referred to as *affect* – is available to, permeates, and partially constitutes aspects of consciousness, and influences the predictive categorization of incoming sensory input. The basis for the predictions of exteroceptive and interoceptive sensory input are concepts: abstract summaries of previous life experiences that can operate largely outside of conscious awareness, but are – at least in part – consciously accessible. These concepts, which include emotion concepts, are learned and refined throughout life and may vary in granularity between individuals depending on a range of biopsychosocial factors. Important in this context is that adverse life experiences, especially during critical developmental periods, can establish concepts with long-lasting consequences –

meaning that the salience/precision signals weighing of the predictions that arise from those concepts can potentially override incoming prediction errors. This is particularly apparent in the finding that in patients with FND, the severity of childhood abuse burden is associated with increased symptom severity and reduced health-related quality of life (Paredes-Echeverri et al. 2022a). In line with the assumption that concepts are formed to explain and deal with momentary sensory input, we speculate that some individuals who formed certain concepts earlier in life to explain the world while experiencing threatening or neglectful environments may develop predictive models with a stronger motor or body-focused component (e.g., fight-or-flight-like motor elements) rather than a self-referential emotional state. While initially adaptive, if carried forward these predictive models may be less helpful in circumstances encountered later in life. Thus, certain individuals might tend to not construct emotions in situation where others would – they might construct bodily/health symptom perceptions instead (i.e., FND symptoms). Notably, this framework also incorporates the idea of how the (micro)structure of the brain – such as the organization and connectivity of cortical layers – supports complex brain functions: predictions originate in the limbic cortices of the default-mode network (precuneus, posterior cingulate, angular gyri, lateral temporal, sub/pregenual anterior cingulate, and medial prefrontal cortices) and are shaped into more fine-grained sensory and motor predictions across the cortical hierarchies by gaining modality-specific attributes and features computed in multimodal and unimodal association areas (see Fig. 3 and caption for more details) (Chanes and Barrett 2016). Sensory input, on the other hand, is initially encoded cortically in the primary sensory areas and then processed in an increasingly multimodal and abstract version, while being compared with the predictions at each step of the processing hierarchy. The resulting prediction errors and predictions are weighted for importance by precision signals originating from the salience network (anterior insula, dorsal anterior cingulate cortex, amygdala, periaqueductal gray, and hypothalamic areas) – a network implicated in the sensing and regulation of the state of the body, and the relationship of this state on the presumed importance of other external and internal events. This offers a window into how the state of the body, in conjunction with previous life experiences, influences predictive processing within salience and default-mode networks, and how alterations in those networks might contribute to the pathophysiology of FND. Overall, this explanation incorporates several themes that have in part been previously discussed in the literature on FND, but not considered in an integrated approach. This includes: (i) explanations of the relationship between chronic bodily dysregulation and the emergence of FND; (ii) a description of how functional neurological symptoms can arise from aberrant emotion construction; (iii) a re-conceptualization of the roles of alexithymia and "panic attack without panic"; (iv) and a delineation of the influence of life experiences on emotion construction.

Functional seizures, a prevalent FND subtype, have been explained within the integrative cognitive model by Brown and Reuber (2016). As an extension of a more general integrative cognitive model for "medically unexplained symptoms" (Brown 2004), the approach focuses on *how* rather than *why* functional seizures arise. Thus,

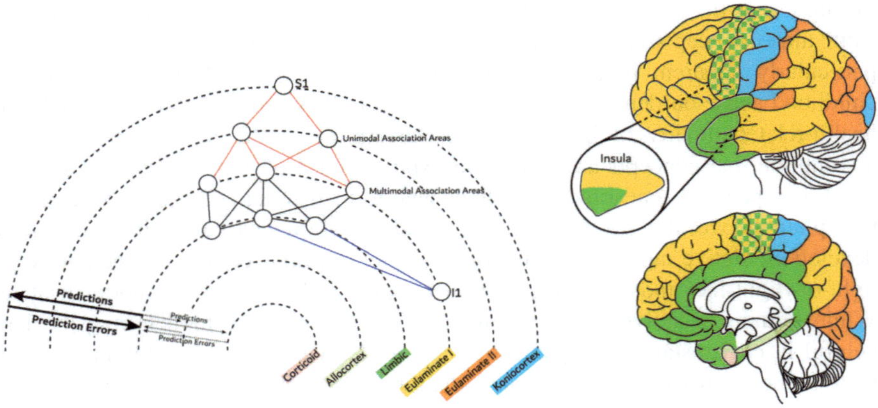

Fig. 3 Relationships between cytoarchitectural complexity and predictive processing. Gray matter regions can be divided based on their lamination profiles, i.e., the number and differentiation of cortical layers from least to most complex. The corticoid regions contain part of the amygdala; allocortex is found among others in hippocampus and part of the amygdala. Limbic cortices entail regions such as the cingulate cortex, ventral anterior insula, posterior orbitofrontal/ventromedial prefrontal cortex, parahippocampus and the temporal pole; there is debate on how to characterize the lamination profiles of the motor cortices, with conflicting evidence supporting an agranular/dysgranular ("limbic like") vs. rudimentary layer IV ("Eulaminate I like") profile. There are two kinds of Eulaminate areas: Eulaminate I entails multimodal association areas such as dorsolateral prefrontal cortex, lateral temporal areas, posterior parietal areas while Eulaminate II areas are unimodal association cortices, e.g., superior parietal lobule. Primary sensory areas such as the primary somatosensory, auditory, and visual areas are the most complexly layered cortical regions (Koniocortex). This cytoarchitectural gradients supports the flow of incoming sensory information from the most detailed representations in the primary sensory cortex (Koniocortex) toward an increasingly abstracted representation across unimodal and multimodal association areas (Eulaminate I/II). Predictions arise at the other end of this axis in areas with limbic structure and increasingly gain granularity while they are compared to the incoming sensory input at every level in-between. Cingulo-insular and parts of amygdala represent the salience network, which weighs predictions and prediction error through salience signaling. Note: This figure only displays the predictive processing of somatosensory and interoceptive signals; other sensory modalities (e.g., visual) have similar processes. S1: primary somatosensory cortex; I1: interoceptive cortex. Figure from Jungilligens et al. (2022a) – reproduced with permission

the model integrates empirical findings on predisposing factors (e.g., chronic stress, traumatic experiences or hard-wired behavioral tendencies), precipitating factors (e.g., elevated arousal, threat perception, and the anticipation of an impending seizure), and perpetuating factors (e.g., the relief resulting from the reduced arousal through the dissociation during the seizure, which might act as reinforcement in an operant learning view). While for the most part not formulated using predictive processing concepts, the integrative cognitive model adopts a constructivist approach. This is most relevant for the idea of a *seizure scaffold*, i.e., a mental representation of the semiology and phenomenology of a seizure. This "rogue representation" includes innate and learned behaviors and experiences that form functional seizure symptoms; through learning/operant conditioning it might become the "go-to" representation when it comes to predicting sensory and motor aspects during heightened physiological arousal states.

Based also on the general integrative cognitive model (Brown 2004), Van den Bergh and colleagues proposed the concept of an *inferential leap* in an attempt to provide a comprehensive explanation for symptom perception across a range of disorders with "medically unexplained symptoms" including FND (Van den Bergh et al. 2017). This framework posits that the interpretation of sensory input is subject to an error-prone system of belief-driven Bayesian hypothesis testing, which in some cases results in *false percepts* or *somatovisceral illusions*. Accordingly, functional neurological and functional somatic symptoms may be only weakly or not at all related to actual bodily processes. Thus, similar to the approach of Edwards, Adams et al. they propose that aberrant prior beliefs influence how individuals perceive incoming sensory input; however, they take a broader perspective by extending this mechanism to any kind of somatic symptom, whether explained or unexplained by a directly identifiable underlying pathology: the authors propose a continuum of symptom perception, ranging from symptoms resulting from clear physical dysfunction at one end, to distorted symptom perception that only partially overlaps with underlying physical dysfunction in the middle, to functional symptoms that have no (or minimal) relationship to incoming physiological data at the other end (Edwards et al. 2012). Therefore, the authors suggest that peripheral physiological input is neither necessary nor sufficient for the experience of functional neurological and functional somatic symptoms. Overall, this approach focusses on sensory rather than motor symptoms.

3 Clinical and Behavioral Evidence for Predictive Processing Difficulties in Functional Neurological Disorder

Evidence for the above models and the role of predictive processing difficulties in FND can be found both in the phenomenology of FND and the related diagnostic domains.

3.1 Phenomenology of Functional Neurological Disorder

The clinical presentation of FND varies considerably between and within subtypes. However, some common elements include: (i) reduced voluntary motor control, which is mostly understood to be related to loss of self-agency perceptions; (ii) the role of emotion, interoception, and affective processes in the pathophysiology; (iii) sensory processing difficulties across external and internal sensory experiences; and (iv) biased attentional processing and aberrant motor learning.

3.1.1 Sense of Self-Agency

An altered sense of self-agency, i.e., the feeling of authorship of one's own behavior, is understood to be play a key role in the pathophysiology of FND, particularly those with functional motor and/or seizure symptoms. Patients with FND report reduced or no subjective control over their symptoms. In predictive processing terms, the subjective loss of voluntary motor control in FND is explained through the physiology underlying voluntary movements. To initiate or modify a movement, the primary motor cortex receives a motor command, while an efferent copy of this command – i.e., a feedforward signal – is issued to the right temporo-parietal junction (among other brain areas). The right temporo-parietal junction represents a key hub implicated in self-agency, among other areas involved in bodily awareness (anterior insula, precuneus, and several prefrontal regions; Nahab et al. 2011; Hallett 2022). Proprioceptive feedback about the ongoing movement is then integrated in the self-agency network, which compares the original plan (the feedforward signal) to the outcome (the feedback signal) (Hallett et al. 2022). In the case of adequate overlap, a sense of self-agency – "I initiated and performed this movement voluntarily" – arises. In FND, this mechanism is commonly considered to be distorted. Thus, patients report feeling no subjective sense of self-agency over the functional neurological symptom – "I did not initiate this movement voluntarily; it just happens to me." Several studies have investigated the sense of self-agency in patients with FND, mostly using paradigms such as the "Libet clock" experiment (participants observing a moving clock hand while reporting the moment they became consciously aware of their intention to perform a specific action) or variations of it. Overall, most studies found reduced self-agency perceptions in patients with FND (Edwards et al. 2011; Kranick et al. 2013), but the effect was found to be more prominent for positive vs. negative functional neurological symptoms (Baek et al. 2017), or identifiable only at trend-level (Jungilligens et al. 2020). Using an electronic glove that controlled a virtual hand at different levels of precision, Nahab et al. found that patients with FND exhibited altered levels of subjective control over the virtual hand compared to healthy controls (Nahab et al. 2017). In a force matching paradigm, patients with functional movement disorder showed lower sensory attenuation, which is a measure of sense of self-agency for movement (Parees et al. 2014a). Other measures such as subliminal and supraliminal priming (Huys et al. 2020) or moving rubber hand illusion paradigms (Marotta et al. 2017) found no differences in explicit sense of self-agency between patients with FND and healthy controls, but hypothesized that the process of conscious motor preparation might be distorted in FND (Huys et al. 2020).

3.1.2 Emotion and Affect

Throughout different historical and contemporary conceptualizations of FND, emotion and emotion processing were proposed to play an important role (Pick et al. 2019). However, only recent advances in affective neuroscience and related disciplines allowed us to rigorously integrate the role of emotion into the pathophysiology of FND (Jungilligens et al. 2022a). This integration hinges on the notion that not only sensory perception and movement are based on predictive mechanisms, but also other processes foundational to the functioning of the body and the brain (such as interoception, allostasis, and affect), and that these brain-mind-body processes also constitute the build blocks of emotions. This includes the predictive regulation of the body through allostasis – the active process of regulating the body, e.g., by managing its energetic supply-and-demand. This fundamental process is key to survival and is thus intertwined with or at the core of other brain functions, such as interoception (the predictive modeling of the internal state of the body) or the predictive categorization of incoming sensory input. Overall, the framework suggests that functional neurological symptoms may occur due to dysfunction at multiple points in the predictive processing emotion construction pathway (Fig. 4). This framework integrates several lines of research relevant to FND, namely the role of interoception, affect and energy management, adverse life experiences, and emotion concept granularity in the pathophysiology. Thus, it allows understanding of those aspects (and potential aberrations) within an overarching predictive processing framework. For example, while the results remain partially inconsistent, some studies investigating interoception in FND populations using heartbeat tracking tasks found decreased interoceptive accuracy at baseline or following induction procedures, which might be interpreted as a distorted interoceptive modeling and model updating (Koreki et al. 2020; Pick et al. 2020; Williams et al. 2021; Ricciardi et al. 2016). Since efficient regulation of the body (allostasis) depends on well-integrated interoceptive information, these findings of distorted interoceptive modeling may help to explain why chronic energy dysregulation in the form of fatigue or hyperarousal is highly prevalent in patients with FND. With an ill-informed interoceptive model, the state of the body cannot be adequately sensed and thus not adaptively managed. Fatigue is the most common symptom reported by patients with FND and is a better predictor of quality of life than symptom severity in motor FND (Butler et al. 2021; Gelauff et al. 2018). The framework also highlights the role of life experiences for predictive processing later in life. Adverse experiences, especially those occurring in critical developmental phases, establish concepts with predictive processing implications that likely carry long-lasting consequences. For example, growing up in a threatening or emotionally abusive environment – as patients with FND report with an odds ratio of 5.6 compared to healthy controls (Ludwig et al. 2018) – might establish concepts related to bodily symptoms and flight-or-flight components instead of self-referential emotion concepts, as these would be more adaptive in dealing with this threatening situation. While this relationship is in line with recent theoretical models (Jungilligens et al. 2022a),

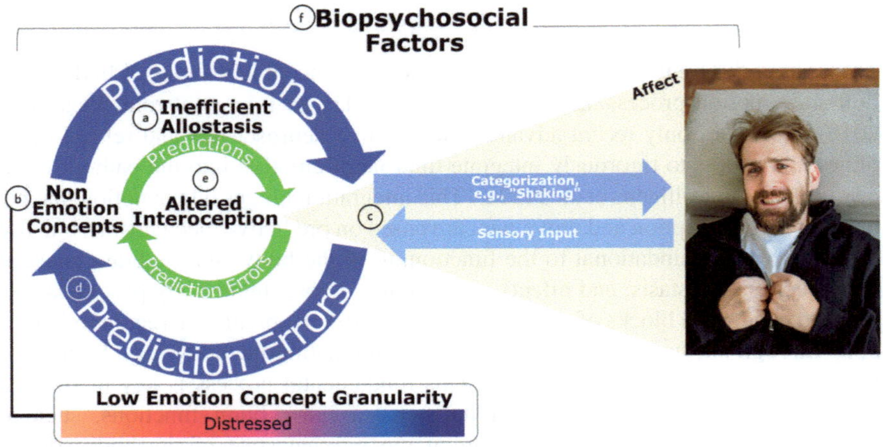

Fig. 4 The theory of constructed emotion applied to functional neurological disorder (FND). The figure highlights points of aberrant emotion construction that might occur in patients with FND. (**a**) Inefficient allostasis reflects chronic energy mismanagement in some people with FND, which can lead to symptoms such as fatigue or states of hyperarousal. (**b**) The limited availability or lack of granular emotion concepts may be a result of adverse life experiences such as emotional neglect or physical threat. These adverse experiences can predispose to the development of bodily and health/illness related non-emotion concepts. (**c**) In cases where incoming sensory input is predicted based on a non-emotion concept, a bodily/illness category is constructed (e.g., "shaking"). (**d**) Aberrant precision signals and integration of prediction error to update internal models can lead to deficits in sensory processing, interoceptive accuracy, biased attention, and impairments in motor learning. (**e**) Dissociation – a common symptom in FND – can be understood as a momentary uncoupling of the brain's interoceptive and allostatic models. (**f**) Deficits at each step of the predictive processing stream can be influenced by biopsychosocial factors. Figure from Jungilligens et al. (2022a) – reproduced with permission, including the picture (public domain, from: https://www.pexels.com)

there is currently a lack of empirical literature. The idea is supported by the finding that in patients with FND, higher adverse life experience burden is related to alexithymia (i.e., a lack of granular emotion concepts), a finding that has been replicated in other cohorts (Farooq and Yousaf 2016; Kaplan et al. 2013; Aust et al. 2013; Zlotnick et al. 2001). Later in life, these concepts may still be at the core of predictive categorization of internal and external sensory input, despite changes in the environment that would make them unnecessary – e.g., a person with FND who is no longer in a stressful environment might still (nonconsciously) predict the experience of such and react accordingly. In this line of thinking, two aspects important to FND – alexithymia (a characteristic described as difficulties identifying/describing feelings and having externally oriented thinking, (Taylor et al. 1991)) and "panic attack without panic" can be reframed as being related to emotion concept granularity or general emotion concept availability. In alexithymia, the availability of granular emotion concepts could be limited, thereby increasing inefficiencies in adaptive contextualization of sensory input. "Panic attack without panic" was described in people with functional seizures to characterize individuals who confirmed the autonomic symptoms commonly associated with panic attacks (such as heart palpitations, hyperventilation, or dizziness) but lacked an associated subjective emotional feeling of fear or panic (Goldstein and Mellers 2006). The phenonemon of "panic attack without panic" may describe an instance where an emotion was not constructed. It is important to highlight that the principles outlined here may be a general predisposing vulnerability for the development of a range of psychopathologies (Shaffer et al. 2022); however, the use of body/illness-focused concepts might be more specific to FND and other functional somatic disorders.

3.1.3 Sensory Processing

Also, the Theory of Constructed Emotion framework applied to FND proposes altered prediction error learning in this population, potentially related to distorted precision signaling from the salience network, which would contribute to deficits in sensory processing. This phenomenon includes alterations in sensitivity and responsiveness to external stimuli (sounds, light, etc.), but might also entail the perception of sensory signals from the body surface (touch, heat/cold) or interoceptive signals. While widely reported clinically by patients with FND, sensory processing difficulties are not well-researched (MacLean et al. 2022). A retrospective chart review study by Ranford et al. (2020) investigated 44 patients with FND, identifying reports of heightened sensory processing (i.e., external sensations such as lights and sounds leading to increased sensory sensitivity or sensory avoidance) and a lowered ability to detect sensory stimuli at other instances (i.e., increased low registration). In a cohort of DSM-IV somatoform disorders, a study using a somatic signal detection task (a paradigm in which participants are asked to report near-threshold non-nociceptive tactile skin stimulation present in some, but not all trials) reported a heightened tendency to experience sensory input, regardless of whether stimuli were actually presented (Katzer et al. 2012). This study was designed to test the "rogue representations" hypothesis referenced in the inferential leap model proposed by Van den Bergh et al. (2017). Thus, this relates to the idea that within predictive processing terms, pathologically precise priors could lead to favoring top-down expectations over the bottom-up sensory input, resulting in reporting distorted sensory experiences. Relatedly, FND as a complex, multifaceted disorder also provides insight into how both interoceptive aspects (i.e., potentially distorted interoceptive modeling and signal integration) and exteroceptive sensations (e.g., strong sensory stimuli such as bright lights or certain sounds) can influence and possibly potentiate each other. Some patients with FND, especially with functional seizures, report that both sensations from the outer world, like an exteroceptive or situational cue, and sensations from within the body (such as heart palpitations) can trigger a seizure.

3.1.4 Attention and Motor Learning

Biased attention/information processing and impairments in motor learning are often found in patients with FND, which can be interpreted to reflect alterations in prediction error learning and subsequent conflicts between top-down predictions and bottom-up information. For example, in a temporal discrimination task of sensory stimuli, patients with FND were found to exhibit a lower rate of information accumulation ("reduced drift rate") (Sadnicka et al. 2020). This implies aberrant allocation of attention; in predictive processing terms, this relates to abnormalities in the weighting of top-down vs. bottom-up signals, potentially through engagement of the salience network. Furthermore, in a cued task of shifting spatial attention, patients with unilateral functional upper limp paresis showed a reduced response to visual and/or tactile targets and an absence of cueing effect on the affected side. This indicates an attentional detection cost and a diametral effect of attending to the symptomatic side (McIntosh et al. 2017). Lastly, motor learning in FND patients was investigated using "the broken escalator phenomenon," a task that implicates walking onto a non-moving platform while anticipating the area to be moving (Lin et al. 2020). Patients with FND were found to exhibit prolonged use of previously learned motor programs (i.e., a tendency to perpetuate motor phenomena despite changing demands), suggesting an impairment in un-learning previously acquired motor programs – which could explain why patients unwillingly perform altered movement patterns.

3.2 Diagnostic Features of FND

Evidence for the role of aberrant predictive processing is also present in the clinical phenomena that are used to diagnose FND. In this context, an important notion is that FND is not a diagnosis of exclusion but has specific and reproducible positive clinical signs (Aybek and Perez 2022). These include, amongst other features, internal inconsistency of symptoms and modifiability (suggestibility, distractibility).

The internal inconsistency of symptoms relates to a phenomenon often observed in FND. Patients are unable to perform a certain movement or report loss of sensation in one situation but can use the same muscle to perform another movement or report sensory input from the same modality in another context. While this is at times incorrectly framed as a sign of feigning or malingering, the inconsistency actually reflects core features of FND (Edwards et al. 2023), as it relates to the decoupling of the brain's internal predictive model about motor control and sensory feedback from the actual sensory input – leading to unpredictable and inconsistent symptoms (Hallett et al. 2022; Espay et al. 2018). For example, the positive sign of entrainment (having the functional tremor frequency change to match the rhythmicity of a repetitive volitional movement performed in another body part) is thought to exemplify the conflict between aberrant (functional) motor programs and "healthy" motor programs, with the latter being able to override abnormal priors, potentially by inducing concordant sensory feedback (Hallett et al. 2022; Huys et al. 2021).

FND symptoms are also modifiable, as best demonstrated through suggestibility (i.e., responsiveness to verbal or situational cues) and distractibility (e.g., through focusing attention away from symptoms). For example, suggestive seizure induction is sometimes used to aid the diagnostic evaluation of a functional vs. epileptic seizure by eliciting an event through verbal and situational cues (Popkirov et al. 2017, 2020). Suggestibility can be understood to be related to a reduction in the metacognition of intention and agency in combination with a tendency to rely on overly precise symptom priors (Wieder et al. 2021; Terhune and Hedman 2017). This may then activate a "rogue representation" (Reuber and Brown 2017) in the form of a symptom model (Bautista et al. 2008). Distractibility, on the other hand, is often used to diagnose functional movement disorder presentations such as tremor – here, the patient is asked to perform a competing cognitive or motor task such as listing the months of the year backward. In many cases, this leads to a weakening or cessation of the functional movement (Bartl et al. 2020). At the core of both techniques is the modification of attentional processes toward or away from certain bodily processes and symptoms. In predictive processing terms, this would pertain to changing the weight of top-down expectations vs. bottom-up sensory input.

Lastly, functional neurological symptoms often develop after physical injury, illness, or surgery, sometimes in form of a "second hit" traumatic experience following early life adversity (Keynejad et al. 2019). A systematic review found physical injury prior to illness onset in 37% of patients (Stone et al. 2009a, b), and a prospective study found that 80% of patients with a functional movement disorder reported such a physical event shortly before illness onset. Notably, 38% of those patients fulfilled criteria of a panic attack related to the event (Parees et al. 2014b). Twenty percent of patients with a sudden onset of functional weakness reported physical injury to the relevant limb, often in conjunction with panic or dissociative symptoms (Stone et al. 2012). Thus, these physical events are understood to provide initial sensory data of bodily experiences in the form of symptoms (i.e., they provide a symptom model); an alternative mechanistic explanation also in line with predictive processing frameworks is that the physical event draws heightened attention to the body. Together with the potentially high allostatic relevance of the respective event (marked through panic symptoms in the respective situation), these sensory experiences are then assigned a high subjective relevance and thus a strong prior – they might become the "go-to" representation or concept, e.g., in states of high arousal and negative affect. This idea is also corroborated by the finding of a higher rate of healthcare professionals in patient samples with FND compared to samples of non-functional neurological disorder, implicating that repeated exposure to or specific knowledge about neurological symptom (in combination with a stressful work environment) might lead to a stronger representation of those "symptoms models" (Vanini et al. 2023).

4 Brain Imaging Findings Consistent with Predictive Processing Abberations in FND

In recent years, neuroimaging studies have enhanced the understanding of neurobiological factors underlying the role of predictive processing FND (Perez et al. 2021). Here, we highlight select studies that are relevant to understanding the role of predictive processing in FND.

Due to the inherent complexity found in FND, it is critical to embed results from these neuroimaging studies into a broader picture of brain function and structure; particularly relevant are emerging insights regarding cortical laminar gradients (cytoarchitectural profiles) in relation to the directionality of predictions and prediction error detection. Briefly summarized, these general principles rest on the finding that brain regions can be divided based on their cytoarchitecture (lamination profiles, i.e., the number and differentiation of cortical layers) from least to most complex (see Fig. 3 and caption). These cytoarchitectural gradients supports a *sensory-fugal axis* (Paquola et al. 2022), which describes the flow of incoming sensory information from the most detailed representations in the primary sensory cortex (Koniocortex) toward an increasingly abstracted representation across unimodal and multimodal association areas (Eulaminate I/II). Predictions arise at the other end of this axis in limbic cortices and gain granularity while they are compared to the incoming sensory input at every level in-between. This means that the most abstracted multimodal representations initiate at the very end of the sensory-fugal axis in regions that overlap with the default-mode network – a domain-general network insulated from direct interaction with the external world and thus able to support integration of external and internal information in a "perceptually-decoupled state" (Paquola et al. 2021). Predictions and prediction errors are weighted for their modeled importance through precision signals originating from the salience network (Seeley 2019). Furthermore, more complex predictive processing formulations also acknowledge the role of the right temporo-parietal junction as a core hub responsible for the subjective feeling of agency over movements through integrated feedback-loops (Hallett et al. 2022). Thus, taken together, regions and networks particularly critical for predictive processing in the brain entail the default-mode network, the salience network, the right temporo-parietal junction, and their respective integration with sensorimotor areas.

4.1 Sense of Self-Agency

In FND cohorts, brain imaging research has further supported a role for the right temporo-parietal junction in mediating impaired action-authorship perceptions. In a landmark neuroimaging study of FND, Voon and colleagues compared brain activity between involuntary functional tremor and voluntarily mimicked tremor (within the same patients) to investigate the lack of sense of self-agency related to the functional tremor (Voon et al. 2010a). They found that the involuntary functional tremor was accompanied by a hypoactivity in the right temporo-parietal junction and a weaker functional connectivity between the right temporo-parietal junction, sensorimotor regions, and limbic cortices such as the ventral anterior cingulate. Baek and colleagues found lower right inferior parietal cortex activity (neighboring the temporo-parietal junction) related to intention to movements (Baek et al. 2017). The subjective degree of control over a virtual hand was related to the temporal dynamics of the activity in the right temporo-parietal junction and an overall dysfunction of the self-agency network (Nahab et al. 2017). In a resting-state functional MRI study, Maurer and colleagues identified reduced functional connectivity between right temporo-parietal junction, sensorimotor regions including the sensorimotor cortex and supplementary motor area, and the right insula when comparing FND patients to healthy controls (Maurer et al. 2016). Importantly, these network changes were observable at resting-state, without a behavioral paradigm to elicit sense of self-agency.

4.2 Emotion and Affect

The role of emotion processing, and associated brain networks, has been investigated in FND using both structural and functional imaging. Thus far, a classical view of emotions – i.e., emotions are reactions to stimuli, coordinated by a dedicated neural circuitry that orchestrates distinctive physiological changes – has guided the design and interpretation of most of these studies (Jungilligens et al. 2022a). Still, these studies also provide valuable insights into the role of emotion and affect in the pathophysiology of FND when interpreted through the lens of predictive processing frameworks. For example, task-based functional MRI studies identified heightened activity and impaired habituation of the amygdala to affectively valenced stimuli, which suggests abnormal salience signaling in patients with FND (Voon et al. 2010b; Aybek et al. 2015). Overall, attending to negatively valenced affective stimuli elicited different responses in patients with FND compared to healthy controls in regions with limbic lamination profiles and promoted abnormal limbic-motor coupling (reviewed in detail in Sojka et al. 2018). Furthermore, emotion regulation abilities assessed using an affective behavioral paradigm were related to the structural connectivity of the right uncinate fasciculus, the bilateral fornix/stria terminalis, and the corpus callosum in a diffusion tensor imaging study (Jungilligens et al. 2021). Closely connected to emotions and affect is interoception. The only study to examine the neural correlates of interoception in FND found that individual differences in interoceptive trait prediction error were associated with lower white matter integrity of fiber tracts originating from the bilateral insula and temporo-parietal junction (Sojka et al. 2021). This supports the notion that changes in the predictive modeling of the body's internal physiological state and the interpretation/categorization of incoming interoceptive sensory data play an important role in FND.

4.3 Sensorimotor Processes

Through neuroimaging studies, the closely-nit relationship between self-agency, emotion/affect, and sensorimotor processes has been highlighted in FND – which is especially relevant to understanding it as a disorder of aberrant prediction and inference. As one example, Voon and colleagues found that affective arousal states related to stronger limbic-motor functional connectivity (Voon et al. 2010b). Similarly, Aybek and colleagues reported increased co-activation in regions involved in motor control during aversive affective activation (Aybek et al. 2014, 2015). Higher connectivity between anterior cingulate cortex and the sensorimotor network correlated with seizure frequency and dissociation scores in those with functional seizures (Li et al. 2015a; van der Kruijs et al. 2014). In general, while only a few studies investigated motor aspects in FND using neuroimaging approaches, enhanced connectivity or activation of motor-network areas have also been found in studies investigating other neural mechanisms (as highlighted above), pointing toward overall aberrations in movement planning and the control of motor output.

4.4 Default-Mode and Salience Network

Several studies with a broader focus have repeatedly found alterations – or significant correlations with disease-related markers – in two brain networks important for predictive processing: the default-mode and salience networks. Alterations in the default-mode network have been linked to FND in general (Begue et al. 2019) and functional seizures specifically; functional connectivity strength within the default-mode network in functional seizure cohorts has been found to relate to dissociation scores (van der Kruijs et al. 2014), and cortical thickness of default-mode network nodes negatively correlated with longer illness duration (Zelinski et al. 2022). Furthermore, a single-photon emission computerized tomography study measured brain activation during functional seizures compared to epileptic seizures and found increases in posterior cingulate and precuneus areas, which are part of the default-mode network (Gallucci-Neto et al. 2021). Gray matter volumes of default-mode network and the salience network nodes were inversely related to symptom severity, illness duration, and trauma burden in patients with functional seizures (Jungilligens et al. 2022b). In a resting-state functional MRI study, Monsa et al. (2018) found increased functional connectivity within the default-mode network, but interestingly also decreases in the interconnectivity between the default-mode and salience networks.

The salience network itself has repeatedly been related to the pathophysiology of FND across structural and functional imaging studies: Functional studies identified aberrant activity (Aybek et al. 2015) and connectivity (Li et al. 2015b; Diez et al. 2019; van der Kruijs et al. 2012, 2014) of the salience network. In a multi-center classification study aimed at distinguishing patients with FND from healthy controls based on resting-state functional MRI activation, cingulo-insular regions were among the most discriminative features allowing to classify a person as belonging to one of either group (Weber et al. 2022b). Given the relevance of life experiences in influencing predictive processing, it is also notable that salience network resting-state connectivity profiles related to childhood maltreatment in patients with FND (Maurer et al. 2016; Diez et al. 2020; Spagnolo et al. 2020); one study showed that childhood physical abuse burden positively correlated with insula-precentral gyrus and amygdala-precentral gyrus connectivity strength – an observation not found in psychiatric controls matched for childhood maltreatment scores (Diez et al. 2020). Besides functional imaging results, structural alterations within the salience network in FND have been reported. While the direction of the effect has been inconsistent, volumetric changes were found compared to healthy controls (Labate et al. 2012, 2021; McSweeney et al. 2018; Ristic et al. 2015; Tatekawa et al. 2020; Weber et al. 2022a), and reduced cingulo-insular volumes have also been found to correlate with symptom severity and illness duration (Jungilligens et al. 2022b; Perez et al. 2017a, b, 2018). Diffusion-weighted imaging studies reported reduced microstructural integrity in white matter tracts of the salience network, such as the uncinate fasciculus, medial forebrain bundle, and stria terminalis/fornix, compared to healthy controls; inverse associations between microstructural integrity in these tracts and physical disability, illness duration, and affective task performance were also observed (Diez et al. 2021; Jungilligens et al. 2021; Hernando et al. 2015; Lee et al. 2015).

5 Findings in Other Modalities

Besides neuroimaging, studies have relied on other modalities of measurement to investigate the pathophysiology of FND.

In children and adolescents with FND, aberrant electroencephalography (EEG) power spectra were found in default-mode network and the premotor/supplementary motor areas, with non-significant trends toward increased power in some frequencies within the salience network (Kozlowska et al. 2018). These findings were interpreted as chronic overactivation of those networks, which are therefore "motoring in idle." The same group found a general heightened cortical arousal in children and adolescents with FND based on an increase in event-related potential components to stimuli (Kozlowska et al. 2017a). Other EEG-based connectivity analyses – while largely relying on relatively small sample sizes – mostly found reduced global connectedness (e.g., between frontal and parietal areas) in patients with FND (Xue et al. 2013; Knyazeva et al. 2011; Barzegaran et al. 2012; Umesh et al. 2017). These findings, particularly the overactivation of default-mode and motor networks, and the reduced global connectivity, have also been understood as reflecting imbalances in the prediction-updating cycle, i.e., an overgeneration of predictions that are not adequately updated (Kozlowska et al. 2018). Another potentially important finding comes from the study of heartbeat evoked potentials, an EEG-signal that reflects the cortical representation and integration of interoceptive signals. Patients with functional seizures showed a reduction in heartbeat evoked potential amplitude in preictal states, pointing to a reduction in the cortical integration of interoceptive information during the acute precipitation of a seizure (Elkommos et al. 2022).

Further evidence for the role of altered interoceptive modeling and related physiological (dys)regulation can be found in peripheral physiology and neuroendocrine markers, such as heart rate and heart rate variability, skin conductance, and cortisol levels. Especially in pediatric FND, elevated baseline heart rates and reduced parasympathetic influence have been found (Kozlowska et al. 2015, 2017b), with results in adults being more heterogeneous, but also potentially pointing toward autonomic dysregulation (review and meta-analysis in Paredes-Echeverri et al. 2022b). With mixed results, studies find some evidence for elevated skin conductance responses (e.g., peri-ictally in functional seizures or in response to an experimental stressor (Mungen et al. 2010; Reinsberger et al. 2012; Pick et al. 2018). Regarding neuroendocrine measures such as cortisol, no clear trend was identified by a recent meta-analysis (Paredes-Echeverri et al. 2022b); notably, though, is a recent finding suggesting a relationship between trauma history and cortisol levels, moderated by emotional neglect burden (Weber et al. 2022a). These findings can be well integrated with predictive processing accounts of FND that include notions of bodily regulation (Jungilligens et al. 2022a).

Lastly, the impact of therapy on functional neurological symptoms – through psychological/behavioral, occupational and physiotherapy, or experimental interventions such as transcranial magnetic stimulation – can offer a window in better understanding changes in predictive processes underlying FND. For example, motor retraining in physiotherapy can be understood as a gradual shifting of expectations/prior about abilities to move a certain limb and better integration of sensory feedback through prediction error learning (Nielsen et al. 2013). This has been investigated by comparing an EEG-component related to motor preparation and anticipatory attention (the contingent negative variation) in patients with functional movement disorder before and after physiotherapy (Teodoro et al. 2020). Results indicate a normalization of this EEG-component, reflecting both the clinical improvement achieved through physiotherapy and the changes in motor preparation and anticipatory attention. On the other hand, directly influencing the sense of self-agency network through transcranial magnetic stimulation of the right temporo-parietal junction led to clinical improvement in a small sample of patients with functional seizures (Peterson et al. 2018).

6 Conclusion

FND is a complex disorder at the interface of brain–mind–body interactions with a pathophysiology in part driven by alterations in predictive processes. Clinical and behavioral evidence for predictive processing difficulties in FND is identifiable in aspects of the phenomenology and the diagnostic features of FND, such as aberrant motor and sensory processes, symptom expectation, self-agency, attention, and illness beliefs. Furthermore, findings from research in neuroimaging and other modalities are consistent with an important role of altered predictive processing for FND.

Acknowledgment

None.

Funding None.

Conflicts/Disclosures of Interest J.J. has received funding from the Mercator Research Center Ruhr, the Ruhr University Bochum Research School, and the Ruhr University Bochum Medical Research Office unrelated to this work.

D.L.P. has received honoraria for continuing medical education lectures on functional neurological disorder, royalties from Springer Nature for a functional movement disorder textbook, is a paid senior editor at *Brain and Behavior*, is on the editorial board of *The Journal of Neuropsychiatry & Clinical Neurosciences* and *Epilepsy & Behavior*, and has received funding from the NIH and Sidney R. Baer Jr. Foundation unrelated to this work.

References

Anderson KE, Gruber-Baldini AL, Vaughan CG, Reich SG, Fishman PS, Weiner WJ, Shulman LM (2007) Impact of psychogenic movement disorders versus Parkinson's on disability, quality of life, and psychopathology. Mov Disord 22:2204–2209

Aust S, Härtwig EA, Heuser I, Bajbouj M (2013) The role of early emotional neglect in alexithymia. Psychol Trauma Theory Res Pract Policy 5:225–232

Aybek S, Nicholson TR, O'daly O, Zelaya F, Kanaan RA, David AS (2015) Emotion-motion interactions in conversion disorder: an FMRI study. PLoS One 10:e0123273

Aybek S, Nicholson TR, Zelaya F, O'daly OG, Craig TJ, David AS, Kanaan RA (2014) Neural correlates of recall of life events in conversion disorder. JAMA Psychiatry 71:52–60

Aybek S, Perez DL (2022) Diagnosis and management of functional neurological disorder. BMJ 376:o64

Baek K, Donamayor N, Morris LS, Strelchuk D, Mitchell S, Mikheenko Y, Yeoh SY, Phillips W, Zandi M, Jenaway A, Walsh C, Voon V (2017) Impaired awareness of motor intention in functional neurological disorder: implications for voluntary and functional movement. Psychol Med 47:1624–1636

Barrett LF (2017) The theory of constructed emotion: an active inference account of interoception and categorization. Soc Cogn Affect Neurosci 12:1–23

Bartl M, Kewitsch R, Hallett M, Tegenthoff M, Paulus W (2020) Diagnosis and therapy of functional tremor a systematic review illustrated by a case report. Neurol Res Pract 2:35

Barzegaran E, Joudaki A, Jalili M, Rossetti AO, Frackowiak RS, Knyazeva MG (2012) Properties of functional brain networks correlate with frequency of psychogenic non-epileptic seizures. Front Hum Neurosci 6:335

Bautista RE, Gonzales-Salazar W, Ochoa JG (2008) Expanding the theory of symptom modeling in patents with psychogenic nonepileptic seizures. Epilepsy Behav 13:407–409

Begue I, Adams C, Stone J, Perez DL (2019) Structural alterations in functional neurological disorder and related conditions: a software and hardware problem? Neuroimage Clin 22:101798

Brown RJ (2004) Psychological mechanisms of medically unexplained symptoms: an integrative conceptual model. Psychol Bull 130:793–812

Brown RJ, Reuber M (2016) Towards an integrative theory of psychogenic non-epileptic seizures (PNES). Clin Psychol Rev 47:55–70

Butler M, Shipston-Sharman O, Seynaeve M, Bao J, Pick S, Bradley-Westguard A, Ilola E, Mildon B, Golder D, Rucker J, Stone J, Nicholson T (2021) International online survey of 1048 individuals with functional neurological disorder. Eur J Neurol 28:3591–3602

Carson A, Lehn A (2016) Epidemiology. Handbook of clinical neurology

Carson A, Stone J, Hibberd C, Murray G, Duncan R, Coleman R, Warlow C, Roberts R, Pelosi A, Cavanagh J, Matthews K, Goldbeck R, Hansen C, Sharpe M (2011) Disability, distress and unemployment in neurology outpatients with symptoms 'unexplained by organic disease'. J Neurol Neurosurg Psychiatry 82:810

Chanes L, Barrett LF (2016) Redefining the role of limbic areas in cortical processing. Trends Cogn Sci 20:96–106

Clark A (2013) Whatever next? Predictive brains, situated agents, and the future of cognitive science. Behav Brain Sci 36:181–204

Diederen KMJ, Fletcher PC (2021) Dopamine, prediction error and beyond. Neuroscientist 27:30–46

Diez I, Larson AG, Nakhate V, Dunn EC, Fricchione GL, Nicholson TR, Sepulcre J, Perez DL (2020) Early-life trauma endophenotypes and brain circuit-gene expression relationships in functional neurological (conversion) disorder. Mol Psychiatry 26. Advance online publication

Diez I, Ortiz-Teran L, Williams B, Jalilianhasanpour R, Ospina JP, Dickerson BC, Keshavan MS, Lafrance WC, Sepulcre J, Perez DL (2019) Corticolimbic fast-tracking: enhanced multimodal integration in functional neurological disorder. J Neurol Neurosurg Psychiatry 90:929–938

Diez I, Williams B, Kubicki MR, Makris N, Perez DL (2021) Reduced limbic microstructural integrity in functional neurological disorder. Psychol Med 51:485–493

Drane DL, Fani N, Hallett M, Khalsa SS, Perez DL, Roberts NA (2020) A framework for understanding the pathophysiology of functional neurological disorder. CNS Spectr:1–7

Ducroizet A, Zimianti I, Golder D, Hearne K, Edwards M, Nielsen G, Coebergh J (2023) Functional neurological disorder: clinical manifestations and comorbidities; an online survey. J Clin Neurosci 110:116–125

Edwards MJ, Adams RA, Brown H, Parees I, Friston KJ (2012) A Bayesian account of 'hysteria'. Brain 135:3495–3512

Edwards MJ, Moretto G, Schwingenschuh P, Katschnig P, Bhatia KP, Haggard P (2011) Abnormal sense of intention preceding voluntary movement in patients with psychogenic tremor. Neuropsychologia 49:2791–2793

Edwards MJ, Yogarajah M, Stone J (2023) Why functional neurological disorder is not feigning or malingering. Nat Rev Neurol 19:246–256

Elkommos S, Martin-Lopez D, Koreki A, Jolliffe C, Mula M, Critchley H, Edwards M, Garfinkel S, Richardson MP, Yogarajah M (2022) 15 Attenuated heart-brain integration predicts functional non-epileptic seizures. J Neurol Neurosurg Psychiatry 93:e3

Espay AJ, Aybek S, Carson A, Edwards MJ, Goldstein LH, Hallett M, Lafaver K, Lafrance WC, Lang AE, Nicholson T, Nielsen G, Reuber M, Voon V, Stone J, Morgante F (2018) Current concepts in diagnosis and treatment of functional neurological disorders. JAMA Neurol 75:1132–1141

Farooq A, Yousaf A (2016) Childhood trauma and alexithymia in patients with conversion disorder. J Coll Physicians Surg Pak 26:606–610

Friston K (2010) The free-energy principle: a unified brain theory? Nat Rev Neurosci 11:127–138

Gallucci-Neto J, Brunoni AR, Ono CR, Fiore LA, Martins Castro LH, Marchetti RL (2021) Ictal SPECT in psychogenic nonepileptic and epileptic seizures. J Acad Consult Liaison Psychiatry 62:29–37

Gelauff JM, Kingma EM, Kalkman JS, Bezemer R, Van Engelen BGM, Stone J, Tijssen MAJ, Rosmalen JGM (2018) Fatigue, not self-rated motor symptom severity, affects quality of life in functional motor disorders. J Neurol 265:1803–1809

Goldstein LH, Mellers JD (2006) Ictal symptoms of anxiety, avoidance behaviour, and dissociation in patients with dissociative seizures. J Neurol Neurosurg Psychiatry 77:616–621

Hallett M (2022) Free will, emotions and agency: pathophysiology of functional movement disorder. In: Lafaver K, Maurer CW, Nicholson TR, Perez DL (eds) Functional movement disorder: an interdisciplinary case-based approach. Springer, Cham

Hallett M, Aybek S, Dworetzky BA, Mcwhirter L, Staab JP, Stone J (2022) Functional neurological disorder: new subtypes and shared mechanisms. Lancet Neurol 21:537–550

Hernando KA, Szaflarski JP, Ver Hoef LW, Lee S, Allendorfer JB (2015) Uncinate fasciculus connectivity in patients with psychogenic nonepileptic seizures: a preliminary diffusion tensor tractography study. Epilepsy Behav 45:68–73

Hohwy J (2020) New directions in predictive processing. Mind & Language 35:209–223

Huys AML, Edwards MJ, Bhatia KP, Haggard P (2020) Modulation of reaction times and sense of agency via subliminal priming in functional movement disorders. Front Neurol 11:989

Huys AML, Haggard P, Bhatia KP, Edwards MJ (2021) Misdirected attentional focus in functional tremor. Brain 144:3436–3450

Jungilligens J, Paredes-Echeverri S, Popkirov S, Barrett LF, Perez DL (2022a) A new science of emotion: implications for functional neurological disorder. Brain 145:2648–2663

Jungilligens J, Popkirov S, Perez DL, Diez I (2022b) Linking gene expression patterns and brain morphometry to trauma and symptom severity in patients with functional seizures. Psychiatry Res Neuroimaging 326:111533

Jungilligens J, Wellmer J, Kowoll A, Schlegel U, Axmacher N, Popkirov S (2021) Microstructural integrity of affective neurocircuitry in patients with dissociative seizures is associated with emotional task performance, illness severity and trauma history. Seizure 84:91–98

Jungilligens J, Wellmer J, Schlegel U, Kessler H, Axmacher N, Popkirov S (2020) Impaired emotional and behavioural awareness and control in patients with dissociative seizures. Psychol Med 50:2731–2739

Kaplan MJ, Dwivedi AK, Privitera MD, Isaacs K, Hughes C, Bowman M (2013) Comparisons of childhood trauma, alexithymia, and defensive styles in patients with psychogenic non-epileptic seizures vs. epilepsy: implications for the etiology of conversion disorder. J Psychosom Res 75: 142–146

Katzer A, Oberfeld D, Hiller W, Gerlach AL, Witthoft M (2012) Tactile perceptual processes and their relationship to somatoform disorders. J Abnorm Psychol 121:530–543

Keynejad RC, Frodl T, Kanaan R, Pariante C, Reuber M, Nicholson TR (2019) Stress and functional neurological disorders: mechanistic insights. J Neurol Neurosurg Psychiatry 90: 813–821

Knyazeva MG, Jalili M, Frackowiak RS, Rossetti AO (2011) Psychogenic seizures and frontal disconnection: EEG synchronisation study. J Neurol Neurosurg Psychiatry 82:505–511

Koreki A, Garfkinel SN, Mula M, Agrawal N, Cope S, Eilon T, Gould Van Praag C, Critchley HD, Edwards M, Yogarajah M (2020) Trait and state interoceptive abnormalities are associated with dissociation and seizure frequency in patients with functional seizures. Epilepsia 61:1156–1165

Kozlowska K, Melkonian D, Spooner CJ, Scher S, Meares R (2017a) Cortical arousal in children and adolescents with functional neurological symptoms during the auditory oddball task. Neuroimage Clin 13:228–236

Kozlowska K, Palmer DM, Brown KJ, Mclean L, Scher S, Gevirtz R, Chudleigh C, Williams LM (2015) Reduction of autonomic regulation in children and adolescents with conversion disorders. Psychosom Med 77:356–370

Kozlowska K, Rampersad R, Cruz C, Shah U, Chudleigh C, Soe S, Gill D, Scher S, Carrive P (2017b) The respiratory control of carbon dioxide in children and adolescents referred for treatment of psychogenic non-epileptic seizures. Eur Child Adolesc Psychiatry 26:1207–1217

Kozlowska K, Spooner CJ, Palmer DM, Harris A, Korgaonkar MS, Scher S, Williams LM (2018) "Motoring in idle": the default mode and somatomotor networks are overactive in children and adolescents with functional neurological symptoms. Neuroimage Clin 18:730–743

Kranick SM, Moore JW, Yusuf N, Martinez VT, Lafaver K, Edwards MJ, Mehta AR, Collins P, Harrison NA, Haggard P, Hallett M, Voon V (2013) Action-effect binding is decreased in motor conversion disorder: implications for sense of agency. Mov Disord 28:1110–1116

Labate A, Cerasa A, Mula M, Mumoli L, Gioia MC, Aguglia U, Quattrone A, Gambardella A (2012) Neuroanatomic correlates of psychogenic nonepileptic seizures: a cortical thickness and VBM study. Epilepsia 53:377–385

Labate A, Martino I, Caligiuri ME, Fortunato F, Bruni A, Segura-Garcia C, Arcuri P, De Fazio P, Cerasa A, Gambardella A (2021) Orbito-frontal thinning together with a somatoform dissociation might be the fingerprint of PNES. Epilepsy Behav 121:108044

Lee S, Allendorfer JB, Gaston TE, Griffis JC, Hernando KA, Knowlton RC, Szaflarski JP, Ver Hoef LW (2015) White matter diffusion abnormalities in patients with psychogenic non-epileptic seizures. Brain Res 1620:169–176

Li R, Li Y, An D, Gong Q, Zhou D, Chen H (2015a) Altered regional activity and inter-regional functional connectivity in psychogenic non-epileptic seizures. Sci Rep 5:11635

Li R, Liu K, Ma X, Li Z, Duan X, An D, Gong Q, Zhou D, Chen H (2015b) Altered functional connectivity patterns of the insular subregions in psychogenic nonepileptic seizures. Brain Topogr 28:636–645

Lin D, Castro P, Edwards A, Sekar A, Edwards MJ, Coebergh J, Bronstein AM, Kaski D (2020) Dissociated motor learning and de-adaptation in patients with functional gait disorders. Brain 143:2594–2606

Ludwig L, Pasman JA, Nicholson T, Aybek S, David AS, Tuck S, Kanaan RA, Roelofs K, Carson A, Stone J (2018) Stressful life events and maltreatment in conversion (functional neurological) disorder: systematic review and meta-analysis of case-control studies. Lancet Psychiatry 5:307–320

Maclean J, Finkelstein SA, Paredes-Echeverri S, Perez DL, Ranford J (2022) Sensory processing difficulties in patients with functional neurological disorder: occupational therapy management strategies and two cases. Semin Pediatr Neurol 41:100951

Maggio J, Alluri PR, Paredes-Echeverri S, Larson AG, Sojka P, Price BH, Aybek S, Perez DL (2020) Briquet syndrome revisited: implications for functional neurological disorder. Brain Commun 2:fcaa156

Marotta A, Bombieri F, Zampini M, Schena F, Dallocchio C, Fiorio M, Tinazzi M (2017) The moving rubber hand illusion reveals that explicit sense of agency for tapping movements is preserved in functional movement disorders. Front Hum Neurosci 11:291

Maurer CW, Lafaver K, Ameli R, Epstein SA, Hallett M, Horovitz SG (2016) Impaired self-agency in functional movement disorders: a resting-state fMRI study. Neurology 87:564–570

Mcintosh RD, Mcwhirter L, Ludwig L, Carson A, Stone J (2017) Attention and sensation in functional motor disorder. Neuropsychologia 106:207–215

Mcsweeney M, Reuber M, Hoggard N, Levita L (2018) Cortical thickness and gyrification patterns in patients with psychogenic non-epileptic seizures. Neurosci Lett 678:124–130

Monsa R, Peer M, Arzy S (2018) Self-reference, emotion inhibition and somatosensory disturbance: preliminary investigation of network perturbations in conversion disorder. Eur J Neurol 25:888–e62

Mungen B, Berilgen MS, Arikanoglu A (2010) Autonomic nervous system functions in interictal and postictal periods of nonepileptic psychogenic seizures and its comparison with epileptic seizures. Seizure 19:269–273

Nahab FB, Kundu P, Gallea C, Kakareka J, Pursley R, Pohida T, Miletta N, Friedman J, Hallett M (2011) The neural processes underlying self-agency. Cereb Cortex 21:48–55

Nahab FB, Kundu P, Maurer C, Shen Q, Hallett M (2017) Impaired sense of agency in functional movement disorders: an fMRI study. PLoS One 12:e0172502

Nielsen G, Stone J, Edwards MJ (2013) Physiotherapy for functional (psychogenic) motor symptoms: a systematic review. J Psychosom Res 75:93–102

Paquola C, Amunts K, Evans A, Smallwood J, Bernhardt B (2022) Closing the mechanistic gap: the value of microarchitecture in understanding cognitive networks. Trends Cogn Sci 26:873–886

Paquola C, Garber M, Frässle S, Royer J, Tavakol S, Rodriguez-Cruces R, Jeffries E, Smallwood J, Bernhardt B (2021) The unique cytoarchitecture and wiring of the human default mode network. bioRxiv. 2021.11.22.469533

Paredes-Echeverri S, Guthrie AJ, Perez DL (2022a) Toward a possible trauma subtype of functional neurological disorder: impact on symptom severity and physical health. Front Psych 13: 1040911

Paredes-Echeverri S, Maggio J, Begue I, Pick S, Nicholson TR, Perez DL (2022b) Autonomic, endocrine, and inflammation profiles in functional neurological disorder: a systematic review and meta-analysis. J Neuropsychiatry Clin Neurosci 34:30–43

Parees I, Brown H, Nuruki A, Adams RA, Davare M, Bhatia KP, Friston K, Edwards MJ (2014a) Loss of sensory attenuation in patients with functional (psychogenic) movement disorders. Brain 137:2916–2921

Parees I, Kojovic M, Pires C, Rubio-Agusti I, Saifee TA, Sadnicka A, Kassavetis P, Macerollo A, Bhatia KP, Carson A, Stone J, Edwards MJ (2014b) Physical precipitating factors in functional movement disorders. J Neurol Sci 338:174–177

Parr T, Friston KJ (2017) Working memory, attention, and salience in active inference. Sci Rep 7: 14678

Perez DL, Matin N, Barsky A, Costumero-Ramos V, Makaretz SJ, Young SS, Sepulcre J, Lafrance WC, Keshavan MS, Dickerson BC (2017a) Cingulo-insular structural alterations associated with psychogenic symptoms, childhood abuse and PTSD in functional neurological disorders. J Neurol Neurosurg Psychiatry 88:491–497

Perez DL, Matin N, Williams B, Tanev K, Makris N, Lafrance WC, Dickerson BC (2018) Cortical thickness alterations linked to somatoform and psychological dissociation in functional neurological disorders. Hum Brain Mapp 39:428–439

Perez DL, Nicholson TR, Asadi-Pooya AA, Begue I, Butler M, Carson AJ, David AS, Deeley Q, Diez I, Edwards MJ, Espay AJ, Gelauff JM, Hallett M, Horovitz SG, Jungilligens J, Kanaan RAA, Tijssen MAJ, Kozlowska K, Lafaver K, Lafrance WC, Lidstone SC, Marapin RS, Maurer CW, Modirrousta M, Reinders A, Sojka P, Staab JP, Stone J, Szaflarski JP, Aybek S (2021) Neuroimaging in functional neurological disorder: state of the field and research agenda. Neuroimage Clin 30:102623

Perez DL, Williams B, Matin N, Lafrance WC, Costumero-Ramos V, Fricchione GL, Sepulcre J, Keshavan MS, Dickerson BC (2017b) Corticolimbic structural alterations linked to health status and trait anxiety in functional neurological disorder. J Neurol Neurosurg Psychiatry 88:1052–1059

Peterson KT, Kosior R, Meek BP, Ng M, Perez DL, Modirrousta M (2018) Right temporoparietal junction transcranial magnetic stimulation in the treatment of psychogenic nonepileptic seizures: a case series. Psychosomatics 59:601–606

Pick S, Goldstein LH, Perez DL, Nicholson TR (2019) Emotional processing in functional neurological disorder: a review, biopsychosocial model and research agenda. J Neurol Neurosurg Psychiatry 90:704–711

Pick S, Mellers JD, Goldstein LH (2016) Emotion and dissociative seizures: a phenomenological analysis of patients' perspectives. Epilepsy Behav 56:5–14

Pick S, Mellers JDC, Goldstein LH (2018) Autonomic and subjective responsivity to emotional images in people with dissociative seizures. J Neuropsychol 12:341–355

Pick S, Rojas-Aguiluz M, Butler M, Mulrenan H, Nicholson TR, Goldstein LH (2020) Dissociation and interoception in functional neurological disorder. Cogn Neuropsychiatry 25:294–311

Popkirov S, Gronheit W, Jungilligens J, Wehner T, Schlegel U, Wellmer J (2020) Suggestive seizure induction for inpatients with suspected psychogenic nonepileptic seizures. Epilepsia 61: 1931–1938

Popkirov S, Jungilligens J, Gronheit W, Wellmer J (2017) Diagnosing psychogenic nonepileptic seizures: video-EEG monitoring, suggestive seizure induction and diagnostic certainty. Epilepsy Behav 73:54–58
Ranford J, Maclean J, Alluri PR, Comeau O, Godena E, Lafrance WC, Hunt A, Stephen CD, Perez DL (2020) Sensory processing difficulties in functional neurological disorder: a possible predisposing vulnerability? Psychosomatics 61:343–352
Reinsberger C, Perez DL, Murphy MM, Dworetzky BA (2012) Pre- and postictal, not ictal, heart rate distinguishes complex partial and psychogenic nonepileptic seizures. Epilepsy Behav 23: 68–70
Reuber M, Brown RJ (2017) Understanding psychogenic nonepileptic seizures-phenomenology, semiology and the integrative cognitive model. Seizure 44:199–205
Ricciardi L, Demartini B, Crucianelli L, Krahe C, Edwards MJ, Fotopoulou A (2016) Interoceptive awareness in patients with functional neurological symptoms. Biol Psychol 113:68–74
Ristic AJ, Dakovic M, Kerr M, Kovacevic M, Parojcic A, Sokic D (2015) Cortical thickness, surface area and folding in patients with psychogenic nonepileptic seizures. Epilepsy Res 112: 84–91
Sadnicka A, Daum C, Meppelink AM, Manohar S, Edwards M (2020) Reduced drift rate: a biomarker of impaired information processing in functional movement disorders. Brain 143: 674–683
Seeley WW (2019) The salience network: a neural system for perceiving and responding to homeostatic demands. J Neurosci 39:9878–9882
Shaffer C, Westlin C, Quigley KS, Whitfield-Gabrieli S, Barrett LF (2022) Allostasis, action, and affect in depression: insights from the theory of constructed emotion. Annu Rev Clin Psychol 18:553–580
Sojka P, Bares M, Kasparek T, Svetlak M (2018) Processing of emotion in functional neurological disorder. Front Psych 9:479
Sojka P, Diez I, Bares M, Perez DL (2021) Individual differences in interoceptive accuracy and prediction error in motor functional neurological disorders: a DTI study. Hum Brain Mapp 42: 1434–1445
Spagnolo PA, Norato G, Maurer CW, Goldman D, Hodgkinson C, Horovitz S, Hallett M (2020) Effects of TPH2 gene variation and childhood trauma on the clinical and circuit-level phenotype of functional movement disorders. J Neurol Neurosurg Psychiatry 91:814–821
Stephen CD, Fung V, Lungu CI, Espay AJ (2021) Assessment of emergency department and inpatient use and costs in adult and pediatric functional neurological disorders. JAMA Neurol 78:88–101
Stone J (2016) Neurologic approaches to hysteria, psychogenic and functional disorders from the late 19th century onwards. In: Handbook of clinical neurology, 1st edn. Elsevier
Stone J, Carson A, Aditya H, Prescott R, Zaubi M, Warlow C, Sharpe M (2009a) The role of physical injury in motor and sensory conversion symptoms: a systematic and narrative review. J Psychosom Res 66:383–390
Stone J, Carson A, Duncan R, Coleman R, Roberts R, Warlow C, Hibberd C, Murray G, Cull R, Pelosi A, Cavanagh J, Matthews K, Goldbeck R, Smyth R, Walker J, Macmahon AD, Sharpe M (2009b) Symptoms 'unexplained by organic disease' in 1144 new neurology out-patients: how often does the diagnosis change at follow-up? Brain 132:2878–2888
Stone J, Sharpe M, Rothwell PM, Warlow CP (2003) The 12 year prognosis of unilateral functional weakness and sensory disturbance. J Neurol Neurosurg Psychiatry 74:591–596
Stone J, Warlow C, Sharpe M (2012) Functional weakness: clues to mechanism from the nature of onset. J Neurol Neurosurg Psychiatry 83:67–69
Tatekawa H, Kerr WT, Savic I, Engel J, Salamon N (2020) Reduced left amygdala volume in patients with dissociative seizures (psychogenic nonepileptic seizures). Seizure 75:43–48
Taylor GJ, Bagby RM, Parker JD (1991) The alexithymia construct. A potential paradigm for psychosomatic medicine. Psychosomatics 32:153–164

Teodoro T, Koreki A, Meppelink AM, Little S, Nielsen G, Macerollo A, Ferreira JJ, Parees I, Lang A, Edwards MJ (2020) Contingent negative variation: a biomarker of abnormal attention in functional movement disorders. Eur J Neurol 27:985–994

Terhune DB, Hedman LRA (2017) Metacognition of agency is reduced in high hypnotic suggestibility. Cognition 168:176–181

Umesh S, Tikka SK, Goyal N, Sinha VK, Nizamie SH (2017) Aberrant gamma band cortical sources and functional connectivity in adolescents with psychogenic non-epileptic seizures: a preliminary report. Psychiatry Res 247:51–54

Van Den Bergh O, Witthoft M, Petersen S, Brown RJ (2017) Symptoms and the body: taking the inferential leap. Neurosci Biobehav Rev 74:185–203

Van Der Kruijs SJ, Bodde NM, Vaessen MJ, Lazeron RH, Vonck K, Boon P, Hofman PA, Backes WH, Aldenkamp AP, Jansen JF (2012) Functional connectivity of dissociation in patients with psychogenic non-epileptic seizures. J Neurol Neurosurg Psychiatry 83:239–247

Van Der Kruijs SJ, Jagannathan SR, Bodde NM, Besseling RM, Lazeron RH, Vonck KE, Boon PA, Cluitmans PJ, Hofman PA, Backes WH, Aldenkamp AP, Jansen JF (2014) Resting-state networks and dissociation in psychogenic non-epileptic seizures. J Psychiatr Res 54:126–133

Voon V, Gallea C, Hattori N, Bruno M, Ekanayake V, Hallett M (2010a) The involuntary nature of conversion disorder. Neurology 74:223–228

Voon V, Brezing C, Gallea C, Ameli R, Roelofs K, Lafrance WC, Hallett M (2010b) Emotional stimuli and motor conversion disorder. Brain 133:1526–1536

Vanini G, Buhler J, Weber S, Steinauer M, Aybek S (2023) Healthcare employment as a risk factor for functional neurological disorder: a case-control study. Eur J Neurol:1–10

Weber S, Buhler J, Vanini G, Loukas S, Bruckmaier R, Aybek S (2022a) Identification of biopsychological trait markers in functional neurological disorders. Brain

Weber S, Heim S, Richiardi J, Van De Ville D, Serranova T, Jech R, Marapin RS, Tijssen MAJ, Aybek S (2022b) Multi-centre classification of functional neurological disorders based on resting-state functional connectivity. Neuroimage Clin 35:103090

Wieder L, Brown R, Thompson T, Terhune DB (2021) Suggestibility in functional neurological disorder: a meta-analysis. J Neurol Neurosurg Psychiatry 92:150–157

Williams IA, Reuber M, Levita L (2021) Interoception and stress in patients with functional neurological symptom disorder. Cogn Neuropsychiatry 26:75–94

Xue Q, Wang ZY, Xiong XC, Tian CY, Wang YP, Xu P (2013) Altered brain connectivity in patients with psychogenic non-epileptic seizures: a scalp electroencephalography study. J Int Med Res 41:1682–1690

Zelinski L, Diez I, Perez DL, Kotz SA, Wellmer J, Schlegel U, Popkirov S, Jungilligens J (2022) Cortical thickness in default mode network hubs correlates with clinical features of dissociative seizures. Epilepsy Behav 128:108605

Zlotnick C, Mattia JI, Zimmerman M (2001) The relationship between posttraumatic stress disorder, childhood trauma and alexithymia in an outpatient sample. J Trauma Stress 14:177–188

The Future of Perceptual Dysregulation in Psychiatric Nosology

Albert R. Powers and Sahib S. Khalsa

Contents

Abstract Like other medical illnesses, psychiatric conditions are defined by the presence of characteristic signs and symptoms that cause distress and dysfunction. Unlike most medical illnesses, however, the pathophysiological mechanisms underlying psychiatric symptoms remain largely unknown. Recent work in theory-driven computational neuroscience has proposed a unified account of how organisms derive estimates of both external and internal states by integrating prior beliefs with sensory input. This framework offers a powerful means of explaining typical perception and action, while also identifying how these processes may go awry to produce a panoply of psychiatric symptoms. Disruptions in this inferential process, which we term *perceptual dysregulation*, span exteroceptive, proprioceptive, and interoceptive

Albert R. Powers and Sahib S. Khalsa contributed equally with all other contributors.

A. R. Powers (✉)
Psychiatry Department, Yale University School of Medicine, New Haven, CT, USA

Yale University Department of Psychology, New Haven, CT, USA

Connecticut Mental Health Center, New Haven, CT, USA
e-mail: albert.powers@yale.edu

S. S. Khalsa (✉)
Laureate Institute for Brain Research, Tulsa, OK, USA

Semel Institute for Neuroscience and Human Behavior, Geffen School of Medicine, University of California Los Angeles, Los Angeles, CA, USA
e-mail: skhalsa@mednet.ucla.edu

Curr Topics Behav Neurosci (2025) 74: 307–322
https://doi.org/10.1007/7854_2025_596
Published Online: 13 August 2025

systems and manifest across conditions ranging from psychosis and autism to anxiety and trauma-related disorders, functional neurological disorder, and dementia. This volume reviews emerging evidence for maladaptive sensory inference across psychiatric and neurological illness and explores how computational models based on Bayesian statistics can illuminate shared mechanisms beneath diverse symptom presentations. We argue that targeting perceptual dysregulation may not only enable new interventions but also catalyze the development of a revised psychiatric nosology grounded in testable models of brain-body interaction.

Keywords Computational Psychiatry · Interoception · Perceptual Inference · Predictive Coding · Psychiatric Disorders · Sensory Processing

1 Introduction: The Bayesian Brain

Since the advent of systems neuroscience, it has been difficult to deny the intuitive appeal of a simple feed-forward model of perception. Beginning in the retina, a stepwise cascade of feature processing seems to call into being an increasingly complete picture of our external world: contrasts become edges, edges become shapes, and shapes, wedded to movement and color, become detailed representations of the contents of our environment (Barlow 1982). Through this organized serial extraction of information, incoming sensory information is seemingly capable of providing everything we need for successful navigation and decision-making in the visual world.

This example of visual perceptual processing is simple and elegant, but wrong. Or, more charitably, it is only a partial description of perception. Responding to the limitations of feed-forward models, contemporary computational neuroscience has proposed the existence of an evolutionarily conserved system for establishing and updating representations of the environment. Using relatively simple message-passing schemes, theorists have demonstrated the ability of this system to form new beliefs and update those beliefs based on new information across all sensory modalities in a manner broadly consistent with the optimal performance of Bayesian statistical systems (Friston and Frith 2015; Clark 2013). Viewing the brain as a prediction-generating organ performing constant inference to update beliefs about the external environment, the body's position in space, and the internal milieu, offers a powerful alternative: rather than passively receiving sensory inputs, the brain constructs percepts by minimizing prediction errors across hierarchical sensory systems. Because this limits the amount of information needed to be processed on a moment-to-moment basis, it is also remarkably efficient (Friston et al. 2014), providing explanations for a wide range of perceptual phenomena that were difficult to account for by appealing only to feedforward information processing. These inferential processes extend across sensory domains: Bayesian reliability weighting explains the dominance of certain modalities in audiovisual illusions (Alais and Burr

2004; Hirst et al. 2019); contextual priors shape not only auditory speech-in-noise perception (Mesgarani et al. 2014) but also interoceptive experiences such as breathlessness or hunger (Smith et al. 2020); and visual depth illusions based on shading (Mumford 1994) have analogs in body-based illusions that alter proprioceptive and interoceptive representations of self-location and ownership (e.g., the rubber hand and full-body illusions) (Aspell et al. 2013). Together, these findings suggest that perception is not a passive decoding of sensory inputs but an active process of hypothesis testing, shaped by prior beliefs and updated through prediction error minimization.

This inferential framework offers a transdiagnostic lens for understanding both normative and pathological perceptual phenomena. Positing that perception follows the expected rules of learning and inference provides a common theoretical framework capable of describing both health and disorders of perception. Deviations of specific learning model parameters from optimality may manifest within specific domains—such as the hypersensitivity to visual detail in body dysmorphic disorder, proprioceptive misintegration in functional neurological disorder, or altered interoceptive precision in trauma and dissociation—and arise from faulty integration across systems. These deviations have been identified as being reliably altered across disorders and tied to expression of specific symptoms, generating a slew of recent advances in computational psychiatry (Browning et al. 2020; Wang and Krystal 2014; Redish and Gordon 2016). This framework carries the possibility of linking symptomatology not only to alterations in learning and inference, but also to upstream abnormalities in sensory processing that could result in symptom development via changes in learning and inference (Powers et al. 2025).

In this volume, we consider perceptual dysregulation, defined as maladaptive learning and inference within or across sensory systems, as a transdiagnostic feature of psychiatric illness. Using the Bayesian brain as a guiding principle for perception, we examine how exteroceptive, interoceptive, and proprioceptive anomalies shape the lived experience of psychiatric symptoms. Our aim is to illuminate how computational principles unify otherwise disparate symptom presentations and how a deeper understanding of perceptual inference may catalyze the evolution of psychiatric nosology itself.

2 Disruptions of Exteroceptive Inference and Self-World Boundaries in Psychopathology

A fundamental task of the nervous system is to infer the causes of sensory inputs—whether they originate from the external world (exteroception), from the body in motion (proprioception), or from within the body (interoception). Psychiatric disorders can be framed as disruptions in these inferential processes. Among the most well-characterized are psychosis and autism spectrum disorders (ASD), which have been associated with altered models of the external world and the self's relation to it.

Psychosis is characterized most saliently by the presence of fixed, false beliefs (delusions) and percepts in the absence of corresponding sensory evidence (hallucinations). Because Bayesian statistics describe the formation and updating of beliefs given new evidence, they lend themselves easily to efforts to explain the latent states driving symptoms like delusions and hallucinations. Accordingly, theories have proliferated in recent years tying delusions to aberrant coding of prediction errors and belief updating, causing inappropriate new belief formation (Sterzer et al. 2018; Adams et al. 2013; Erdmann and Mathys 2022; Corlett et al. 2010) and hallucinations to hyper-precise perceptual priors (Kelley and Corlett 2016; Corlett et al. 2019). Evidence for this theoretical framework has also grown: a tendency to learn inappropriately appears to correspond to delusional propensity in psychosis, schizotypy, the clinical high risk state for psychosis, and in healthy volunteers under the influence of the psychotomimetic drug ketamine (Corlett et al. 2006, 2007; Rossi-Goldthorpe et al. 2024). Hallucinations, on the other hand, have been repeatedly linked to a tendency to overweight existing beliefs more than new evidence during perception (Corlett et al. 2019; Powers et al. 2017; Alderson-Day et al. 2017; Kafadar et al. 2020, 2022; Benrimoh et al. 2024).

Despite this evidence, it remains unclear whether the functioning of sensory systems themselves is implicated in symptom formation. Recent theoretical efforts have attempted to describe how such a state of affairs might develop (Powers et al. 2025; Sheldon et al. 2022). In this formulation, a tendency to over-rely on priors during perception is conceptualized as a secondary effect, a compensation against earlier disruptions in sensory systems caused by excessive noise and resulting in inappropriate belief formation (Powers et al. 2025). This excessive noise, then, would result in learning abnormalities leading directly to symptom formation and also to distinct sensory system dysfunction. Two contributions to this series describe consistent sensory deficits and aberrant multisensory integration, an early form of inference and belief updating based on sensory reliability (Alais and Burr 2004).

Many years of work have demonstrated specific sensory abnormalities in schizophrenia and other forms of psychosis, with multimodal deficits ranging from impaired processing of basic acoustic features of simple and complex sounds (Billingberg and Jonsson 1965; Hari et al. 1984) and disruption of auditory memory traces (Strous et al. 1995) to basic detection of visual stimuli (Dondé et al. 2019) and visual facial processing (Ruocco et al. 2014). These sensory disruptions also extend beyond auditory and visual domains to sensory modalities less commonly tied to hallucinations. Deficits in tactile, proprioceptive, and interoceptive processing have been reported in psychosis, including impaired body ownership, reduced proprioceptive precision, and blunted or misattributed visceral sensations (Javitt et al. 1999; Ardizzi et al. 2016; Melvin et al. 2021; Jeganathan et al. 2024; Critchley et al. 2023), implicating disruptions in internal sensory systems alongside traditional exteroceptive pathways and suggesting a broader impairment in constructing an embodied model of the self. These deficits appear to occur very early, in pre-attentive phases of sensory processing, indicating disruption of sensory information in pathways proximal to complex cortical processing (Rabinowicz et al. 2000). This provides compelling evidence that the observed discrimination deficits are not due to the broad cognitive deficits of schizophrenia but may be a more fundamental process that could give rise to other manifestations of the illness. The downstream effects of basic sensory deficits are likely to be seen in multisensory interactions, which Vassall and Wallace show to be disrupted in psychosis (Vassall and Wallace 2025). Consistent with the broad disruptions highlighted by Javitt in unimodal processing (Javitt 2025), multisensory disruptions in psychosis seem to describe a broad smearing of information across stimulus types and modalities, with a tendency to over-integrate information across longer time scales (Zvyagintsev et al. 2017; Haß et al. 2017; Vanes et al. 2016; Martin et al. 2013) and more readily than unaffected healthy comparison participants (Peled et al. 2000). Although it remains unclear whether unimodal deficits fully account for the disruptions described in multisensory integration, the findings are consistent with the likelihood of early, large-scale disruptions in sensory processing that may contribute meaningfully to the emergence of psychotic symptoms (Powers et al. 2025).

While sensory abnormalities in psychosis often reflect a broad degradation of signal precision and temporal binding, ASD-related sensory differences may be more circumscribed and context-dependent. Theories of ASD posit impairments in integrating sensory signals across time and modality, which are particularly in socially salient contexts (Gambra et al. 2024; Happé and Frith 2006). These include difficulties with audiovisual speech integration, tactile defensiveness, and altered proprioceptive feedback during social interaction. Such impairments may reflect differences in the weighting of sensory precision and attention allocation, rather than global sensory deficits. Integration abnormalities similar to those seen in psychosis are also present in ASD, characterized by over-integration of stimuli over time (Foss-Feig et al. 2010). Interestingly, these may be more specific than the deficits seen in psychosis: ASD integration difficulties appear to impact more complex stimuli, including speech (Stevenson et al. 2014). These distinct patterns of sensory processing suggest differences in underlying pathophysiology and contribute differentially to core social communication difficulties seen in ASD (Manjaly et al. 2007). Understanding how these integration deficits give rise to specific symptoms offers a pathway toward targeted interventions aimed at recalibrating sensory precision, improving multisensory integration and ultimately supporting social function.

Although we have thus far only touched on alterations in inference that pertain to understanding some state of the world outside oneself, it bears noting that certain sensory deficits impact our understanding of our own bodies. As Diaz-Fong and Feusner (2024) demonstrate, similar deficits described in visual processing in ASD may also be present in body dysmorphic disorder (BDD) (Diaz-Fong and Feusner 2024). These include general enhancements of local processing to the detriment of global integration (Li et al. 2015a, b) as well as specific visual attentional biases (Feusner et al. 2010) that may map more directly onto symptomatology (Buhlmann et al. 2014). But what if these visual biases were to interact with atypical proprioceptive and interoceptive input, such as altered heartbeat perception or body position sense, to reinforce a distorted self-image? Although there are currently no empirical demonstrations of this concept, several consistent theoretical propositions have recently emerged, including ones relying on predictive processing and hierarchical models of brain function (Jenkinson and Rossell 2024; Choquette and Khalsa 2025). As summarized by Reilly et al. (2024), similar disruptions are observed in anorexia nervosa and bulimia nervosa, where overestimation of body size has been linked to altered processing of cardiac signals (Saramandi et al. 2024), along with aberrant visual processing of local features (Alexi et al. 2019). As is also seen in BDD, these perceptual factors may exist independently of the cognitive and affective factors that drive specific eating disorder phenomenology (Cornelissen et al. 2019).

These findings underscore a broader principle: interactions between low-level perceptual biases and higher-order attentional or contextual factors can give rise to disorder-specific symptomatology. Building on this idea, recent models propose that baseline tendencies in sensory processing, such as enhanced local visual detail processing, may interact with attentional modulation and sociocultural expectations to shape perceptual experiences in ways that reinforce maladaptive self-representations. This intersection between intrinsic sensory processing styles and flexible top-down inference offers an entry point for targeted interventions, ranging from perceptual retraining to neuromodulatory techniques. More broadly, as we elaborate throughout this volume, perceptual dysregulation—whether in the form of heightened priors, degraded sensory precision, or impaired multisensory integration—provides a powerful unifying lens for understanding how dysfunction in discrete sensory systems can propagate into the diverse and complex symptom profiles seen across psychiatric diagnoses.

3 Interoceptive Dysregulation: Altered Estimates of Bodily and Emotional States

One of the most profound insights of Bayesian-influenced theories of psychiatric illness is that we infer not only the contents of our outer environments but also the inner states of our own bodies. This is because, just as the brain must infer the causes of visual or auditory sensations, it must also interpret ambiguous and noisy interoceptive signals to generate models of affective, emotional, and physiological states (Paulus and Stein 2010; Seth and Critchley 2013), as direct neural access to these states does not exist.

Disruptions in interoceptive inference have significant transdiagnostic implications in psychiatry, extending beyond simple emotion recognition deficits into altered perception of physiological cues such as cardiovascular variability, respiratory rhythm, and gastrointestinal sensations (Khalsa et al. 2018). As detailed by Mehta and colleagues (2025), applying phenotypic descriptions of individuals using computational measures may facilitate the identification of diagnostic subgroups with symptoms due to pathway-specific abnormalities, or be helpful in predicting treatment responders. This could be applied to specific symptoms within a disorder, such as fatigue in major depression (Stephan et al. 2016), or mechanistically span broad symptom categories (Petzschner et al. 2017). The implications of this insight are profound and may extend even into a formalized understanding of what it means for an organism to find an experience rewarding (Weber et al. 2024). It also provides a useful framework for extending the work described above in exteroception into other, more internalizing psychiatric disorders.

Because interoceptive inference so broadly applies to an understanding of internal states, identifying alterations in these processes has the potential to touch upon symptoms that cross traditional diagnostic boundaries. Dissociation provides a particularly vivid illustration of how disruptions in interoceptive inference can alter one's sense of self. As described by Woelk and colleagues (Woelk and Garfinkel 2024), dissociative symptoms, ranging from depersonalization to emotional numbing, are seen not only in dissociative disorders but also in conditions such as substance use, ASD, psychosis, and even transiently in the general population. These symptoms, often described as a perceptual uncoupling from one's own body, may be reasonably thought to involve disruptions of interoceptive inference. This is because factors that are characteristically disrupted in dissociation, including a sense of agency (Stern et al. 2020), recognition of self-generated sensory information (Haggard et al. 2002), and emotional dysregulation (Simeon and Abugel 2008), have been linked to disruptions in interceptive inference. Just as described above in psychosis, low-level disruptions in autonomic regulation (Owens et al. 2015) and bodily reactivity to emotional stimuli (Dewe et al. 2016) may result in a down-weighting of interoceptive information and subsequent distancing from the information most informative for one's own bodily state. This pathophysiological pathway may also help to explain why the experience of adverse life events could lead to dissociation. As outlined by Fani and colleagues (2024), traumatic experiences can induce chronic autonomic dysregulation and heightened visceral reactivity (D'Andrea et al. 2013; Sack et al. 2012), with even brief exposure to aversive emotional stimuli, making interoceptive input more volatile and less reliable. Similar interoceptive disturbances have been identified in other trauma-linked conditions, including chronic pain and eating disorders (Dunne et al. 2021; Brown et al. 2017), suggesting a shared pathophysiological mechanism that cuts across diagnostic boundaries.

Interoceptive inference disruptions can result not only from environmental exposures and genetic predispositions, but also from degeneration of neural circuits critical to maintaining allostatic stability. This is most saliently seen in the case of neurodegenerative disease, where the pathways involved in maintaining allostasis are impacted early and may result in some of the symptoms of these illnesses. As argued by Franco-O'Byrne and colleagues (2024), dementia subtypes, particularly frontotemporal dementia (FTD), show pronounced early disruptions in circuits underpinning interoceptive inference (Franco-O'Byrne et al. 2024; Migeot et al. 2022). The resulting breakdown in top-down predictions and bottom-up signal fidelity may create a growing disconnect between predicted and actual bodily states, manifesting as both autonomic instability and maladaptive social-emotional behavior—hallmarks of allostatic interoceptive overload (Franco-O'Byrne et al. 2024). Thus, a predictive coding perspective unifies diverse behavioral and physiological symptoms in neurodegeneration, offering novel therapeutic targets oriented around maintaining interoceptive precision and stabilizing predictive processing (Franco-O'Byrne et al. 2024; Birba et al. 2022).

4 Proprioceptive Dysregulation: Disrupted Body Representation and Motor Control

While the previous sections have focused on perceptual inference, how the brain constructs models of the world and the body from sensory inputs, Bayesian modeling approaches based on the Active Inference framework extend these ideas into action (Parr et al. 2022). That is, the brain not only updates its internal models to minimize prediction error, but also acts upon the world to fulfill predictions and reduce uncertainty (Pezzulo et al. 2024). A classic example is motor control: when the brain strongly predicts a limb's position, any mismatch with proprioceptive input creates a prediction error that can be minimized by moving the limb accordingly (Parr and Friston 2017). This principle has been fruitfully applied to functional neurological disorder (FND), as discussed by Jungilligens and Perez (2024). In FND, symptoms such as limb weakness, tremor, or non-epileptic seizures may emerge from aberrant precision weighting of proprioceptive signals and prior expectations (Jungilligens and Perez 2024). Individuals with FND demonstrate a slower rate of evidence accumulation (Sadnicka et al. 2020) and impaired motor program learning (Lin et al. 2020), suggesting that incoming sensory signals are treated as unreliable, leading the brain to lean excessively on top-down priors. This disrupted precision may then produce altered movement seen in FND. Notably, top-down priors may, in turn, be shaped by prior bodily experiences, particularly in individuals with a history of trauma. Psychogenic non-motor symptoms may arise from a similar process, whereby disruptions of interoceptive input lead to a downweighting of this information and a relative hyper-precision of priors around internal bodily states. As the risk for FND is elevated in those with past traumatic experiences concordant with the endorsed symptom (i.e., past epileptic seizures in those with psychogenic seizures, past injury in those with pain-related FND), hyper-precise priors could conceivably cause ongoing inference of bodily states consistent with the previous traumatic events. Central to this application is the understanding that the body's processes are regulated by allostasis, a predictive form of regulation relying on interoceptive and proprioceptive input to infer the body's physiological state (Jungilligens et al. 2022).

5 Gaps and Next Steps

Bayesian frameworks like those highlighted in this collection are immensely powerful because they provide a common method for linking normative processing to pathophysiology (Browning et al. 2020). However, this explanatory power may also be something of a liability: if these frameworks are maximally flexible, they may also be difficult to falsify, limiting their ability to generate new knowledge with any degree of certainty.

Central to most if not all of the contributions above is the distinction between top-down and bottom-up input. In most cases, symptoms and disease states are attributed to disruptions in the precision afforded to these two sources of information. However, using only behavior or even basic neurophysiological measures, it can be difficult to disentangle the influence of prior precision from the influence of likelihood or sensory precision. Take, for example, the case of hallucinations. While we and others have demonstrated that hallucination propensity and severity are related to the relative precision of priors compared to that of sensory evidence (Powers et al. 2017; Kafadar et al. 2022), this is only a relative measure: it can just as easily be explained as decreased precision of incoming sensory evidence as absolute hyper-precision of priors. These forms of perceptual dysregulation are not exclusive to exteroception. They are equally likely—perhaps even more fundamentally—operant in the domain of interoception, where inferential challenges may be even more acute due to the ambiguity of signals, their affective salience, and the absence of direct access to external referents. The same computational tools used to model prediction error minimization in visual cortex could be extended to interoceptive regions such as the posterior and anterior insula, where feedforward visceral signals interface with top-down predictions to construct bodily self-awareness (Paulus et al. 2019; Petzschner et al. 2021; Adamic et al. 2024).

Haarsma and Kok offer insight into how these factors may be disentangled in the service of falsifiability and replicability. Using the fact that cortical information processing follows a conserved motif that allows for feedforward and feedback message passing (Friston 2005), they make the case that differences in activation of specific brain areas may be attributed to bottom-up versus top-down input with the use of high-field MRI (Haarsma et al. 2022). Because of the high spatial resolution made possible by the use of very strong static electromagnetic fields, it becomes feasible to identify voxels that not only correspond to functional activation of a specific cortical region, but to determine which layers of cortex are specifically activated. The authors and others have already demonstrated the utility of this approach in validating predictive-coding-based cortical architectures and identifying the influence of priors in visual cortex (Thomas et al. 2024; Haarsma et al. 2023). It is not too difficult to imagine the extension of this concept to layer-specific processing of interoceptive input in primary visceral cortex such as the insula. To date, however, no published work has examined laminar-specific interoceptive signal flow in humans. Developing methods to identify distinct cortical layers of the insula and map feedforward versus feedback signal components could offer a transformative step toward evaluating whether the predictive coding architecture described in visual cortex extends to the viscera. Such advances would provide a rigorous framework for disentangling whether interoceptive symptoms in psychiatric conditions stem from aberrant priors, degraded afferent input, or their interaction.

Another remaining challenge stems directly from a poverty of data. The chapters in this volume apply theories of information processing to the expression of symptoms across neuropsychiatric illness. To do so, they necessarily consider how alterations in typical processes might feasibly produce perception and behavior that resembles symptomatology. As noted, these efforts have been fruitful, identifying states that appear to be just proximal to symptom expression and are linked to symptom severity (Kafadar et al. 2022). However, these observations provide hints at only the very end of a pathway toward symptom emergence. The pathophysiology of psychiatric illness likely comprises a series of insults and compensatory responses within a system attempting to maintain allostasis (Powers et al. 2025).

Seen in this light, the production and sharing of large-scale longitudinal data becomes crucial. Large, international, harmonized efforts such as the Accelerating Medicines Partnership-Schizophrenia (AMP-SCZ) provide potential sources of data (Brady et al. 2023) such as these. However, not all data sets are equally capable of answering questions posed within computational frameworks. Additionally, few (if any) large-scale longitudinal studies include measures of interoception, such as heartbeat perception, respiratory load tasks, gastric or nociceptive probes, or layer-specific insular imaging. Ongoing and future projects should consider incorporating behavioral, electrophysiological, and functional imaging measures capable of directly testing these emerging computationally oriented hypotheses in a developing sample so that the signatures of symptom emergence might be seen most clearly and these ideas put directly to the test.

If our collective challenge remains one of identifying pathophysiological pathways, it becomes necessary to identify how these pathways might differ from one another in ways that feasibly map onto the reality of patients' experiences. Stated differently, while classical frameworks identify five exteroceptive modalities, this vastly underrepresents the true sensory complexity of the human body. Interoception alone comprises multiple parallel input channels—cardiovascular, respiratory, gastrointestinal, urogenital, nociceptive, and others—each with distinct cortical pathways and computational demands. If psychiatric symptoms arise from disruptions in perceptual inference, we must ask how dysfunction in such a high-dimensional sensory space maps onto specific symptom profiles. Why do similar disruptions in signal precision or integration yield hallucinations in one individual, dissociation in another, and affective flattening in a third? Understanding this mapping requires not only richer phenotyping but new models of how distributed sensory perturbations shape the emergence of psychiatric categories. Doing so will require much more work than is obvious now, as we recognize early successes and begin to turn the pathophysiological clock back toward discovery of distal causes. The degree to which there is fine-grained mapping between neural abnormality, information processing alteration, and individual experience may determine the feasibility of using theoretical insights to inform new nosological schemes and personalized treatment approaches.

Beyond finding solutions to these challenges, the early translation of computationally oriented findings into new biomarkers and therapeutic approaches should be a goal that simultaneously addresses major clinical challenges and demonstrates the utility of these approaches in advancing treatment (Browning et al. 2020). Some of the findings described here may begin to exhibit the sensitivity and specificity required for development into meaningful potential psychiatric biomarkers

(Abi-Dargham et al. 2023), but few have been tested in a broad enough patient population over enough time to establish whether this is, in fact, the case.

Ultimately, this volume advances the hypothesis that disruptions in perceptual inference lie at the heart of a diversity of neuropsychiatric illnesses. This may or may not prove to be correct. But the development and testing of interventions that explicitly target perceptual mechanisms, whether through perceptual learning, feedback modulation, or sensory retraining, offers not only the promise of novel treatments, but also the opportunity to establish causal links between latent computational processes and observable symptoms. In doing so, these efforts may help lay the groundwork for a revised nosology—one rooted not in symptom checklists or clusters, but in empirically grounded models of sensory inference, brain-body interaction, and bodily regulation.

Acknowledgments ARP was supported by a K23 Career Development Award, one R21, and two R01s from the National Institute of Mental Health (K23MH115252-01A1; 5R21MH122940-02; R01MH129721; R01MH131768), by a Career Award for Medical Scientists from the Burroughs-Wellcome Fund, a Carol and Eugene Ludwig Award for Early Career Research, and by the Yale Department of Psychiatry and the Yale School of Medicine. The work described in this chapter was funded in part by the State of Connecticut, Department of Mental Health and Addiction Services, but this publication does not express the views of the Department of Mental Health and Addiction Services or the State of Connecticut. SSK was supported by grants from the National Institute of Mental Health (R01MH127225), the Binational Science Foundation (BSF2023143), by the Louis Jolyon West Innovation Chair, the UCLA Semel Institute for Neuroscience and Human Behavior, and by the UCLA Departments of Psychiatry and Medicine. The views and opinions expressed are those of the authors.

References

Abi-Dargham A, Moeller SJ, Ali F, DeLorenzo C, Domschke K, Horga G, Jutla A, Kotov R, Paulus MP, Rubio JM, Sanacora G, Veenstra-VanderWeele J, Krystal JH (2023) Candidate biomarkers in psychiatric disorders: state of the field. World Psychiatry 22:236–262

Adamic EM, Teed AR, Avery J, de la Cruz F, Khalsa S (2024) Hemispheric divergence of interoceptive processing across psychiatric disorders. elife 13:13

Adams RA, Stephan KE, Brown HR, Frith CD, Friston KJ (2013) The computational anatomy of psychosis. Front Psych 4:47

Alais D, Burr D (2004) The ventriloquist effect results from near-optimal bimodal integration. Curr Biol 14:257–262

Alderson-Day B, Lima CF, Evans S, Krishnan S, Shanmugalingam P, Fernyhough C, Scott SK (2017) Distinct processing of ambiguous speech in people with non-clinical auditory verbal hallucinations. Brain 140:2475–2489

Alexi J, Palermo R, Rieger E, Bell J (2019) Evidence for a perceptual mechanism relating body size misperception and eating disorder symptoms. Eat Weight Disord 24:615–621

Ardizzi M, Ambrosecchia M, Buratta L, Ferri F, Peciccia M, Donnari S, Mazzeschi C, Gallese V (2016) Interoception and positive symptoms in schizophrenia. Front Hum Neurosci 10:379

Aspell JE, Heydrich L, Marillier G, Lavanchy T, Herbelin B, Blanke O (2013) Turning body and self inside out: visualized heartbeats Alter bodily self-consciousness and tactile perception. Psychol Sci 24:2445–2453

Barlow HB (1982) David Hubel and Torsten Wiesel: their contributions towards understanding the primary visual cortex. Trends Neurosci 5:145–152

Benrimoh D, Fisher VL, Seabury R, Sibarium E, Mourgues C, Chen D, Powers A (2024) Evidence for reduced sensory precision and increased reliance on priors in hallucination-prone individuals in a general population sample. Schizophr Bull 50:349–362

Billingberg O, Jonsson CO (1965) The ability of schizophrenic patients to interpret intonation. Acta Psychiatr Scand 41:218–226

Birba A, Santamaría-García H, Prado P, Cruzat J, Ballesteros AS, Legaz A, Fittipaldi S, Duran-Aniotz C, Slachevsky A, Santibañez R, Sigman M, García AM, Whelan R, Moguilner S, Ibáñez A (2022) Allostatic-interoceptive overload in frontotemporal dementia. Biol Psychiatry 92:54–67

Brady LS, Larrauri CA, AMP SCZ Steering Committee (2023) Accelerating medicines partnership schizophrenia (AMP SCZ): developing tools to enable early intervention in the psychosis high risk state. World Psychiatry 22:42–43

Brown TA, Berner LA, Jones MD, Reilly EE, Cusack A, Anderson LK, Kaye WH, Wierenga CE (2017) Psychometric evaluation and norms for the multidimensional assessment of interoceptive awareness (MAIA) in a clinical eating disorders sample. Eur Eat Disord Rev 25:411–416

Browning M, Carter C, Chatham C, den Ouden H, Gillan C, Baker J, Goldstein R, Lawson R, Rindskopf D, Roiser J (2020) Realizing the clinical potential of computational psychiatry: report from the Banbury Center Meeting, February 2019. Biol Psychiatry 88:e5–e10

Buhlmann U, Rupf L, Gleiss MJL, Zschenderlein K, Kathmann N (2014) Seeing "changes" that aren't there: Facial and object discrimination in body dysmorphic disorder. Compr Psychiatry 55:468–474

Choquette EM, Khalsa SS (2025) Sensing the self: the role of the insula and Interoception in body image. Curr Top Behav Neurosci

Clark A (2013) Whatever next? Predictive brains, situated agents, and the future of cognitive science. Behav Brain Sci 36:181–204

Corlett PR, Honey GD, Aitken MR, Dickinson A, Shanks DR, Absalom AR, Lee M, Pomarol-Clotet E, Murray GK, McKenna PJ, Robbins TW, Bullmore ET, Fletcher PC (2006) Frontal responses during learning predict vulnerability to the psychotogenic effects of ketamine: linking cognition, brain activity, and psychosis. Arch Gen Psychiatry 63:611–621

Corlett PR, Murray GK, Honey GD, Aitken MR, Shanks DR, Robbins TW, Bullmore ET, Dickinson A, Fletcher PC (2007) Disrupted prediction-error signal in psychosis: evidence for an associative account of delusions. Brain 130:2387–2400

Corlett PR, Taylor JR, Wang XJ, Fletcher PC, Krystal JH (2010) Toward a neurobiology of delusions. Prog Neurobiol 92:345–369

Corlett PR, Horga G, Fletcher PC, Alderson-Day B, Schmack K, Powers AR (2019) Hallucinations and strong priors. Trends Cogn Sci 23:114

Cornelissen KK, Widdrington H, McCarty K, Pollet TV, Tovée MJ, Cornelissen PL (2019) Are attitudinal and perceptual body image the same or different? Evidence from high-level adaptation. Body Image 31:35–47

Critchley HD, Sherrill SP, Ewing DL, van Praag CG, Habash-Bailey H, Quadt L, Eccles JA, Meeten F, Jones A-M, Garfinkel SN (2023) Cardiac interoception in patients accessing secondary mental health services: a transdiagnostic study. Auton Neurosci 245:103072

D'Andrea W, Pole N, DePierro J, Freed S, Wallace DB (2013) Heterogeneity of defensive responses after exposure to trauma: blunted autonomic reactivity in response to startling sounds. Int J Psychophysiol 90:80–89

Dewe H, Watson DG, Braithwaite JJ (2016) Uncomfortably numb: new evidence for suppressed emotional reactivity in response to body-threats in those predisposed to sub-clinical dissociative experiences. Cogn Neuropsychiatry 21:377–401

Diaz-Fong JP, Feusner JD (2024) Visual perceptual processing abnormalities in body dysmorphic disorder. Curr Top Behav Neurosci

Dondé C, Avissar M, Weber MM, Javitt DC (2019) A century of sensory processing dysfunction in schizophrenia. Eur Psychiatry 59:77–79

Dunne J, Flores M, Gawande R, Schuman-Olivier Z (2021) Losing trust in body sensations: interoceptive awareness and depression symptom severity among primary care patients. J Affect Disord 282:1210–1219

Erdmann T, Mathys C (2022) A generative framework for the study of delusions. Schizophr Res 245:42–49

Fani N, Fulton T, Botzanowski B (2024) The neurophysiology of interoceptive disruptions in trauma-exposed populations. Curr Top Behav Neurosci

Feusner JD, Moller H, Altstein L, Sugar C, Bookheimer S, Yoon J, Hembacher E (2010) Inverted face processing in body dysmorphic disorder. J Psychiatr Res 44:1088–1094

Foss-Feig JH, Kwakye LD, Cascio CJ, Burnette CP, Kadivar H, Stone WL, Wallace MT (2010) An extended multisensory temporal binding window in autism spectrum disorders. Exp Brain Res 203:381–389

Franco-O'Byrne D, Santamaría-García H, Migeot J, Ibáñez A (2024) Emerging theories of Allostatic-interoceptive overload in neurodegeneration. Curr Top Behav Neurosci

Friston K (2005) A theory of cortical responses. Philos Trans R Soc Lond B Biol Sci 360:815–836

Friston K, Frith CA (2015) Duet for one. Conscious Cogn 36:390–405

Friston KJ, Stephan KE, Montague R, Dolan RJ (2014) Computational psychiatry: the brain as a phantastic organ. Lancet Psychiatry 1:148–158

Gambra L, Magallon S, Crespo-Eguílaz N (2024) Weak central coherence in neurodevelopmental disorders: a comparative study. Front Psychol 15:1348074

Haarsma J, Kok P, Browning M (2022) The promise of layer-specific neuroimaging for testing predictive coding theories of psychosis. Schizophr Res 245:68–76

Haarsma J, Deveci N, Corbin N, Callaghan MF, Kok P (2023) Expectation cues and false percepts generate stimulus-specific activity in distinct layers of the early visual cortex. J Neurosci 43: 7946–7957

Haggard P, Clark S, Kalogeras J (2002) Voluntary action and conscious awareness. Nat Neurosci 5: 382–385

Happé F, Frith U (2006) The weak coherence account: detail-focused cognitive style in autism spectrum disorders. J Autism Dev Disord 36:5–25

Hari R, Hämäläinen M, Ilmoniemi R, Kaukoranta E, Reinikainen K, Salminen J, Alho K, Näätänen R, Sams M (1984) Responses of the primary auditory cortex to pitch changes in a sequence of tone pips: neuromagnetic recordings in man. Neurosci Lett 50:127–132

Haß K, Sinke C, Reese T, Roy M, Wiswede D, Dillo W, Oranje B, Szycik GR (2017) Enlarged temporal integration window in schizophrenia indicated by the double-flash illusion. Cogn Neuropsychiatry 22:145–158

Hirst RJ, Setti A, Kenny RA, Newell FN (2019) Age-related sensory decline mediates the sound-induced flash illusion: evidence for reliability weighting models of multisensory perception. Sci Rep 9:19347

Javitt DC (2025) Disturbances in auditory and visual perceptual function in schizophrenia: patterns, causes, and consequences. Curr Top Behav Neurosci

Javitt DC, Liederman E, Cienfuegos A, Shelley AM (1999) Panmodal processing imprecision as a basis for dysfunction of transient memory storage systems in schizophrenia. Schizophr Bull 25: 763–775

Jeganathan J, Campbell MEJ, Legrand N, Allen M, Breakspear M (2024) Aberrant cardiac interoception in psychosis. Schizophr Bull 51:208–216

Jenkinson PM, Rossell SL (2024) Disturbed Interoception in body dysmorphic disorder: a framework for future research. Aust N Z J Psychiatry 58:300–307

Jungilligens J, Perez DL (2024) Predictive processing and the pathophysiology of functional neurological disorder. Curr Top Behav Neurosci

Jungilligens J, Paredes-Echeverri S, Popkirov S, Barrett LF, Perez DL (2022) A new science of emotion: implications for functional neurological disorder. Brain 145:2648–2663

Kafadar E, Mittal VA, Strauss GP, Chapman HC, Ellman LM, Bansal S, Gold JM, Alderson-Day B, Evans S, Moffatt J, Silverstein SM, Walker EF, Woods SW, Corlett PR, Powers AR (2020) Modeling perception and behavior in individuals at clinical high risk for psychosis: support for the predictive processing framework. Schizophr Res 226:167–175
Kafadar E, Fisher VL, Quagan B, Hammer A, Jaeger H, Mourgues C, Thomas R, Chen L, Imtiaz A, Sibarium E, Negreira AM, Sarisik E, Polisetty V, Benrimoh D, Sheldon AD, Lim C, Mathys C, Powers AR (2022) Conditioned hallucinations and prior overweighting are state-sensitive markers of hallucination susceptibility. Biol Psychiatry 92:772–780
Khalsa SS, Adolphs R, Cameron OG, Critchley HD, Davenport PW, Feinstein JS, Feusner JD, Garfinkel SN, Lane RD, Mehling WE, Meuret AE, Nemeroff CB, Oppenheimer S, Petzschner FH, Pollatos O, Rhudy JL, Schramm LP, Simmons WK, Stein MB, Stephan KE, Van den Bergh O, Van Diest I, von Leupoldt A, Paulus MP (2018) Interoception Summit 2016 participants. Interoception and mental health: a roadmap. Biol Psychiatry Cogn Neurosci Neuroimaging 3:501–513
Li W, Lai TM, Bohon C, Loo SK, McCurdy D, Strober M, Bookheimer S, Feusner J (2015a) Anorexia nervosa and body dysmorphic disorder are associated with abnormalities in processing visual information. Psychol Med 45:2111–2122
Li W, Lai TM, Loo SK, Strober M, Mohammad-Rezazadeh I, Khalsa S, Feusner J (2015b) Aberrant early visual neural activity and brain-behavior relationships in anorexia nervosa and body dysmorphic disorder. Front Hum Neurosci 9:301
Lin D, Castro P, Edwards A, Sekar A, Edwards MJ, Coebergh J, Bronstein AM, Kaski D (2020) Dissociated motor learning and de-adaptation in patients with functional gait disorders. Brain 143:2594–2606
Manjaly ZM, Bruning N, Neufang S, Stephan KE, Brieber S, Marshall JC, Kamp-Becker I, Remschmidt H, Herpertz-Dahlmann B, Konrad K, Fink GR (2007) Neurophysiological correlates of relatively enhanced local visual search in autistic adolescents. NeuroImage 35:283–291
Martin B, Giersch A, Huron C, van Wassenhove V (2013) Temporal event structure and timing in schizophrenia: preserved binding in a longer "now". Neuropsychologia 51:358–371
Mehta M, Paulus MP, Smith R (2025) Computational approaches for uncovering interoceptive mechanisms in psychiatric disorders and their biological basis. Curr Top Behav Neurosci
Melvin K, Crossley J, Cromby J (2021) The feeling, embodiment and emotion of hallucinations in first episode psychosis: a prospective phenomenological visual-ecological study using novel Multimodal Unusual Sensory Experience (MUSE) maps. EClinicalMedicine 41:101153
Mesgarani N, David SV, Fritz JB, Shamma SA (2014) Mechanisms of noise robust representation of speech in primary auditory cortex. Proc Natl Acad Sci USA 111:6792–6797
Migeot JA, Duran-Aniotz CA, Signorelli CM, Piguet O, Ibáñez A (2022) A predictive coding framework of allostatic-interoceptive overload in frontotemporal dementia. Trends Neurosci 45: 838–853
Mumford D (1994) Neuronal architectures for pattern-theoretic problems. In: Large-scale neuronal theories of the brain
Owens AP, David AS, Low DA, Mathias CJ, Sierra-Siegert M (2015) Abnormal cardiovascular sympathetic and parasympathetic responses to physical and emotional stimuli in depersonalization disorder. Front Neurosci 9:89
Parr T, Friston KJ (2017) Uncertainty, Epistemics and active inference. J R Soc Interface 14: 20170376
Parr T, Pezzulo G, Friston KJ (2022) Active inference: The free energy principle in mind, brain, and behavior. MIT Press
Paulus MP, Stein MB (2010) Interoception in anxiety and depression. Brain Struct Funct 214:451–463
Paulus MP, Feinstein JS, Khalsa SS (2019) An active inference approach to interoceptive psychopathology. Annu Rev Clin Psychol 15:97–122
Peled A, Ritsner M, Hirschmann S, Geva AB, Modai I (2000) Touch feel illusion in schizophrenic patients. Biol Psychiatry 48:1105–1108

Petzschner FH, Weber LAE, Gard T, Stephan KE (2017) Computational psychosomatics and computational psychiatry: toward a joint framework for differential diagnosis. Biol Psychiatry 82:421–430

Petzschner FH, Garfinkel SN, Paulus MP, Koch C, Khalsa SS (2021) Computational models of interoception and body regulation. Trends Neurosci 44:63–76

Pezzulo G, Parr T, Friston K (2024) Active inference as a theory of sentient behavior. Biol Psychol 186:108741

Powers AR III, Kelley M, Corlett PR (2016) Hallucinations as top-down effects on perception. Biol Psychiatry 1:393–400

Powers AR, Mathys C, Corlett PR (2017) Pavlovian conditioning-induced hallucinations result from overweighting of perceptual priors. Science 357:596–600

Powers A, Angelos PA, Bond A, Farina E, Fredericks C, Gandhi J, Greenwald M, Hernandez-Busot G, Hosein G, Kelley M, Mourgues C, Palmer W, Rodriguez-Sanchez J, Seabury R, Toribio S, Vin R, Weleff J, Woods S, Benrimoh D (2025) A computational account of the development and evolution of psychotic symptoms. Biol Psychiatry 97:117–127

Rabinowicz EF, Silipo G, Goldman R, Javitt DC (2000) Auditory sensory dysfunction in schizophrenia: imprecision or distractibility? Arch Gen Psychiatry 57:1149–1155

Redish AD, Gordon JA (2016) Chapter 7 - How could we get nosology from computation? In: Computational psychiatry: new perspectives on mental illness. The MIT Press: Cambridge, MA, pp xii, 408 p

Reilly EE, Brown TA, Frank GKW (2024) Perceptual dysfunction in eating disorders. Curr Top Behav Neurosci

Rossi-Goldthorpe R, Silverstein SM, Gold JM, Schiffman J, Waltz JA, Williams TF, Powers AR, Woods SW, Zinbarg RE, Mittal VA, Ellman LM, Strauss GP, Walker EF, Levin JA, Castiello S, Kenney J, Corlett PR (2024) Different learning aberrations relate to delusion-like beliefs with different contents. Brain 147:2854–2866

Ruocco AC, Reilly JL, Rubin LH, Daros AR, Gershon ES, Tamminga CA, Pearlson GD, Hill SK, Keshavan MS, Gur RC, Sweeney JA (2014) Emotion recognition deficits in schizophrenia-Spectrum disorders and psychotic bipolar disorder: findings from the Bipolar-Schizophrenia Network on Intermediate Phenotypes (B-SNIP) study. Schizophr Res 158:105–112

Sack M, Cillien M, Hopper JW (2012) Acute dissociation and cardiac reactivity to script-driven imagery in trauma-related disorders. Eur J Psychotraumatol 3:3

Sadnicka A, Daum C, Meppelink A-M, Manohar S, Edwards M (2020) Reduced drift rate: a biomarker of impaired information processing in functional movement disorders. Brain 143: 674–683

Saramandi A, Crucianelli L, Koukoutsakis A, Nisticò V, Mavromara L, Goeta D, Boido G, Gonidakis F, Demartini B, Bertelli S, Gambini O, Jenkinson PM, Fotopoulou A (2024) Updating prospective self-efficacy beliefs about cardiac interoception in anorexia nervosa: an experimental and computational study. Comput Psychiatr 8:92–118

Seth AK, Critchley HD (2013) Extending predictive processing to the body: emotion as interoceptive inference. Behav Brain Sci 36:227

Sheldon AD, Kafadar E, Fisher V, Greenwald MS, Aitken F, Negreira AM, Woods SW, Powers AR (2022) Perceptual pathways to hallucinogenesis. Schizophr Res 245:77–89

Simeon D, Abugel J (2008) Feeling unreal: depersonalization disorder and the loss of the self. Oxford University Press

Smith R, Kuplicki R, Feinstein J, Forthman KL, Stewart JL, Paulus MP, Tulsa 1000 investigators, Khalsa SS (2020) A Bayesian computational model reveals a failure to adapt interoceptive precision estimates across depression, anxiety, eating, and substance use disorders. PLoS Comput Biol 16:e1008484

Stephan KE, Manjaly ZM, Mathys CD, Weber LAE, Paliwal S, Gard T, Tittgemeyer M, Fleming SM, Haker H, Seth AK, Petzschner FH (2016) Allostatic self-efficacy: a metacognitive theory of dyshomeostasis-induced fatigue and depression. Front Hum Neurosci 10:550

Stern Y, Koren D, Moebus R, Panishev G, Salomon R (2020) Assessing the relationship between sense of agency, the bodily-self and stress: four virtual-reality experiments in healthy individuals. J Clin Med 9:9

Sterzer P, Adams RA, Fletcher P, Frith C, Lawrie SM, Muckli L, Petrovic P, Uhlhaas P, Voss M, Corlett PR (2018) The predictive coding account of psychosis. Biol Psychiatry 84:634–643

Stevenson RA, Siemann JK, Woynaroski TG, Schneider BC, Eberly HE, Camarata SM, Wallace MT (2014) Brief report: arrested development of audiovisual speech perception in autism Spectrum disorders. J Autism Dev Disord 44:1470–1477

Strous RD, Cowan N, Ritter W, Javitt DC (1995) Auditory sensory ("echoic") memory dysfunction in schizophrenia. Am J Psychiatry 152:1517–1519

Thomas ER, Haarsma J, Nicholson J, Yon D, Kok P, Press C (2024) Predictions and errors are distinctly represented across V1 layers. Curr Biol 34:2265–2271.e4

Vanes LD, White TP, Wigton RL, Joyce D, Collier T, Shergill SS (2016) Reduced susceptibility to the sound-induced flash fusion illusion in schizophrenia. Psychiatry Res 245:58–65

Vassall SG, Wallace MT (2025) Sensory and multisensory processing changes and their contributions to autism and schizophrenia. Curr Top Behav Neurosci

Wang XJ, Krystal JH (2014) Computational psychiatry. Neuron 84:638–654

Weber L, Yee D, Small D, Petzschner FH (2024) Rethinking reinforcement learning: the interoceptive origin of reward

Woelk SP, Garfinkel SN (2024) Dissociative symptoms and interoceptive integration. Curr Top Behav Neurosci

Zvyagintsev M, Parisi C, Mathiak K (2017) Temporal processing deficit leads to impaired multisensory binding in schizophrenia. Cogn Neuropsychiatry 22:361–372

MIX
Papier aus verantwortungsvollen Quellen
Paper from responsible sources
FSC® C105338

If you have any concerns about our products,
you can contact us on
ProductSafety@springernature.com

In case Publisher is established outside the EU,
the EU authorized representative is:
Springer Nature Customer Service Center GmbH
Europaplatz 3, 69115 Heidelberg, Germany

Printed by Libri Plureos GmbH
in Hamburg, Germany